Decarbonizing through Sustainable Hydrogen and its Derivatives

Decarbonizing through Sustainable Hydrogen and its Derivatives

Edited by

Asmae Berrada
LERMA Lab, International University of Rabat, Sala Al Jadida, Morocco

Rachid El Mrabet
Senior Expert and General Director, INNOREM Group, Morocco

ELSEVIER

Elsevier
Radarweg 29, PO Box 211, 1000 AE Amsterdam, Netherlands
50 Hampshire Street, 5th Floor, Cambridge, MA 02139, United States

ISBN: 978-0-443-27632-3

For Information on all Elsevier publications visit our website at https://www.elsevier.com/books-and-journals

Publisher: Priscilla Braglia
Acquisitions Editor: Ambika Kapoor
Editorial Project Manager: Juwairiya Ali
Production Project Manager: Maria Shalini A
Cover Designer: Raman Kumar

Typeset by Aptara, New Delhi, India

Contents

Part II
Hydrogen derivatives

Part III
Green and sustainable hydrogen economy

Contributors

Jamil Abdelmajid, Higher School of Technology of Fez (EST Fez), Sidi Mohamed Ben Abdellah University (USMBA), Fez, Morocco

Youssef Achour, LERMA Lab, College of Engineering and Architecture, International University of Rabat, Sala Al Jadida, Morocco

Mustafa M. Amin, Department of Chemical Engineering, King Fahd University of Petroleum and Minerals, Dhahran, Saudi Arabia

Abdellah Benzaouak, ENSAM, Laboratory of Spectroscopy, Molecular Modeling, Nanomaterials, Water and Environment, Environmental Materials Team, Mohammed V University in Rabat, Rabat, Morocco

Asmae Berrada, LERMA Lab, College of Engineering and Architecture, International University of Rabat, Sala Al Jadida, Morocco

Salma Boujmiraz, LRIT, Faculty of Science, Mohammed V University in Rabat, Rabat, Morocco

Khalid Bouziane, LERMA Lab, College of Engineering and Architecture, International University of Rabat, Sala Al Jadida, Morocco

Abdellatif El Ghazi, TIC Lab, College of Engineering and Architecture, International University of Rabat, Sala Al Jadida, Morocco

Asmae El Mokrini, BEAR Lab, Rabat Business School, International University of Rabat, Sale, Morocco

Rachid El Mrabet, INNOREM Group, Rabat, Morocco

Anisa Emrani, ENSAS/Laboratory of Materials, Processes, Environment and Quality (LMPEQ), Cadi Ayyad University (UCA), Safi, Morocco

Bashir Hashim, Department of Chemical Engineering, King Fahd University of Petroleum and Minerals, Dhahran, Saudi Arabia

Mohammad M. Hossain, Department of Chemical Engineering, King Fahd University of Petroleum and Minerals, Dhahran, Saudi Arabia; Interdisciplinary Research Center for Refining and Advanced Chemicals, King Fahd University of Petroleum and Minerals, Dhahran, Saudi Arabia

Rasheed M. Ibrahim, Department of Chemical Engineering, King Fahd University of Petroleum and Minerals, Dhahran, Saudi Arabia

Umar M. Jafar, Department of Chemical Engineering, King Fahd University of Petroleum and Minerals, Dhahran, Saudi Arabia

Samson Olaitan Jeje, Department of Chemical, Metallurgical and Materials Engineering, Faculty of Engineering and Built Environment, Tshwane University of Technology, Pretoria, South Africa

Khadija Jihar, LERMA Lab, College of Engineering and Architecture, International University of Rabat, Sala Al Jadida, Morocco

Ayesha Kaleem, School of Design and the Built Environment, Curtin University, Bentley, WA, Australia

Ilham Kirm, Laboratory of Chemistry-Biology Applied to the Environment, URL-CNRST-13, Applied Materials & Catalysis, Department of Chemistry, Faculty of Science, Moulay Ismail University (UMI), Meknes, Morocco

Umar H. Nuhu, Department of Chemical Engineering, King Fahd University of Petroleum and Minerals, Dhahran, Saudi Arabia

Mohammed Sahab, Higher School of Technology of Fez (EST Fez), Sidi Mohamed Ben Abdellah University (USMBA), Fez, Morocco

Mxolisi Brendon Shongwe, Department of Chemical, Metallurgical and Materials Engineering, Faculty of Engineering and Built Environment, Tshwane University of Technology, Pretoria, South Africa

Eseoghene Helen Umukoro, Department of Chemistry, Obafemi Awolowo University, Ile Ife, Nigeria

Atiq Zaman, School of Design and the Built Environment, Curtin University, Bentley, WA, Australia

Abdellah Zouheir, LERMA Lab, College of Engineering and Architecture, International University of Rabat, Sala Al Jadida, Morocco

Part I

Direct use of hydrogen

Chapter 1

Potential of green hydrogen energy for decarbonization

Asmae Berrada[a] and Rachid El Mrabet[b]
[a] *LERMA Lab, College of Engineering and Architecture, International University of Rabat, Sala Al Jadida, Morocco,* [b] *INNOREM Group, Rabat, Morocco*

1.1 Introduction

Energy is transported from its source (primary sources) to its end-use (consumption) by energy carriers, which include heat, electricity, and solid, liquid, and gaseous fuels. Energy carriers' carbon footprints must be reduced or removed in order for future net-zero energy systems to function. Low-carbon electricity, such as that produced by solar or wind power, and the electrification of end uses, such as transportation and heating, are anticipated to be key components of energy transitions. However, to decarbonize applications across a variety of sectors that are difficult to electrify, low greenhouse gas (GHG) liquid or gaseous fuels, such as hydrogen, ammonia, biofuels, and synthetic hydrocarbons, will be needed. Depending on technological advancements and implementation, any of these fuels may play a significant role in decarbonization (IPCC, 2023). Interest in hydrogen stems from its special qualities, which limit its production, storage, transportation, and use while making it a great substitute for carbon-intensive energy sources such as fossil fuels.

Hydrogen is considered a means of decarbonizing sectors where direct electrification or other low-carbon alternatives may not be practical or affordable. Although there has been interest in hydrogen since the 1970s, earlier efforts to develop this promising energy carrier, most recently in the early 2000s, were never successful in making hydrogen a viable substitute for traditional fossil fuels. Some contend that things are different now, and in some significant ways, they are. Climate concerns are the main driver of today's interest in hydrogen. The idea of creating hydrogen from renewable electricity has also gained fresh life due to the extraordinary expansion of renewable energy production and the sharp drop in the price of solar and wind energy. Further, the idea of generating hydrogen from renewable electricity has gained new momentum due to the

Decarbonizing through Sustainable Hydrogen and its Derivatives.
DOI: https://doi.org/10.1016/B978-0-443-27632-3.00001-3

unparalleled expansion of renewable energy production and the sharp drop in the cost of solar and wind power.

The physical and chemical characteristics of hydrogen, particularly its low volumetric energy density, which makes it challenging to store and transport, have not altered. These and other characteristics have historically made it difficult for hydrogen to find new uses outside of the industrial processes for which it has been employed.

About 2% of global carbon dioxide (CO_2) emissions come from direct CO_2 emissions linked to current hydrogen production methods, including roughly 680 million tons per year (MT/y) from industrial processes and 240–380 MT/y from refining (International Energy Agency, 2022a). This makes hydrogen production a critical target for decarbonization efforts. Therefore current hydrogen usage offers a compelling decarbonization opportunity. Water electrolysis and methane reforming with carbon capture and storage (CCS) are the two primary routes that need to be prioritized in order to successfully decarbonize hydrogen generation at scale.

Water electrolysis is the process of separating water into its components using an electrical current. The resulting hydrogen can be regarded as low-carbon as long as the energy is produced in a low-carbon way (the electricity source may need to fulfill certain criteria or a predetermined carbon intensity threshold in order to be deemed "clean" or "low-carbon" for regulatory or policy purposes). Water electrolysis currently makes up a very small portion of the world's hydrogen supply and is far more costly than traditional, carbon-intensive methods of producing hydrogen.

Natural gas is used as a feedstock for methane reforming with CCS. This production process needs CCS in order to be low-carbon. This hydrogen generation pathway's life-cycle carbon emissions consist of direct CO_2 emissions from the reforming unit itself, indirect CO_2 emissions from the use of electricity, and upstream methane emissions from the production and transmission of natural gas. In order to be considered low-carbon, these emissions must be reduced as much as is practical by the use of low-carbon electricity, high rates of carbon capture, and low methane leakage. For instance, these direct emissions must be reduced by 90% or more through the use of carbon capture technology in order to be considered low-carbon under European rules.

This chapter acknowledges the decarbonization difficulty given by current hydrogen generation while offering a workable framework for hydrogen to effectively address climate change. By adopting this perspective, engineers, planners, and policymakers may more effectively pinpoint high-potential and economically viable uses for clean hydrogen.

1.2 Hydrogen production technologies

As a nonprimary energy source, hydrogen is generated from both renewable and nonrenewable sources. Once produced, it can be stored for later use, transformed into other chemicals and fuels, or utilized to replace carbon-intensive industrial operations such as steel manufacturing. It can also be used as fuel to generate heat or power in a variety of applications, including combustion engines, turbines, or fuel cells.

Fossil fuels, renewable energy sources, or other nonrenewable energy sources can all be used to produce hydrogen. Depending on the feedstock and technological pathway utilized for its production, hydrogen is usually labeled with a color nomenclature. Each of these factors has a unique efficiency, technology readiness level (TRL), cost, and environmental impact. The production routes' carbon intensity can be used to identify the technology that will best support decarbonization initiatives. However, resource availability, local supply chain peculiarities, and country-specific energy system characteristics can also affect emission intensity within pathways (Hydrogen Council, 2021; World Energy Council, 2021). Furthermore, other factors such as technology maturity, resource depletion, ecosystem quality, and human health can also be considered when assessing the impacts of each pathway (Al-Qahtani et al., 2021).

The most widely used method of producing hydrogen at the moment is gray hydrogen. By using steam methane reforming (SMR) or autothermal reforming (ATR), gray hydrogen is produced from natural gas, or methane, but the GHGs that are produced during this process are not captured (Nikolaidis & Poullikkas, 2017).

Gasification, the process of combining black coal or lignite (brown coal) with oxygen and steam at high temperatures, produces both black and brown hydrogen (Vidas & Castro, 2021). Although reforming and gasification are currently the most cost-effective and technologically advanced methods for producing hydrogen on a large scale, their impact as a decarbonization tool is diminished because they produce large amounts of GHG emissions, either as a byproduct or as an energy input (Osman et al., 2022).

Natural gas is used to produce **blue hydrogen**, which combines carbon capture, utilization, and storage (CCUS) to separate CO_2 from the product stream. As a result, "low-carbon hydrogen" is another term for blue hydrogen. The cost of developing and running CCUS systems and concerns about upstream CO_2 emissions, during storage or transportation, are some obstacles to the adoption of blue hydrogen applications (Griffiths et al., 2021; Hydrogen Forecast to 2050, 2022). Blue hydrogen can lower emissions, but it does not solve the current energy system's reliance on fossil fuels.

Turquoise hydrogen is obtained by thermal or catalytic decomposition of natural gas (methane pyrolysis); since the process occurs in the absence of oxygen, the products are hydrogen and solid carbon instead of CO_2 emissions. The latter has potential market applications and can be stored more easily than CO_2, reducing capital and operational costs. However, methane pyrolysis is not a mature technology, with only pilot plants in early development stages. Despite these limitations, both turquoise and blue hydrogen are expected to have an expanding role in a net-zero future, particularly as cost-effective transition technologies, whereas the infrastructure and end-use applications for the other types of hydrogen are deployed (Parkinson et al., 2018).

Methane pyrolysis, the thermal or catalytic breakdown of natural gas, produces turquoise hydrogen. Since oxygen is not present during the process, the end products are solid carbon and hydrogen rather than CO_2 emissions. In addition to having potential market uses, the latter is easier to store than CO_2, which lowers operating and capital expenses. However, only pilot plants are currently in the early stages of development, making methane pyrolysis an immature technology. Notwithstanding these drawbacks, it is anticipated that both blue and turquoise hydrogen will play a larger role in a net-zero future, especially as affordable transition technologies are implemented alongside the infrastructure and end-use uses for the other forms of hydrogen (Parkinson et al., 2018). The transition to a fossil fuel-free energy future and mitigation measures may be hampered by methane leakage and carbon lock-in hazards, as turquoise hydrogen, like blue hydrogen, depends on the extraction of fossil fuels (Kemfert et al., 2022).

Another large-scale option for delivering low-carbon energy is nuclear power, which can also provide electricity for the electrolysis of pink hydrogen. Pink hydrogen raises the same issues as nuclear, including the need for significant investment, cost overruns, radioactive waste management, and differing degrees of public and political support (Davis et al., 2018).

Green hydrogen is created through water electrolysis using renewable energy sources such as solar, wind, and hydropower. Utilizing electricity, electrolysis separates water molecules into hydrogen and oxygen, yielding the purest hydrogen possible, which is advantageous for subsequent uses in numerous supply chains. Alkaline electrolysis (AE), polymer electrolyte membrane (PEM), anion exchange membrane (AEM), and solid oxide electrolysis (SOE) are the four main electrolyzer technologies that currently show promise for large-scale green hydrogen production. These technologies differ in their maturity levels and efficiencies.

Water electrolysis, which is frequently regarded as the most promising production pathway for low-carbon hydrogen, has the potential to become a cost-effective option for hydrogen production when combined with inexpensive renewable energy (Nikolaidis & Poullikkas, 2017). Accessibility and affordability of freshwater are crucial in this setting (Osman et al., 2022); freshwater and renewable energy may not always be available at the same time, and conflicting uses may call into question the fairness of the distribution and allocation of energy and water. Green hydrogen is regarded as one of the cleanest and most environmentally friendly ways to produce hydrogen because it is made by electrolyzing water using renewable energy sources like solar or wind power, which produces no GHG emissions and only oxygen as a byproduct. Based on manufacturing processes, Fig. 1.1 summarizes the hydrogen color scheme and offers information on TRL, carbon intensity, and efficiency.

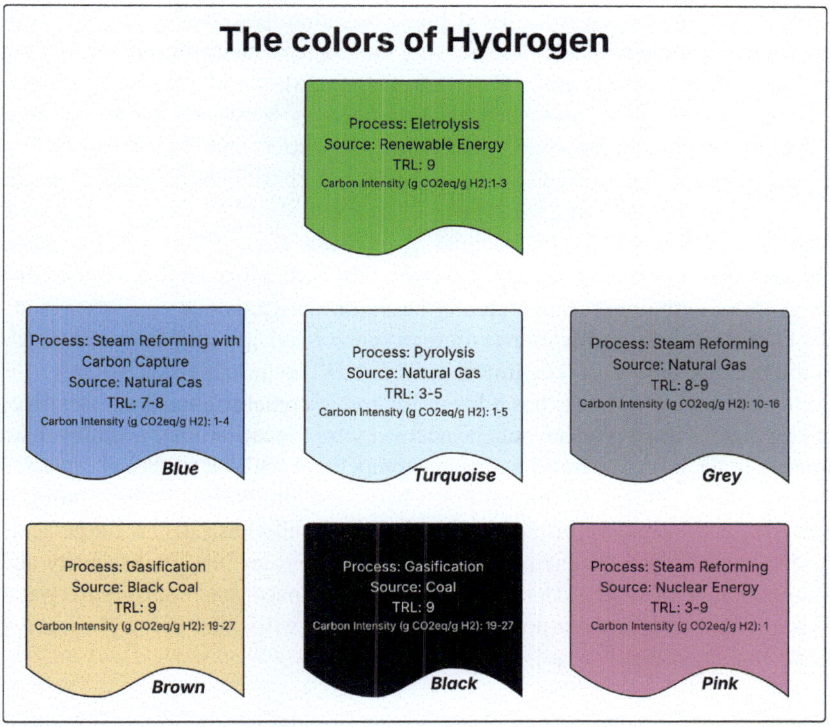

FIGURE 1.1 Hydrogen colors, carbon intensity, and technology readiness level (TRL). Hydrogen color scheme.

Electrolysis can help with decarbonization efforts and has low GHG emissions (related to the life cycle of materials), depending on the electrical source. However, electrolyzers are currently less economic than hydrogen produced from natural gas using coal gasification or SMR due to their high electricity consumption. All electrolyzer technologies must advance in terms of efficiency, life span, capital cost, grid balancing options, system integration, compactness, and hydrogen production rate in order to become a more cost-effective choice (Nilsson et al., 2021). Additionally, green hydrogen is not affected by variations in the price of fossil fuels, which might make it more competitive and unaffected by shifts in geopolitics. Additionally, expenditures are needed to standardize large-scale applications, boost manufacturing capabilities, promote technological research, and deploy electrolyzers on a larger scale (Hydrogen Forecast to 2050, 2022).

Grid-connected electrolyzers have the potential to help countries gradually decarbonize their hydrogen supply, which currently has a comparatively high proportion of renewable energy sources in their power mixes. Investing in the full decarbonization of the electrical supply is crucial for the production of green hydrogen in the future. In order to assist nations in gradually ceasing the production of hydrogen from fossil fuels, including blue hydrogen pathways, investments should be made in increasing the usage of electrolyzers, even if the electricity is not entirely green (Berrada et al., 2025).

Green hydrogen is also sometimes defined as hydrogen generated from biomass, which includes wood, crops, and crop wastes, and other forms of biological waste. Although GHG is released during the gasification process, the carbon impact is thought to be small because biomass absorbs CO_2 during growth. When paired with CCUS, this technology pathway could result in net negative emissions (Al-Qahtani et al., 2021). Some of the difficulties include the fact that the process's efficiency varies greatly based on the feedstock (Iribarren et al., 2014), that biomass gasification is more expensive and complex than coal gasification because biomass pretreatment is required (Thomas, 2018), and that yields are lower because biomass has a lower hydrogen content. Therefore incentives and regulatory frameworks would be necessary for the technology's scaling up in order to make biomass-based hydrogen competitive with fossil fuel alternatives (Nilsson et al., 2021). Additional green hydrogen alternatives include biomass pyrolysis and biogas reforming (Nikolaidis & Poullikkas, 2017). Large-scale production may present environmental and social issues like food security and competition with food systems, deforestation, impacts on biodiversity, water consumption, and fertilizer pollution, as is the case with other uses of bioenergy (Clarke et al., 2022).

1.3 Clean hydrogen applications

The current global consumption of hydrogen is around 95 MT/y, nearly all of which is used as a feedstock (not as fuel) in steel manufacturing (5 MT/y), ammonia production (33 MT/y), methanol production (16 MT/y), and refining (41 MT/y) (Roy et al., 2025). Clean hydrogen would need to significantly increase its use in fuel applications in order to contribute significantly to decarbonization.

Low-carbon hydrogen works best in situations where there are not any other viable decarbonization solutions, considering the aforementioned physical property and economic constraints. Ideally, low-carbon hydrogen will be directed to the best end applications by market forces influenced by stricter decarbonization regulations. Economically and environmentally efficient results, however, should be promoted by industrial policy, or at the very least, not discouraged. As investors and policymakers consider challenging trade-offs in order to meet decarbonization goals over the coming decades, the following sections address a strategy for ranking clean hydrogen uses.

Given proven industrial uses of hydrogen and related infrastructure, switching to low-carbon hydrogen from current carbon-intensive hydrogen usage provides a no-regrets option to cut carbon emissions. In situations when there are no viable hydrogen substitutes, this strategy can also guarantee the long-term sustainability of production.

Refineries use hydrogen as a vital feedstock in their operations and generate a wide range of products that are essential to the operation of the modern economy. It is difficult and costly to replace many of these products, and they will probably continue to be used in our economy in the future. For instance, in transportation fuels, hydrogen is utilized to eliminate sulfur, nitrogen, oxygen, olefins, and heavy metals. These hydrotreatment processes are required to guarantee that petroleum products meet safety, legal, and technical standards. Additionally, hydrogen contributes to the creation of a range of nonfuel products, including lubricants and anode-grade coke, which is essential for the manufacturing of steel and aluminum, as well as to the increase of product yields from hydrocracking operations. A decarbonization opportunity might be created by substituting 26 MT/y of unabated hydrogen production in refineries with low-carbon hydrogen, which would reduce CO_2 emissions by 240–380 MT/y (which is equal to UK emissions) (Cao et al., 2025). Ammonia is a crucial component of nitrogen fertilizers, which are vital to ensuring a safe food supply for people all over the world. In fact, fertilizer production uses 70% of the world's ammonia supply (Zhu et al., 2025). Other significant applications for ammonia include synthetic fibers, mining explosives, and specialist applications. Around 70% of ammonia is manufactured using natural gas, with the remaining portion being produced using coal, mostly in China. The current global ammonia production rate of 190 MT/y is predicted to result in 450 MT/y of CO_2 emissions (Wakim, 2023). A key ingredient in ammonia synthesis is hydrogen, which reacts with atmospheric nitrogen to produce ammonia. Decarbonizing the hydrogen used to produce ammonia should be a priority because ammonia is essential to the contemporary, worldwide industrial-scale agriculture system (Achour et al., 2025a).

Methanol is a vital industrial chemical that is used in the manufacturing of plastics (which convert methanol to olefins), adhesives, latex paints, various resins, and sealants, as well as formaldehyde (an intermediate chemical used to make polyurethane and other resins) and acetic acid (an intermediate chemical used to make adhesives) (Chou et al., 2025). Additionally, methanol and its derivatives are added to fuel to enhance combustion characteristics. Since natural gas is used to manufacture 60% of methanol and coal is used to produce the remaining 50%, primarily in China, the current global methanol production rate of 98 MT/y is anticipated to contribute 130 MT/y to global CO_2 emissions (Wakim, 2023). An intermediary substance in the synthesis of methanol is hydrogen. Decarbonising the hydrogen used to produce methanol should be high on the list of uses for low-carbon hydrogen because of the significance of methanol for numerous industrial applications (Achomo et al., 2025).

One strategy to lessen the climatic effects of steel production is to use low-carbon hydrogen in already-existing direct reduction of iron (DRI) applications. Hydrogen is utilized in the DRI-electric arc furnace (DRI-EAF) process to remove oxygen from DR-grade iron ore. This process uses hydrogen from a synthetic gas, primarily H_2+CO.

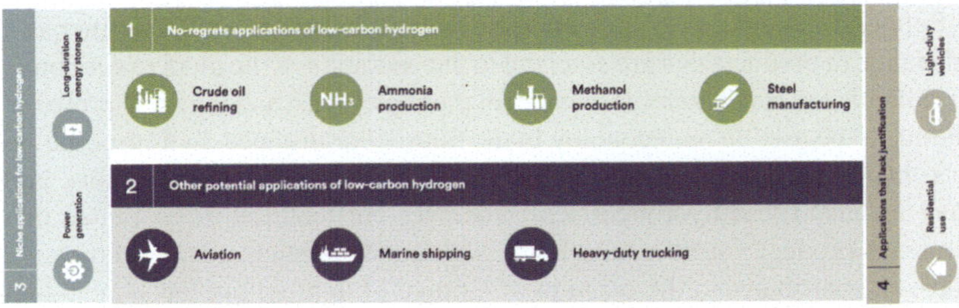

FIGURE 1.2 Hydrogen applications. No-regret applications.

One strategy to lessen the climatic effects of steel production is to use low-carbon hydrogen in already-existing DRI applications. Intriguingly, the Middle East and North Africa (MENA) region exhibits significant decarbonization potential in this area: although producing only 3% of the world's crude steel, it produces 46% (55 MT/y) of all DRI-based steel.

1.3.1 Potential applications of low-carbon hydrogen

Although hydrogen is expected to be a significant player in the global energy transition, it is vital to carefully examine the conditions under which it may help achieve a truly sustainable and cost-effective decarbonization in various regions of the world. There are three important factors to take into account while discussing hydrogen expansion. Efficiency losses occur during the synthesis of green hydrogen, resulting in less usable energy being available after the process. Additional efficiency losses may arise depending on the fuel's use (e.g., in combustion engine automobiles or gas turbines to generate electricity). Electricity from renewable sources is typically more economical and energy-efficient when used directly rather than converted to hydrogen.

Power-to-gas technology, like green hydrogen, can be economical in isolated areas with significant renewable energy potential. However, these areas must also have access to plentiful water supplies for this technology to be sustainable.

Setting priorities is therefore necessary when implementing hydrogen applications (International Renewable Energy Agency, 2022). When hydrogen is the only or most economical choice for decarbonization, future deployment should concentrate on no-regret applications (Fig. 1.2). Similarly, end uses should be avoided when there are better options.

1.4 Transportation

The decarbonization of transportation will be difficult, especially for forms of transportation that are difficult to electrify. Hydrogen will be necessary to produce fuel for these segments. Low-carbon hydrogen will be necessary for the aviation industry's decarbonization, in order to improve biomass-based sustainable aviation fuels (bio-SAF), create jet fuel from hydrogen and collected carbon (synthetic SAF), and maybe power aircraft that run directly on hydrogen fuel. The benefit of these fuels is their compatibility with current engines and infrastructure, which is why they are frequently referred to as "drop-in" fuels. In order to create them, biogenic oils and fats must undergo hydrogen treatment in order to yield straight-chain paraffinic hydrocarbons devoid of sulfur, oxygen, and aromatics (Corcoran et al., 2025). Reducing related life-cycle emissions may be possible by using low-carbon hydrogen in the production of transportation fuels derived from biomass. However, land-use and supply chain constraints on biomass feedstocks will necessitate the development of alternative fuel options, such as synthetic fuels (also known as "e-fuels") made from nonbiogenic feedstocks using a combination of hydrogen, electricity, and CO_2. However, producing synthetic gasoline is currently difficult from a technological and financial standpoint.

Hydrogen's suitability for different means of transportation differs greatly. Although technologies are still in the research and development stage, hydrogen and its derivatives (ammonia and synthetic fuels) are anticipated to have the greatest influence on the decarbonization of long-haul aviation and maritime shipping (Jaramillo & Newman et al., 2022). Green ammonia has drawn special attention for use in maritime ships and has the potential to cut life-cycle emissions by 70%–80% when compared to fuel oil (Achour et al., 2025b). The long lifespan of most vessels and the requirement for additional development of safe storage and handling procedures may hinder the implementation of new systems (Staffell et al., 2019). Smaller short-haul aircraft are more likely to be electrified, while synthetic fuels are anticipated to compete with biofuels for long-haul flights (Jaramillo & Newman et al., 2022). In the short to medium future, hydrotreatment of vegetable oils (HVOs) to produce biofuel is another possible method of replacing fossil fuels because hydrogen may be utilized to make green diesel and jet fuel (Esposito et al., 2022).

Using hydrogen fuel cell electric vehicles in conjunction with battery electric vehicles can help decarbonize the heavy-duty trucking industry. Heavy-duty fuel cell electric vehicles (FCEVs) can travel long distances without making several extra refueling stops, can be refueled in about the same amount of time, and have powertrains that are only marginally heavier than diesel trucks. As a result, when fully loaded, FCEVs can transport almost all of the cargo that diesel trucks can. Heavy-duty FCEVs would provide the same operational flexibility as diesel vehicles and avoid the time penalties associated with charging battery electric vehicles, increasing the percentage of the truck fleet that can be decarbonized if they prove to be a viable alternative to diesel on long-haul routes. However, there are additional considerations that will also affect FCEVs, such as the total cost of ownership and well-to-wheels life-cycle emissions, which can be higher than those of a fully electrified approach. Although hydrogen can be produced off-site, there is still a problem with the cost of trucking it to various filling stations. An alternative method of reducing electric grid loads is to source low-carbon hydrogen from natural gas reforming with CCS and rigorous upstream methane emissions management. However, this approach might result in somewhat smaller emissions savings. The need for transportation to deliver hydrogen to refueling stations may also rise due to the centralized structure of natural gas reforming facilities.

Hydrogen FCEVs, a technology that has advanced via research and development but is still immature for many forms of transportation beyond light-duty vehicles, are essential to land transport applications (Jaramillo & Newman et al., 2022). Battery electric vehicles are anticipated to continue to be the most cost-effective decarbonization solution for short-distance and light-duty vehicles (such as passenger cars and urban buses) for the foreseeable future. Trucks and other heavy-duty, long-distance vehicles will be more affected by hydrogen, as it will compete with electric haulage systems in this market (Jaramillo & Newman et al., 2022). Decarbonization using hydrogen would primarily target long-range and low-frequency networks, such as those for freight in rural or remote locations, while several nations already have highly electrified rail lines (Cabeza et al., 2022). Lastly, there is a high probability that fuel cells will be used to decarbonize additional off-road vehicle types, including tractors, forklifts, and logistics hub equipment (Staffell et al., 2019).

Compared to battery electric vehicles (BEVs), hydrogen FCEVs consume 2.0–2.5 times the amount of energy and cost more per mile to operate. While some consumers may value the longer range and shorter filling time that light-duty fuel cell electric cars currently offer over BEVs, advancements in battery technology will probably make these benefits less significant when it comes to selecting FCEVs.

Low-carbon ammonia is a promising substitute for conventional marine fuel in maritime transportation. However, before widespread use of ammonia can be approved for a possible role in decarbonizing marine transportation on the open seas, the health, safety, and environmental issues related to bunkering, storing, and burning ammonia in marine engine rooms must be carefully investigated at a system level. Methanol is another possible marine fuel that can lower life-cycle carbon emissions. In fact, many contemporary cargo ships are being constructed with dual fuel capabilities to handle methanol in the future and marine diesel oil today. Finding "sustainable" carbon atoms for the methanol synthesis process is necessary to reduce life-cycle emissions since, unlike ammonia, methanol releases carbon dioxide as it burns.

1.5 Power generation

In the energy sector, hydrogen is an attractive alternative to natural gas because it can replace it with no emissions of carbon dioxide. However, there are many technological and infrastructure obstacles in this application that go beyond just running turbines on hydrogen. Geologic storage of hydrogen and specialized transmission and distribution pipelines, for instance, would probably be required to provide the fuel volumes required for power production. The carbon intensity of any hydrogen used should be the main focus due to associated expenses. Hydrogen from electrolysis with clean energy or hydrogen from natural gas with CCS and stringent upstream methane emissions management.

Unfortunately, the principle of round-trip efficiency (RTE), which is the proportion of energy input to a system that can subsequently be recovered for productive purposes, makes using electrolytic hydrogen powered by low-carbon electricity to decarbonize the power system unlikely to be any more appealing. In practical terms, 76% of the electricity used to produce hydrogen is wasted because it cannot be recovered (Achour et al., 2023). In other words, four units of clean electricity will be diverted from further grid decarbonization to deliver one unit of clean electricity in a grid that is not yet fully decarbonized (especially one where full grid decarbonization still faces significant obstacles). This means that three units of clean electricity that could be used to serve other direct electricity end uses will be lost. In a decarbonized power system, hydrogen produced by electrolysis may be helpful as a long-duration energy storage method for grid balancing during periods when demand for renewable energy would otherwise outstrip supply and would have to be reduced. This role is likely to be relevant only in fully decarbonized grids. Nevertheless, an evidence-based analysis will be required to assess all aspects of the power system design, evaluate options for long-term energy storage, and optimize total system cost and decarbonization.

Demand for electricity is predicted to climb as end-use electrification arises, and electrical systems will be impacted by the growth of intermittent renewable energy sources such as wind and solar. Maintaining costs and reliability will require balancing supply and demand, especially when there are significant seasonal demand imbalances, using energy storage capacity or large-scale peak power plants. The integration and expansion of solar and wind-powered electrical systems can be facilitated by hydrogen technology (Clarke et al., 2022).

Electrolysis-based methods for producing hydrogen can utilize surplus renewable energy. When demand peaks, the hydrogen generated can be stored for later conversion to electricity (re-electrification) using either fuel cells or combined-cycle gas turbines reconfigured to run on hydrogen (Staffell et al., 2019).

For flexible power-to-hydrogen (and vice versa) plants and long-term hydrogen storage, the strategic positioning of renewable energy sources, storage facilities, hydrogen, and electricity networks will determine their economic viability. According to (Clarke et al., 2022), the characteristics of the distribution and storage infrastructure that will be constructed will be determined by the distance between the producers and consumers of energy carriers, such as hydrogen, ammonia, and synthetic fuels.

1.6 Residential heating

The major strategies for decarbonizing heating in commercial and residential buildings include district heating, electrification, demand reduction, on-site renewable energy sources, and, to a lesser extent, alternative gaseous fuels like biogas. Although hydrogen could also contribute to the decarbonization of buildings, most analyses predict that this contribution will be little because providing heat from hydrogen would be more expensive than other commercially available methods, such as electric heat pumps (Cabeza et al., 2022). Gas-driven heat pumps, fuel cell combined heat and power (CHP), and hydrogen boilers are examples of low-carbon heating options. Applications are more cost-effective for district energy networks and large-scale commercial or residential complexes (International Energy Agency, 2019). In addition to upgrading major appliances, switching to direct hydrogen consumption in buildings would necessitate broad power-to-hydrogen-to-power operations (such as methanation), which are less efficient, complex, and expensive (Staffell et al., 2019). As a short-term fix, blending hydrogen with natural gas has also been suggested; however, this is ineffective (20% blends by volume reduce emissions by 7%), prolongs the life of fossil fuel assets, and replaces strategies with greater potential for decarbonization (Staffell et al., 2019).

When it comes to home heating options, alternative options are more efficient and environmentally friendly than hydrogen, according to a number of independent studies. Furthermore, hydrogen is commonly used in industrial settings, but its use in residential settings poses serious safety risks due to its high leakage potential (without a visible flame or odor) and its six times greater flammability range than natural gas. Hydrogen is not expected to be the favored decarbonization method for home usage because of these factors.

1.7 Industry

Low-carbon hydrogen is anticipated to have the greatest effect on decarbonization in heavy industrial sectors, where gray hydrogen is a crucial feedstock that cannot be substituted with other chemicals. These industries include methanol production, a common chemical feedstock and solvent in many industries; ammonia production through the Haber-Bosch process, which would help decarbonize fertilizer production and other industrial applications (such as the manufacturing of explosives and plastics); and oil and gas refining, where hydrogen is used to remove impurities (hydrotreating and desulphurization) and upgrade oil fractions (hydrocracking) (International Renewable Energy Agency, 2022). Another industry where path dependence and technology lock-in might impede energy transitions is iron and steel production, where quick adoption of low-carbon hydrogen can be crucial to the sectoral transformation.

DRI, which can be accomplished with hydrogen (and carbon monoxide), can be put into electric arc furnaces (EAFs) to make steel with possibly zero emissions. Although EAFs are now in operation, their market share is only 24% due to the fact that they rely on steel scraps for secondary steel production (International Renewable Energy Agency, 2022). At the moment, blast furnace-basic oxygen furnaces (BF-BOFs) produce 71% of the world's steel, where hydrogen can partially replace coke as a reducing agent (International Renewable Energy Agency, 2022). However, given that it can only reduce process emissions by 21%, this substitution is not a sustainable route to decarbonization (Fan & Friedmann, 2021). In the long run, the decarbonization of the iron and steel industry is incompatible with new plants that use BF-BOF technology.

In many industrial areas where high-temperature heat (>400°C) is needed, such as cement, glass, and ceramics, hydrogen and hydrogen derivatives can take the place of fossil fuels. It is simpler to electrify applications using heat that is between mid- and low-temperature. The viability of using hydrogen for high-temperature heat, however, is debatable because it would necessitate redesigning furnaces and burners, utilizing new materials to stop equipment from corroding and becoming brittle, and resolving safety concerns related to handling hydrogen and monitoring combustion (Staffell et al., 2019). The aforementioned problems are exacerbated by the immaturity of these technologies, unpredictable expenses, and the sluggish system turnover. Hydrogen is still a desirable alternative for large-scale processes that are difficult to electrify, such as cement kilns and steam crackers, as well as for geographically dispersed industrial sectors, even though biofuels and electricity-based heating may be more competitive than hydrogen for high-temperature heating (International Energy Agency, 2020).

1.8 Recent trends affecting the use of hydrogen

The increase in interest in hydrogen as an alternative energy carrier has once again brought it into the spotlight as a key component of meeting international climate targets and preventing the worst impacts of climate change. Investments in hydrogen production and end-use applications are being influenced by several trends. Energy supply diversification is necessary in light of growing costs and worries about energy security. The invasion of Ukraine by Russia has affected the price of petroleum and agricultural items globally and increased inflation. Countries that import fossil fuels on a net basis will probably be the most impacted (Cárdenas et al., 2022). By using its abundant renewable resources to create green hydrogen and improve local energy security, the area may diversify its energy supply. In order to do this, countries must encourage the production and demand for hydrogen while obtaining foreign financial and technical assistance for the implementation of technology that is not easily available in the area.

Today, there is an increased interest in low-carbon technologies and public awareness of climate change. Research demonstrates that concern and awareness of climate change have grown on a worldwide scale (Milfont et al., 2021). If people are motivated to change and receive interventions that assist them in overcoming institutional and market barriers, public opinion can have a significant impact on mitigation efforts by influencing their willingness to adopt and implement new policies and technologies, incur higher financial costs, or change energy-related behavior (e.g., demand-side measures) (Creutzig et al., 2022). Deploying hydrogen-based applications that are motivated or impacted by cultural factors and user preferences requires a shift in public perception (Cvetković et al., 2021).

Several studies show that fossil-fuel-based systems emit large amounts of methane. The IEA claims that national emissions inventories have significantly underestimated global methane emissions from the energy sector (International Energy Agency, 2022b). Methane emissions from fossil fuel value chains are 20%–60% higher than previously thought, according to recent studies (Hmiel et al., 2020). Therefore building more infrastructure for fossil fuels, especially natural gas, which is sometimes referred to as a bridge or transition technology, may cause carbon lock-in and impede efforts to reduce carbon emissions (Kemfert et al., 2022). Therefore investments should be directed to increasing the use of renewable energy sources and utilizing alternative carriers, such as hydrogen, in industries that are difficult to electrify. Furthermore, since many of the expertise and resources needed for its implementation may be transferred from the oil and gas sector, hydrogen can help achieve a fair and gradual energy transition. For instance, the energy transition's overall costs could be reduced by converting existing natural gas infrastructure, such as pipelines, heaters, and turbines, to hydrogen use (Cvetković et al., 2021). Several countries have identified these emission sources and created specialized policy tools to address them.

Green hydrogen production costs are expected to be lower, and investments in the sector will increase. Costs of producing hydrogen by electrolysis are expected to keep declining (Glachant & dos Reis, 2021). The electrolyzer CAPEX and the levelized cost of energy are decreasing, as a result of improved mixtures and less expensive renewable production. Costs will be further reduced by the implementation or development of larger hydrogen projects, improved electrolyzer efficiency, and optimized integrated energy system designs. When these conditions are met, green hydrogen will be able to compete and even outperform gray or blue hydrogen production pathways in locations characterized by high renewable potential and low capital costs. Furthermore, rising gas prices and possible carbon taxes may make blue hydrogen less affordable, casting doubt on policy support for development and implementation. Uncertainties about methane leakage, carbon capture rates, and the lack of progress in reaching suggested targets over time are among other criticisms leveled toward blue hydrogen (Flora & Jaller-Makarewicz, 2022).

1.9 Future hydrogen demand and supply

Expanding the use of hydrogen and hydrogen-based fuels in both new and existing applications is necessary to achieve net-zero emissions by 2050. According to various assessments, the amount of hydrogen produced in 2050 will be between 30 and 100 EJ, or three to almost ten times the current world output. In these cases, nearly all of the hydrogen produced in 2050 will be low-carbon, with electrolysis from renewable sources often accounting for two-thirds of this amount. The remaining portion will come from fossil fuels with CCUS. This would necessitate a storage capacity of 1–3 Gt H_2/year and an electrolyzer installation capacity of 3000–5000 GW. The estimated electrolyzer demand would require between 4800 and 5400 GW of installed renewable capacity to supply green hydrogen. These figures are more than three times the installed capacity of wind and solar photovoltaic (PV) worldwide as of right now (IRENASTAT Online Data Query Tool, 2022). This emphasizes how crucial it is to make concurrent investments and efforts to increase the capacity of renewable energy.

According to more optimistic projections, hydrogen will account for almost 20% of worldwide final energy consumption, while more realistic projections place it at around 3%. The sector with the highest need for hydrogen is frequently the industrial sector, which requires 8–40 EJ of hydrogen and its derivatives. Hydrogen's current applications as a feedstock for chemical production (44%) and as a reagent for DRI in steelmaking (29%) are anticipated to account for the majority of industrial demand (Hydrogen demand in industry, 2020). The timeline of its deployment will primarily depend on policy measures and cost progression. These are applications where hydrogen is thought to be the sole practical decarbonization alternative. It is also anticipated that new applications would be important, especially for heating in the cement sector (6.4%). The industrial sector, which accounts for 6.1% of total energy consumption, will be especially affected by on-site electricity use for hydrogen production. The only application where future hydrogen demand will decline is oil refining, as consumption is expected to decline significantly as climate ambition rises.

Derivatives such as ammonia and e-fuels will probably be produced using about one-third of the entire amount of hydrogen produced. In the transportation sector, where direct electrification is less feasible and high energy content is required, hydrogen and hydrogen-based fuels may replace fossil fuels. This is where the use of derivatives will have the most impact. Since hydrogen has a higher energy content than the majority of modern batteries, this is the case for heavy-duty and long-haul freight transportation. According to the IEA, ammonia could be used as a fuel to meet half of the demand for maritime shipping, while synthetic fuels could account for one-third of the aviation industry's ultimate energy consumption. Due to the greater efficiency and cost-effectiveness of electric cars, FCEVs, despite being a mature technology, are not expected to be widely used for light-duty vehicles. By 2050 they may only account for 9.3% of the worldwide fleet. It is anticipated that by the middle of the century, 31% of heavy-duty trucks would be fueled by hydrogen fuel cells.

As hydrogen's application for energy storage and grid management increases, it may also contribute to meeting 2.4%–5% of future electricity demand. According to the IEA's Net Zero Emissions (NZE) scenario, 1700 TWh of energy will be generated from hydrogen overall by 2050, with a final demand of 12 EJ in the power sector. With estimates of 26 and 7.7 EJ, respectively, BloombergNEF and the Hydrogen Council predict a sizable portion of hydrogen consumption for power generation. Lastly, under all scenarios, hydrogen will contribute less than 3% of the sector's total energy requirement, making it the least useful fuel for residential and commercial structures.

Current project pipelines show that the goal of decarbonization scenarios is not in line with the expected deployment of technologies for hydrogen production and end-use applications. An additional 3.5 GW of installed capacity per year by 2030 would result from the completion of all stated electrolyzer projects by 2022. While this represents a massive increase over current production levels, it would need to be accelerated rapidly to meet the required levels of production. To achieve the scale-up requirements, electrolyzer manufacturing capacity would need to be increased to 130–160 GW/year by 2050.

1.10 Conclusion

In the decarbonization effort to reduce global warming, hydrogen is considered a key player. By 2050 net-zero estimates predict that it will account for between 3% and 20% of total energy usage. Significant expenditures are required to lower production costs and enhance end-use applications in order for hydrogen technologies to achieve this objective. To ensure that hydrogen's future use does not depend on fossil fuels, it is imperative to increase the renewable energy generation capacity.

Although it has occasionally been portrayed as a crucial element for decarbonization, low-carbon hydrogen may be a crucial tool for lowering GHG emissions from specific industries due to its physical and financial constraints. Furthermore, because hydrogen production is expensive and energy-intensive, it is not a primary energy source and does not contribute to energy security. In fact, the energy requirements are what currently limit the supply of low-carbon hydrogen, even more so than the availability of electrolysis equipment or natural gas reformers. In order to manufacture a wide range of essential chemicals, as a feedstock for fertilizers needed to feed the world's population, and as fuel for important sectors of the transportation industry, hydrogen is and will remain a crucial chemical that needs to be produced on an industrial scale. Where and to what extent switching to low-carbon hydrogen can be an energy-efficient and cost-effective strategy for reducing GHG emissions, taking into account the availability of alternative decarbonization pathways, is currently the main topic of discussion, rather than whether it will be technically possible to greatly expand low-carbon hydrogen production and use. While determining how to prioritize clean hydrogen and when establishing policies pertaining to the production and use of low-carbon hydrogen, policymakers will need to be careful. This will be essential to determining the right path for quick and efficient decarbonization of the entire economy.

The iron and steel industry's high emissions are drawing attention from all over the world. Although it is currently not cost-competitive with existing manufacturing methods, the technology to completely decarbonize steel production with hydrogen is anticipated to alter in the near future. The first step in decarbonizing the industry and preventing dangerous carbon lock-in is to replace EAF with BF-BOF units. The industry can advance by optimizing scrap recycling and becoming ready for the DRI reduction process.

Europe will soon see the first hydrogen-based steel factories go into production, offering technological know-how. As consumer demand for low-carbon products rises, early adopters can also gain advantages in international trade. To reduce the risk of this endeavor, specific actions such as carbon contracts for difference or strong certification programs could serve as a foundation for the development of green steel markets.

At the moment, the majority of green hydrogen capacity expansions are directly connected to their industrial offtakers. Investment in independent electrolyzer projects will be required to meet the growing demand. The cost of electrolyzers is rapidly decreasing, and replacing the present need for hydrogen with green hydrogen, along with the anticipated expansion of additional applications, will necessitate much more production capacity than what is presently planned.

Another intriguing prospect is the chemical sector, since green ammonia is required to produce low-carbon fertilizers, and methanol and ammonia are new clean fuels. Given that hydrogen is already utilized in the production process and that replacing fossil fuel-based hydrogen can accelerate the region's low-carbon hydrogen production, this industry presents a short-term opportunity.

As a cost-effective alternative to diesel, hydrogen can also be used to power heavy-duty mining vehicles. Due to their large power and range requirements, it is the most attractive application for land transport. With local hydrogen production, fuel purchase and transportation costs can be avoided, and additional ventilation savings can strengthen the case for Electric Fuel Cell Vehicles (EFCV) in the industry. Creating simultaneous projects for on-site renewable electricity and hydrogen generation capacity offers the chance to completely bypass fuel logistics expenditures. With a carbon price of about \$60 USD/t CO_2, this application is cost-competitive, making policy interventions essential to its effective deployment.

References

Achomo, M. A., Muthukumar, P., & Peela, N. R. (2025). , Thermodynamic, energy and exergy analysis of steam reforming of methanol for hydrogen production. *International Journal of Hydrogen Energy, 149*, 150040. doi:10.1016/j.ijhydene.2025.150040.

Achour, Y., Berrada, A., Arechkik, A., & El Mrabet, R. (2023). Techno-economic assessment of hydrogen production from three different solar photovoltaic technologies. *International Journal of Hydrogen Energy, 48*(83), 32261–32276. doi:10.1016/j.ijhydene.2023.05.017.

Achour, Y., El Mokrini, A., & Berrada, A. (2025a). Techno-economic assessment of large-scale power-to-ammonia plant: Integrating hybrid energy storage system (gravity and battery) and advanced day-ahead forecasting algorithms. *International Journal of Hydrogen Energy, 136*, 202–216. doi:10.1016/j.ijhydene.2025.04.102.

Achour, Y., El Mokrini, A., El Mrabet, R., & Berrada, A. (2025b). Optimizing power-to-ammonia plant: Sizing, operation, and production forecasting using deep learning approach. *Renewable Energy, 240*, 122234. doi:10.1016/j.renene.2024.122234.

Al-Qahtani, A., Parkinson, B., Hellgardt, K., Shah, N., & Guillen-Gosalbez, G. (2021). Uncovering the true cost of hydrogen production routes using life cycle monetisation. *Applied Energy, 281*, 115958. doi:10.1016/j.apenergy.2020.115958.

Berrada, A., Sanjari, M. J., & El Mrabet, R. (2025). Techno-economic assessment of hydrogen production: Comparative analysis of electrolyser technologies in a hybrid PV/wind system. *International Journal of Hydrogen Energy, 141*, 193–211. https://www.sciencedirect.com/science/journal/03603199. doi:10.1016/j.ijhydene.2025.05.384.

Cabeza, L. F., Bai, Q., & Bertoldi (2022). Chapter 9 - Buildings. IPCC, 2022. *Climate Change 2022: Mitigation of climate change. Contribution of Working Group III to the Sixth Assessment Report of the Intergovernmental Panel on Climate Change*. doi:10.1017/9781009157926.011.

Cao, Y., Bai, Y., & Du, J. (2025). Numerical simulation of hydrogen production from co-gasification of biomass and refuse-derived fuel using ASPEN Plus. *International Journal of Hydrogen Energy, 165*, 150928.

Cárdenas, M., Jobet, J. C., Palacios, L., Rivera Rivota, D., (2022). Implications of the Russian war in Ukraine on Latin America's energy sector. Center on Global Energy Policy, https://www.energypolicy.columbia.edu/research/global-energy-dialogue/implications-russian-war-ukraine-latin-america-s-ener-gy-sector.

Chou, Y.-C., Gómez-Hernández, A., Ochi, K. K., Agba, C., Horton, M., & Erickson, P. A. (2025). Analysis of carbon monoxide and hydrogen yield in steam methanol reformation with passive flow disturbance. *International Journal of Hydrogen Energy, 157*, 150281. doi:10.1016/j.ijhydene.2025.150281.

Clarke, Wei, Y.-M., & Navarro, A. (2022). Chapter 6—Energy systems. *IPCC, 2022: Climate change 2022: Mitigation of climate change*. doi:10.1017/9781009157926.008.

Corcoran, C. K., Davíðsdóttir, B., Cook, D., & Jóhannsdóttir, L. (2025). Drivers and barriers for the development and deployment of hydrogen-based fuels for the maritime sector: A case study from Iceland. *International Journal of Hydrogen Energy, 162*, 150367. doi:10.1016/j.ijhydene.2025.150367.

Creutzig, F., Roy, J., & Devine-Wright (2022). Chapter 5—Demand, services and social aspects of mitigation. IPCC, 2022. *Climate change 2022: Mitigation of climate change. Contribution of Working Group III to the Sixth Assessment Report of the Intergovernmental Panel on Climate Change*. doi:10.1017/9781009157926.007.

Cvetković, S. M., Radoičić, T. K. đć, Kijevčanin, M., & Novaković, J. G. ć (2021). Life Cycle Energy Assessment of biohydrogen production via biogas steam reforming: Case study of biogas plant on a farm in Serbia. *International Journal of Hydrogen Energy, 46*(27), 14130–14137. doi:10.1016/j.ijhydene.2021.01.181.

Davis, S. J., Lewis, N. S., Shaner, M., Aggarwal, S., Arent, D., Azevedo, I. L., Benson, S. M., Bradley, T., Brouwer, J., Chiang, Y. M., Clack, C. T. M., Cohen, A., Doig, S., Edmonds, J., Fennell, P., Field, C. B., Hannegan, B., Hodge, B. M., Hoffert, M. I., … Caldeira, K. (2018). Net-zero emissions energy systems. *Science, 360*(6396). http://science.sciencemag.org/content/360/6396/eaas9793/tab-pdf. doi:10.1126/science.aas9793.

Fan, Z., & Friedmann, S. J. (2021). Low-carbon production of iron and steel: Technology options, economic assessment, and policy. *Joule, 5*(4), 829–862. https://www.journals.elsevier.com/joule. doi:10.1016/j.joule.2021.02.018.

Esposito, G., Robinson, B., & Savage, J. (2022). Decarbonising heavy duty vehicles and machinery (A. Eastlake, Ed.). Zemo Partnership. https://www.zemo.org.uk/assets/reports/Decarbonising_Heavy_Duty_Vehicles_and_Machinery_Zemo_Nov2022.pdf.

Flora, A., Jaller-Makarewicz, A. M., Russia sanctions and gas price crisis reveal danger of investing in "blue" hydrogen. Institute for Energy Economics and Financial Analysis (2022), https://ieefa.org/resources/russia-sanctions-and-gas-price-crisis-reveal-danger-investing-blue-hydrogen.

Glachant, J.-M., & dos Reis, P. C. (2021). A Snapshot of clean hydrogen costs in 2030 and 2050. *Florence School of Regulation [Preprint]*, (12). https://fsr.eui.eu/publications/?han-dle=1814/70971.

Griffiths, S., Sovacool, B. K., Kim, J., Bazilian, M., & Uratani, J. M. (2021). Industrial decarbonization via hydrogen: A critical and systematic review of developments, socio-technical systems and policy options. *Energy Research & Social Science, 80*, 102208. doi:10.1016/j.erss.2021.102208.

Hmiel, B., Petrenko, V. V., Dyonisius, M. N., Buizert, C., Smith, A. M., Place, P. F., Harth, C., Beaudette, R., Hua, Q., Yang, B., Vimont, I., Michel, S. E., Severinghaus, J. P., Etheridge, D., Bromley, T., Schmitt, J., Faïn, X., Weiss, R. F., & Dlugokencky, E. (2020). Preindustrial 14CH4 indicates greater anthropogenic fossil CH4 emissions. *Nature, 578*(7795), 409–412. doi:10.1038/s41586-020-1991-8.

Hydrogen Council, Hydrogen de-carbonization pathways. Hydrogen Council (2021), https://hydrogenchttps/hydrogencouncil.com/en/hy-drogen-decarbonization-pathways/.

Hydrogen demand in industry, 2020, Charts – data & statistics. International Energy Agency. https://www.iea.org/data-and-statistics/charts/hydro-gen-demand-in-industry-2020.

Hydrogen Forecast to 2050. ETN (2022), https://www.dnv.com/focus-areas/hydrogen/forecast-to-2050.html.

Intergovernmental Panel on Climate Change (IPCC), Technical summary (pp. 51–148). Cambridge University Press (CUP), (2023). doi:10.1017/9781009157926.002.

International Energy Agency, The future of hydrogen: Seizing today's opportunities. International Energy Agency (2019), https://iea.blob.core.windows.net/assets/9e3a3493-b9a6-4b7d-b499-7ca48e357561/The_Future_of_Hydrogen.pdf.

International Energy Agency, Levelised costs of hydrogen production by technology in Latin America, 2020 and mid-term potential, charts – data & statistics. International Energy Agency (2020), https://www.iea.org/da-ta-and-statistics/charts/levelised-costs-of-hydro-gen-production-by-technology-in-latin-ameri-ca-2020-and-mid-term-potential.

International Energy Agency, Global hydrogen review 2022. International Energy Agency. (2022a), https://www.iea.org/reports/global-hydro-gen-review-2022.

International Energy Agency, World energy outlook 2022, International Energy Agency, 2022b. https://www.iea.org/.

International Renewable Energy Agency, Green hydrogen for industry: A guide to policy making. International Renewable Energy Agency (2022), https://www.irena.org/pub-lications/2022/Mar/Green-Hydrogen-for-Industry.

IRENASTAT Online Data Query Tool, Installed renewable electricity capacity (MW) by region/country/area, technology, and year, IRENASTAT Online Data Query Tool. (2022), https://pxweb.irena.org/pxweb/en/IRENASTAT.

Iribarren, D., Susmozas, A., Petrakopoulou, F., & Dufour, J. (2014). Environmental and exergetic evaluation of hydrogen production via lignocellulosic biomass gasification. *Journal of Cleaner Production, 69*, 165–175. doi:10.1016/j.jclepro.2014.01.068.

Jaramillo, K. , R., & Newman, P. (2022). Chapter 10—Transport. IPCC, 2022. *Climate change 2022: Mitigation of climate change. Contribution of Working Group III to the Sixth Assessment Report of the Intergovernmental Panel on Climate Change*. doi:10.1017/9781009157926.012.

Kemfert, C., Präger, F., Braunger, I., Hoffart, F. M., & Brauers, H. (2022). The expansion of natural gas infrastructure puts energy transitions at risk. *Nature Energy, 7*(7), 582–587. www.nature.com/nenergy/. doi:10.1038/s41560-022-01060-3.

Milfont, T. L., Zubielevitch, E., Milojev, P., & Sibley, C. G. (2021). Ten-year panel data confirm generation gap but climate beliefs increase at similar rates across ages. *Nature Communications, 12*(1). http://www.nature.com/ncomms/index.html. doi:10.1038/s41467-021-24245-y.

Nikolaidis, P., & Poullikkas, A. (2017). A comparative overview of hydrogen production processes. *Renewable and Sustainable Energy Reviews, 67*, 597–611. doi:10.1016/j.rser.2016.09.044.

Nilsson, A., De Vivero, G., Legarreta, P. L., Day, T., (2021). Green hydrogen applications in Mongolia - technology potential and policy options. NewClimate Institute doi:10.13140/RG.2.2.15300.50561.

Osman, A. I., Mehta, N., Elgarahy, A. M., Hefny, M., Al-Hinai, A., Al-Muhtaseb, A. H., & Rooney, D. W. (2022). Hydrogen production, storage, utilisation and environmental impacts: A review. *Environmental Chemistry Letters, 20*(1), 153–188. https://www.springer.com/journal/10311. doi:10.1007/s10311-021-01322-8.

Parkinson, B., Tabatabaei, M., Upham, D. C., Ballinger, B., Greig, C., Smart, S., & McFarland, E. (2018). Hydrogen production using methane: Techno-economics of decarbonizing fuels and chemicals. *International Journal of Hydrogen Energy, 43*(5), 2540–2555. doi:10.1016/J.IJHYDENE.2017.12.081.

Roy, R., Antonini, G., Hayibo, K. S., Rahman, M. M., Khan, S., Tian, W., Boutilier, M. S. H., Zhang, W., Zheng, Y., Bassi, A., & Pearce, J. M. (2025). Comparative techno-environmental analysis of grey, blue, green/yellow and pale-blue hydrogen production. *International Journal of Hydrogen Energy, 116*, 200–210. doi:10.1016/j.ijhydene.2025.03.104.

Staffell, I., Scamman, D., Velazquez Abad, A., Balcombe, P., Dodds, P. E., Ekins, P., Shah, N., & Ward, K. R. (2019). The role of hydrogen and fuel cells in the global energy system. *Energy & Environmental Science, 12*(2), 463–491. doi:10.1039/c8ee01157e.

Thomas, H., (2018). Policy briefing: Options for producing low-carbon hydrogen at scale. IGEM House, https://royalsociety.org/⟨/media/policy/projects/hydrogen-pro-duction/energy-briefing-green-hydrogen.pdf.

Vidas, L., & Castro, R. (2021). Recent developments on hydrogen production technologies: State-of-the-art review with a focus on green-electrolysis. *Applied Sciences, 11*(23), 11363. doi:10.3390/app112311363.

Wakim, G., (2023). Hydrogen for decarbonization: A realistic assessment. Clean Air Task Force. https://www.catf.us/resource/hydrogen-decarbonization-realisticassessment/.

World Energy Council, (2021). National hydrogen strategies, Working Paper. World Energy Council, https://www.worldenergy.org/assets/downloads/Working_Paper_-_National_Hydrogen_Strate-gies_-_September_2021.pdf.

Zhu, L., Pan, X., Alahakoon, Y., Zhou, X., Jia, P., Chaerusani, V., Kang, D.-J., Wang, C., Xu, G., Abudula, A., & Guan, G. (2025). Optimizing hydrogen-rich gas production from miscanthus biochar through steam gasification in a batch-feeding fluidized-bed reactor. *International Journal of Hydrogen Energy, 166*, 150889. doi:10.1016/j.ijhydene.2025.150889.

Chapter 2

Decarbonizing the building and residential sector

Ayesha Kaleem and Atiq Zaman
School of Design and the Built Environment, Curtin University, Bentley, WA, Australia

2.1 Introduction

The urgent challenge of climate change has driven global efforts to reduce greenhouse gas (GHG) emissions, with the Paris Agreement serving as a landmark commitment to limit global warming to well below 2°C, ideally 1.5°C, above preindustrial levels (Viñuales, 2016). This international accord emphasizes the need for rapid decarbonization by transitioning away from fossil fuels and scaling up renewable energy sources. However, despite these commitments, fossil fuels remain deeply embedded in the global energy landscape. Over the past half-century, the world's increasing reliance on fossil fuels to meet rising energy demands has become a significant vulnerability for national economies, accompanied by severe environmental consequences. The continuous rise in carbon dioxide (CO_2) levels has led to a sustained increase in global surface temperatures, resulting in widespread disruptions to weather patterns, rising sea levels, ecosystem collapse, and a growing frequency of extreme weather events (Bindoff, 2019; Venegas et al., 2023).

As GHG emissions originate from diverse sectors, they are driven by different activities and socioeconomic processes. The pie chart in Fig. 2.1 shows the share of global emissions by sector (Emissions Gap Report 2024, 2024). The power sector is the largest emitter at 26.3%, mainly due to fossil fuel-based electricity generation. Transport follows at 15.2%, while industry and agriculture each contribute 11.1%. Fuel production accounts for 10.1%, and industrial processes like cement manufacturing add 9.1%. Land use changes and forestry contribute 7%, the building sector accounts for 6.1%, and waste management represents 4%. These sectors vary in their emissions sources, from energy production to agricultural practices and waste disposal.

The energy sector, responsible for nearly 75% of global GHG emissions, is both the largest contributor to climate change and the key to mitigating its impacts (Global Energy Review, 2022). Transforming this sector is essential to

Decarbonizing through Sustainable Hydrogen and its Derivatives.
DOI: https://doi.org/10.1016/B978-0-443-27632-3.00004-9

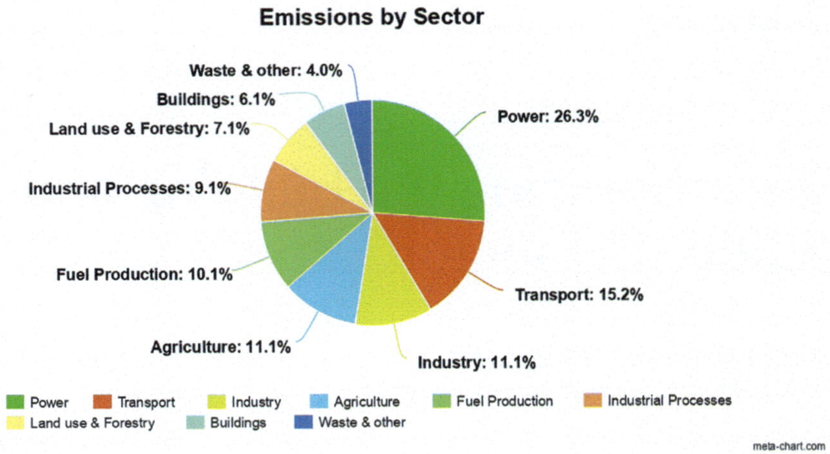

FIGURE 2.1 The pie chart visually represents the percentage contribution of different sectors to global emissions. The share of global emissions by sector.

preventing the most severe consequences of climate change, widely considered one of humanity's greatest challenges (International Energy Agency, 2021). However, reducing emissions from industries, buildings, and transport remains particularly challenging due to their reliance on energy-intensive processes and existing infrastructure. The integration of renewable energy in these sectors can help reduce global temperatures by lowering GHG emissions and decreasing reliance on fossil fuels. By prioritizing renewable energy expansion in the residential sector, fossil fuel-based systems can be replaced with sustainable alternatives, allowing renewable energy to not only support decarbonization efforts but also strengthen energy security while promoting long-term environmental and economic sustainability. This energy transition is driven by the urgent need to reduce carbon emissions and mitigate climate change.

The adoption of solar photovoltaic (PV) technology in the residential and building sector has expanded rapidly in recent years, providing a cost-effective and environmentally friendly alternative for electricity generation. Many residential users are now incorporating solar panels along with battery storage systems and electric vehicles to manage their energy consumption better and reduce dependence on the central power grid (Kaleem et al., 2024). However, this shift toward decentralized energy generation places added pressure on the grid, particularly during periods of low solar output, such as in winter when sunlight is scarce. A possible way to address this challenge is by utilizing solar energy to produce hydrogen, which is especially suitable for localized distribution. Green hydrogen not only generates minimal GHG emissions during operation but also aligns well with existing energy needs.

The power-to-hydrogen-to-power (P2H2P) process is one approach that enables the storage of excess renewable electricity through hydrogen production.

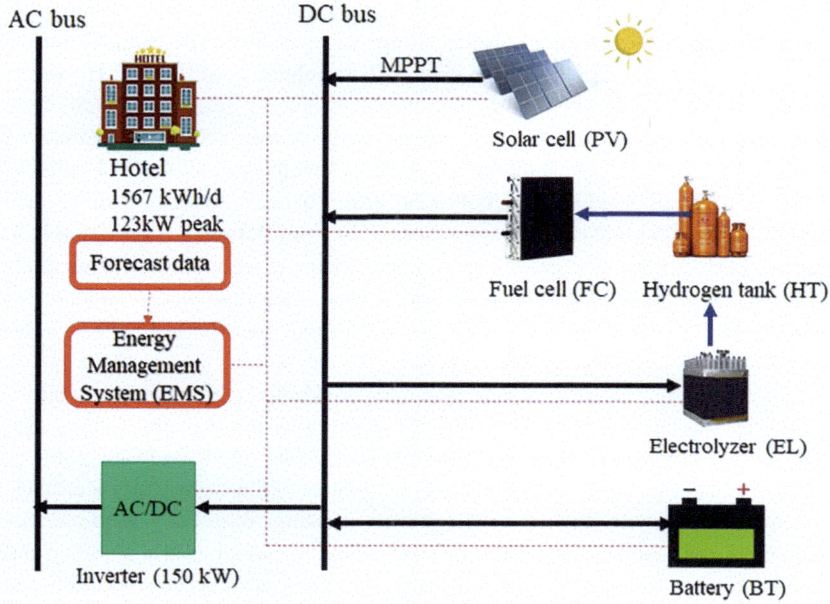

FIGURE 2.2 P2H2P-based stand-alone microgrid integrating PV, battery, electrolyzer, hydrogen storage, and fuel cell for optimized energy management in a hotel. *PV*, Photovoltaic; *P2H2P*, power-to-hydrogen-to-power. (*Reproduced from Xie et al. (2021), licensed under CC BY-NC-ND 4.0.*).

This stored hydrogen can later be utilized to generate electricity during periods of insufficient renewable output or increased demand, helping to maintain a stable and reliable energy system. An example of a P2H2P-based stand-alone microgrid is shown in Fig. 2.2. In this system, the electricity produced by the PV array is supplied as direct current (DC) power, which can either be consumed immediately or stored in a battery (BT) to meet short-term energy requirements. Any surplus electricity is directed to an electrolyzer (EL), which produces hydrogen stored in a dedicated hydrogen tank (HT). When the combined output from the PV system and the battery is unable to meet demand, the stored hydrogen is converted back into electricity using a fuel cell (FC). An inverter (AC/DC) ensures that the generated electricity is compatible with alternating current (AC) loads, such as those found in buildings or facilities. The overall system is regulated by an energy management system (EMS), which optimizes energy generation, storage, and distribution based on predicted consumption patterns to ensure efficient and continuous operation.

Beyond grid applications, green hydrogen also offers alternative applications for powering remote locations. Many rural communities and islands lack direct access to electricity networks, making grid expansion costly and impractical. Consequently, they often rely on diesel generators, which come with high fuel costs and significant carbon emissions. In such cases, hydrogen-based microgrids present a cleaner and more sustainable off-grid energy solution, providing both environmental benefits and energy independence (Barzola-Monteses & Espinoza-Andaluz, 2019; Groppi et al., 2018; Schöne et al., 2022). However, ensuring economic viability and operational efficiency remains a challenge. Since off-grid systems cannot rely on the main power network, minimizing the wear and tear of key components, such as electrolyzers and fuel cells, is crucial. A well-balanced system that incorporates battery storage for short-term energy buffering, alongside hydrogen storage for long-term reliability, can help optimize performance and extend the lifespan of these technologies. This dual-storage strategy allows for efficient energy management by leveraging batteries for immediate demand and hydrogen storage for extended periods of low solar generation.

This chapter explores the role of green hydrogen in decarbonizing the building and residential sector, focusing on its integration with renewable energy systems. As homeowners increasingly adopt solar PV and other renewables due to cost-effectiveness, energy independence, resilience, and government incentives, the demand for decentralized energy solutions is rising. While this trend reduces reliance on fossil fuels, it also introduces challenges related to grid stability and surplus energy management. This chapter examines how green hydrogen can serve as a viable solution by utilizing excess renewable energy for storage and later use in households, thereby supporting grid stability while advancing the transition to a low-carbon built environment.

2.2 Surplus renewable energy and grid stability

New capacity installations in the global renewable electricity sector are predominantly driven by wind and solar PV technologies. In 2023 global solar PV capacity reached **1624 gigawatts (GW)** with **447 GW** of new installations, while wind power capacity exceeded **1 terawatt (TW),** adding over **100 GW** compared to the previous year (Clean Energy Council, 2024; Statista. (n.d.), 2024). Despite this remarkable growth, the inherent variability of renewable energy sources presents significant challenges, particularly when peak electricity generation does not align with peak demand. This misalignment often leads to periods of surplus renewable energy generation, where electricity production exceeds grid demand. This phenomenon, referred to as the "low-load situation," strains grid infrastructure and complicates grid stability. A well-known example is the **Duck Curve challenge**, where midday peaks in solar generation are followed by steep demand ramps as solar output declines in the evening. Fig. 2.3 illustrates the "Duck Curve" pattern observed in the South West Interconnected System (SWIS), highlighting daily electricity demand variations influenced by solar energy generation (Synergy, 2021). The lines represent electricity demand throughout the day, forming the characteristic "Duck Curve" shape. The dip in the middle of the day occurs when solar energy production is at its highest,

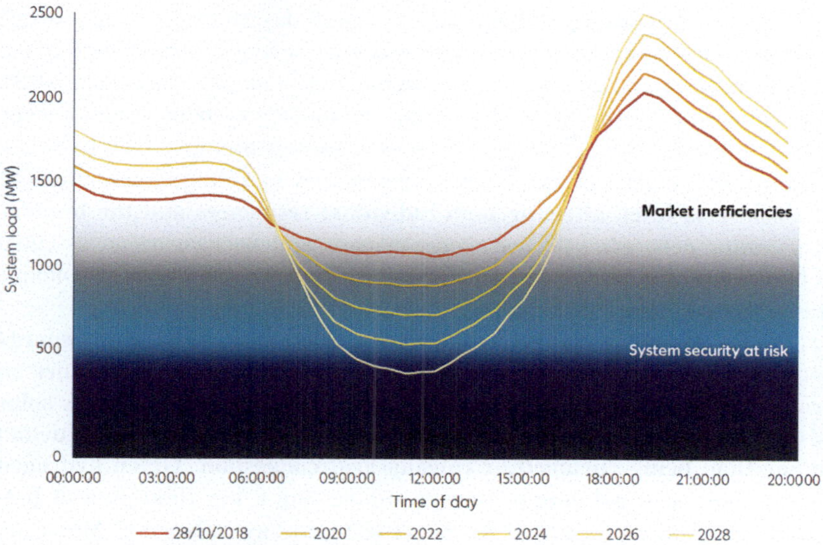

FIGURE 2.3 The Duck Curve in SWIS: Impact of solar energy on daily electricity demand. *SWIS*, South West Interconnected System.

reducing the need for conventional power generation. As the sun sets and solar generation declines, demand for electricity from the grid increases rapidly, resulting in a steep rise in the curve. This fluctuation poses challenges for grid stability, requiring careful energy management and storage solutions.

To manage these fluctuations, grid operators often curtail excess renewable energy, resulting in inefficiencies, economic losses, and lower financial returns for renewable energy projects. This not only discourages future investments but also increases the cost of integrating renewables into the grid. These challenges underscore the **urgent need for advanced energy storage and grid management solutions** to maximize the utilization of renewable energy while maintaining grid stability.

2.3 Hybrid renewable energy systems for grid stability

In addition to stand-alone storage solutions, **hybrid renewable energy systems (HRES)** are being developed to integrate multiple generation and storage technologies, ensuring a **more balanced and resilient energy supply for residential and building sectors**. These systems enhance household **energy reliability, improve efficiency, and reduce dependence on fossil fuels** by combining renewable sources such as **rooftop solar PV, small-scale wind, and battery storage** with energy management technologies (Baruah et al., 2021; Khan et al., 2022).

However, **integrating HRES into residential settings** presents several challenges, including the variability of renewable energy sources, high initial investment costs, and the complexity of system integration (Basnet et al., 2023). **For homeowners adopting rooftop solar and battery systems, excess energy production can create challenges related to storage and grid feed-in limits. Uncontrolled export of surplus power can lead to voltage fluctuations at the local grid level, affecting both individual residences and the broader distribution network.** Effective energy management mechanisms are essential to **balance power generation and demand**, ensuring stability and economic feasibility (Al Busaidi et al., 2014).

By using **batteries and green hydrogen storage within residential energy systems,** HRES can effectively manage surplus energy**, reducing reliance on grid imports and stabilizing local power distribution. When excess solar power is generated in homes, it can be directed toward hydrogen production rather than being curtailed or causing grid congestion. Green hydrogen also enables seasonal energy storage, addressing a key limitation of batteries, which are best suited for short-term energy balancing.** Moreover, **the integration of hydrogen with household microgrids can enable energy independence, particularly in remote or off-grid communities.** By combining technologies such as **solar PV, hydrogen storage, and smart EMSs, homeowners and building operators can create self-sufficient, low-carbon energy solutions** (Mohammed et al., 2014).

In addition to stabilizing residential energy systems, green hydrogen provides environmental benefits by replacing fossil fuel-based backup power with hydrogen fuel cells, contributing to decarbonized built environments. Furthermore, green hydrogen enhances household energy security by allowing residents to generate and store their own energy, reducing dependence on grid power and minimizing the impact of power outages.

2.4 The case for green hydrogen in residential settings

The unpredictable and fluctuating output of solar panels installed in the residential and building sector makes it difficult to maintain a continuous balance between electricity supply and demand. Traditionally, power generation and consumption cooccur to ensure grid stability. However, since solar and wind energy do not generate electricity on demand, they introduce volatility that can cause substantial voltage flickers, frequency deviations, reduced minimum fault currents, and increased harmonic distortions (Alsayegh et al., 2008). To address these challenges and enhance grid flexibility, a range of solutions can be implemented. These include advanced energy storage systems to store surplus electricity, demand-side management to optimize consumption patterns, grid network enhancements, improved control mechanisms for variable renewable energy sources, and modifications to system operation protocols (Ecosys, 2014).

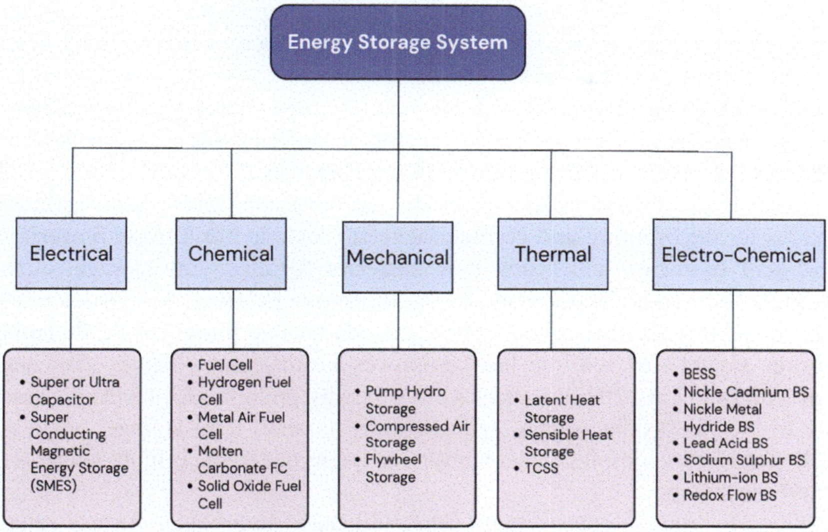

FIGURE 2.4 Classification of energy storage systems based on mechanical, chemical, electrical, electrochemical, and thermal storage technologies.

By integrating these measures, power systems can better accommodate the variability of renewables while ensuring a stable and reliable energy supply.

Energy storage systems are essential for supporting both small and large-scale use of renewable energy by helping to manage peak electricity demand, maintain voltage stability, and enhance power quality (Rana et al., 2023; Shafiullah et al., 2013). By compensating for the natural fluctuations in renewable energy output, storage solutions help maintain a consistent and reliable power supply. For effective energy management, it is important to have large-capacity storage options that can capture surplus electricity during periods of high production and release it when demand rises. Selecting the appropriate type of storage technology is crucial to improving grid stability and addressing the variability of renewable energy sources.

Various storage methods are available, each designed to support efficient power system operation based on specific energy needs and grid conditions. Fig. 2.4 categorizes energy storage systems (ESS) into five primary types, each serving different energy management needs (Rana et al., 2023). The first is mechanical storage, which relies on methods such as hydroelectric pumping, compressed air systems, and flywheels to physically hold and release energy. Chemical storage, on the other hand, uses processes such as FCs to store energy in chemical compounds, making this method especially useful in hydrogen-related applications. Electrical storage solutions, including devices such as supercapacitors and magnetic systems, are well-suited for delivering quick bursts of power during peak demand. Electrochemical storage is widely recognized for its application in various battery systems, including lithium-ion, lead-acid, and flow batteries, which are commonly used for storing renewable energy and supporting grid operations. Lastly, thermal storage focuses on capturing heat, whether through sensible heat, latent heat, or thermochemical processes, to be used later in industrial settings or solar-based energy systems. Altogether, these storage methods play a vital role in creating a more flexible and reliable grid while maximizing the use of renewable energy sources.

Among all of these storage methods, the most applicable energy storage options for the **building and residential sector** include **lithium-ion batteries, lead-acid batteries**, and **redox flow batteries** for electricity storage, often integrated with solar PV (Go et al., 2023). **Hydrogen FCs** (e.g., solid oxide FCs) such as solid oxide fuel cells (SOFCs) provide backup power, while **thermal storage** (latent and sensible heat) improves heating and cooling efficiency (Ali Yildirim et al., 2023). Batteries are generally preferred for residential use due to their cost-effectiveness and efficiency in short-term storage, whereas hydrogen storage, despite higher conversion losses and costs, is more viable for long-term storage.

Green hydrogen storage systems can be utilized to decarbonize the residential sector (Maestre et al., 2022), yet it faces significant barriers that hinder their widespread implementation. High upfront costs for electric appliances and home retrofits, such as upgrading electrical panels or rewiring, pose a financial challenge for many households (Economics, 2022). Additionally, the existing electricity grid infrastructure in some regions is not equipped to handle the increased demand associated with extensive electrification (Blonsky et al., 2019; Elmallah et al., 2022). This limitation raises concerns about grid reliability and necessitates costly upgrades. Cultural preferences also play a role, particularly in the case of cooking, where many households favor gas due to its perceived precision and familiarity (Gordon et al., 2023; Sandoval et al., 2025). In extreme climates, the efficiency of electric heat pumps may decline, reducing their appeal as a dependable solution. Moreover, electricity prices are often perceived as higher than natural gas, deterring some consumers from switching. Other barriers include the "split incentive" in rental properties, where landlords are reluctant to invest in electrification because tenants pay the utility bills, and inconsistent policy support that fails to provide adequate financial incentives for homeowners to transition (Melvin, 2018).

Given these challenges, green hydrogen emerges as a complementary solution that can address many of the limitations of direct electrification. Unlike full electrification, green hydrogen can be integrated into existing gas infrastructure with minimal retrofitting (Ammermann, Hoff, Atanasiu, Tisler, & Kaufmann, 2015), making it a practical option for households that rely on natural gas for heating, cooking, and hot water (Quarton & Samsatli, 2018). Hydrogen-ready appliances, such as stoves, boilers, and water heaters, can provide a low-carbon alternative for households that are hesitant to adopt fully electric systems. Its versatility and compatibility with existing systems reduce the need for significant behavioral changes or costly retrofits, addressing cultural and financial barriers simultaneously (Schiro et al., 2020).

Beyond its immediate benefits, green hydrogen also enhances the resilience of energy systems in both urban and rural communities. As energy demand fluctuates seasonally, long-term and large-capacity storage solutions become essential. Community-based green hydrogen systems offer greater flexibility than traditional battery storage by efficiently coordinating multiple storage resources and addressing uncertainty, particularly during extreme energy demand events. These systems combine the high efficiency of battery storage with the high energy density and low leakage rate of hydrogen storage, ensuring a stable energy supply (Wu & Zhong, 2025).

From a technical perspective, the development of community green hydrogen systems must prioritize safety and ease of deployment. Hydrogen's potential risks, such as leakage and spontaneous ignition, necessitate strict safety measures to ensure secure integration. Additionally, designing compact and easily installable hydrogen equipment is essential for minimizing disruptions in urban environments while enhancing portability and accessibility. By addressing both short-term electrification challenges and long-term energy security, green hydrogen offers a scalable and sustainable solution to decarbonizing the residential sector.

2.5 Decarbonizing building materials and reducing operational emissions

The **construction and building sector** is one of the major sources of global GHG emissions, contributing around **39% of total emissions** (Adams, Richardson, Drinkwater, & Gamboa, 2019). Despite this significant impact, it has received **comparatively less funding** for climate-focused initiatives than other industries. Efforts to reduce emissions have primarily centered on **operational carbon**, which includes energy use for heating, cooling, and lighting (UNEP, 2023). These emissions are projected to decline over time. However, **embodied carbon** emissions, which are generated during the production, transportation, and installation of materials like **cement, steel, and aluminum**, have not seen the same level of progress in mitigation strategies.

Achieving **net-zero embodied carbon** in building materials is possible by advancing **technologies** that reduce emissions in material production. Transitioning from traditional materials to **low-carbon alternatives** such as **green steel, sustainable cement, and eco-friendly bricks** can significantly cut emissions. Decarbonizing the built environment requires a dual approach: reducing emissions from the production of materials and minimizing operational emissions throughout a building's lifecycle.

Steel is one of the most widely used materials in the world, contributing 11% of total CO_2 emissions due to its carbon-intensive production process, which heavily depends on coal-based blast furnaces (Geschwindt, 2024). A more sustainable alternative is green hydrogen, which can replace coking coal in direct reduced iron (DRI) processes, resulting in steel production with water as a byproduct instead of CO_2. Initiatives like HYBRIT in Sweden and H_2 Green Steel have already demonstrated the viability of hydrogen-based steelmaking (Stegra, 2025).

Similarly, cement production is responsible for approximately **7% of global CO_2 emissions** (Adams, Richardson, Drinkwater, & Gamboa, 2019), mainly due to the **calcination of limestone** and the combustion of fossil fuels in kilns. **Hydrogen can decarbonize cement manufacturing by replacing natural gas in high-temperature kiln operations**, reducing direct emissions. Additionally, **alternative binders such as geopolymers, carbon-cured concrete, and recycled aggregates** can further lower the carbon footprint of construction materials (Scrivener et al., 2018).

In the same way, hydrogen-processed bricks and building blocks can also be utilized, as traditional brick manufacturing involves firing clay at high temperatures, typically using **natural gas or coal-fired kilns**, contributing to CO_2 emissions. The **HyBric project** explored the use of **hydrogen as a replacement for natural gas** in firing clay bricks while ensuring their **structural integrity remained unchanged**. The trials showed promising results, with a **carbon emission reduction of 81%–84%** across three test runs (Michelmersh Brick Holdings PLC. (n.d.)). That means **hydrogen can be used as a clean combustion fuel** in brick kilns, which can eliminate fossil fuel-related emissions. Moreover, **alkali-activated bricks and autoclaved aerated concrete (AAC)** offer alternative low-carbon options (Michelini et al., 2023).

The transition to **hydrogen-based building materials** can reduce GHG emissions significantly, but it faces challenges such as **high production costs**, the need for **policy support**, and the **scaling of hydrogen infrastructure** for both industrial and residential use. Additionally, **building codes and material standards** must adapt to encourage the adoption of low-carbon materials. However, **rising investments, technological advancements, and stricter carbon regulations** are accelerating this shift. By integrating **hydrogen-based materials and energy systems**, the construction sector can **significantly lower emissions** and contribute to a **more sustainable built environment.**

2.6 Possible integration of green hydrogen technologies in the building and residential sector

Hydrogen is a versatile energy carrier with numerous applications across diverse energy systems, making it a key contender in addressing climate change (Griffiths et al., 2021; Scita et al., 2020). It is increasingly explored as a low-carbon replacement for natural gas in residential heating, particularly in regions like Europe, where hydrogen adoption is a key energy transition pathway.

Countries with high natural gas dependence, such as the UK and the Netherlands, are well-positioned to integrate hydrogen into their heating infrastructure (Arapostathis et al., 2014; Roberts & Geels, 2019; Sovacool, 2016).

Many countries are actively exploring the integration of green hydrogen into their natural gas distribution systems as a step toward reducing carbon emissions and transitioning to cleaner energy sources. **Blending hydrogen with conventional gas allows gradual decarbonization without requiring immediate large-scale changes to appliances or distribution systems. However, current blending is limited by regulatory constraints, as existing natural gas standards do not fully accommodate hydrogen's unique properties.** Research is ongoing to establish safe blending limits and update technical guidelines for long-term integration.

Beyond gas networks, green hydrogen supports decentralized energy applications, particularly in residential and commercial buildings. **Its long-term storage capability makes it an effective complement to intermittent renewables such as solar and wind. By capturing surplus renewable electricity, hydrogen enhances grid stability and reduces fossil fuel reliance.**

Hydrogen storage solutions continue to evolve, including metal hydride-based systems for mobile and off-grid applications. Companies, such as Endua, have developed hydrogen-powered storage solutions such as the "Power Bank," **demonstrating hydrogen's potential for clean, portable power.** Stored hydrogen can be converted into electricity or heat using fuel cells, hydrogen boilers, and micro-CHP systems, **allowing households to reduce their carbon footprint while meeting energy demands. Additionally, hydrogen refueling for vehicles at home is emerging as a new application, further integrating hydrogen into daily energy use.** Intelligent EMSs optimize hydrogen utilization for peak demand periods.

Rather than replacing conventional energy sources, **small-scale hydrogen systems act as a complementary solution, enhancing resilience and sustainability in modern energy grids. Their modular design enables scalability for different building types and energy needs.** Although initial costs, particularly for storage infrastructure, remain high, **commercial sectors may benefit from long-term savings and environmental advantages.**

Hybrid configurations combining solar PV, battery storage, and hydrogen fuel cells can create self-sufficient energy models. As hydrogen technology advances and costs decline, its role in residential energy systems is expected to grow. **By integrating hydrogen with renewables, a more efficient and sustainable energy framework can be developed, supporting global decarbonization goals.**

Studies indicate that integrating **hybrid hydrogen and battery storage systems** with eco-friendly buildings can help achieve **energy independence in residential spaces.** A case study focusing on a single-family home powered entirely by rooftop PV panels explored different storage technology options. The findings suggest that by 2030 hydrogen storage could enable complete energy

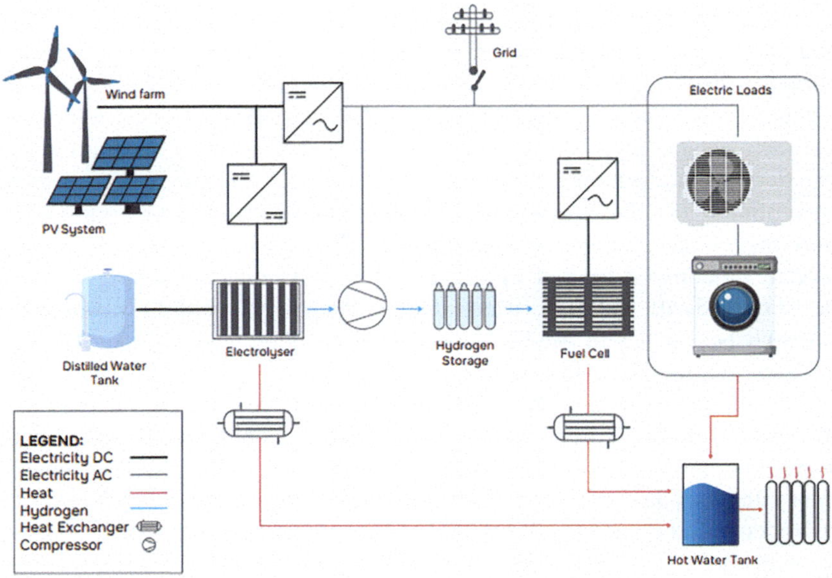

FIGURE 2.5 Model of a hydrogen energy system (power to power and power to gas) integrating renewable energy and fuel cells for residential use.

independence, with a 52% annual cost advantage over traditional grid electricity. This demonstrates the potential economic viability of hydrogen storage, especially when compared to lithium-ion battery systems (Kaleem et al., 2024). By combining sustainability, affordability, and technological advancements, such integrated energy systems offer a promising pathway for decarbonizing residential energy consumption.

Recent studies have explored the feasibility of incorporating green hydrogen into power-to-gas-to-power (P2G2P) systems for residential use, both in individual homes and centralized setups serving multiple households. There are two primary configurations for integrating green hydrogen into residential energy systems, shown in Fig. 2.5. The first is a stand-alone setup that combines PV panels, a fuel cell, and an electrolyzer, but it lacks dedicated energy storage. This limitation makes it vulnerable to fluctuations in sunlight and periods of high energy demand. The second approach is a hybrid system that includes a lithium-ion (Li-ion) battery, allowing excess energy to be stored and utilized when solar generation is insufficient. While this hybrid model enhances flexibility and reliability, it also introduces additional costs, complexity, and integration challenges. Choosing between these configurations requires balancing simplicity, efficiency, and economic feasibility to achieve optimal performance.

Despite its potential, the power-to-gas strategy faces challenges such as high infrastructure costs, storage limitations, and regulatory barriers. Hydrogen production costs depend on location, electrolyzer efficiency, and market conditions, but advancements and economies of scale are expected to reduce costs. Expanding infrastructure such as pipelines and storage, along with supportive policies and investments, will be key to scaling hydrogen adoption in the residential and building sector.

2.7 Geospatial analysis for measuring the green hydrogen production potential

2.7.1 Case study

Australia's commitment to hydrogen energy development was formalized in 2019 with the launch of its National Hydrogen Strategy. This initiative brought together various government agencies, private sector entities, and research institutions to accelerate advancements in hydrogen technologies. Concurrently, state governments developed individual hydrogen roadmaps to support the establishment of hydrogen production, storage, and distribution infrastructure (Dawood et al., 2020). Western Australia, in particular, is well-positioned due to its vast renewable energy resources, including some of the highest solar radiation levels globally. The state receives an estimated 58 million petajoules of solar energy annually, offering a strong foundation for large-scale hydrogen production (Bahadori & Nwaoha, 2013).

The widespread adoption of rooftop solar PV systems further supports Western Australia's renewable energy landscape. Currently, more than 30% of residential buildings in the state are equipped with solar panels, generating approximately 1300 MW of capacity (Australian Energy Market Operator, 2019). Projections suggest this figure could double in the next decade due to technological improvements and policy incentives promoting renewable energy expansion.

Given the immense potential for solar energy generation, selecting optimal sites for green hydrogen production is critical for maximizing efficiency and sustainability. However, site selection involves more than just resource availability; factors such as proximity to infrastructure, accessibility to water resources, and integration with urban development must also be considered. Geographic Information Systems (GIS) play a pivotal role in this process by facilitating spatial analysis and data integration. By using GIS methodologies, planners can systematically evaluate geographic, environmental, and economic factors to identify suitable locations for hydrogen production facilities.

Advanced GIS techniques enhance the assessment of renewable energy potential, ensuring a data-driven approach to hydrogen site selection. By analyzing multiple spatial datasets, GIS enables a comprehensive evaluation of critical parameters such as resource distribution, infrastructure accessibility, and environmental impact. This analytical framework aids decision-making by refining site selection and optimizing resource allocation.

To determine the most viable locations for green hydrogen production in Western Australia, a site suitability analysis was conducted using the Analytical

Hierarchy Process (AHP) within ArcGIS Pro. This evaluation ranked potential sites based on key criteria weighted according to their significance in hydrogen production. Spatial data in both vector and raster formats representing land use, road networks, electricity transmission infrastructure, and topographical features were analyzed to ensure a robust site selection framework. This assessment incorporated insights from expert analyses, scientific studies, and industry reports. Our study explored the feasibility of utilizing 60% surplus solar energy from residential PV systems for green hydrogen production. We estimated Western Australia's green hydrogen production potential using solar installation data from the Australian Government's Clean Energy Regulator (Regulator, 2024). As of 2024 a total of 513,870 small-scale solar PV systems have been deployed across the state (Regulator, 2024). These installations provide a substantial renewable energy resource that can be utilized for hydrogen production via electrolysis. Two cases were developed to estimate hydrogen production: households with battery storage, which generate a surplus of 1,517,718.24 MWh/year, and those without batteries, which contribute a surplus of 3,794,481.6 MWh/year. This highlights Western Australia's significant solar PV capacity, which can be utilized for green hydrogen production while contributing to grid stability.

To evaluate different implementation models, three primary production approaches were considered. The **decentralized production** model involves generating hydrogen within residential areas by utilizing surplus solar energy. This approach offers a high level of resilience and requires minimal infrastructure investment. However, its effectiveness is constrained by the limited energy availability in localized residential settings. The **semicentralized production** approach aggregates surplus solar energy from multiple residential zones, directing it toward medium-sized electrolysis plants. This model provides a balance between scalability and infrastructure efficiency, allowing for greater hydrogen output while maintaining a reasonable level of flexibility in deployment. The **centralized production** model focuses on large-scale electrolysis plants that utilize extensive surplus solar energy to generate hydrogen for industrial applications. This method benefits from economies of scale, making large-scale production more cost-effective. However, it also requires significant infrastructure investments and faces challenges related to energy transmission and integration into the existing grid.

Fig. 2.6 illustrates the suitability maps for green hydrogen production plants under three different scenarios. Spatial analysis revealed that high-suitability sites for hydrogen production tend to be near existing settlements with essential utilities available. For larger industrial-scale production, optimal locations are farther from residential zones but remain connected through key infrastructure networks. As the energy landscape continues to evolve, the growing integration of distributed solar generation presents new challenges for maintaining grid stability. Solar energy generation peaks during daylight hours, often resulting in excess electricity that can destabilize the grid if not managed properly.

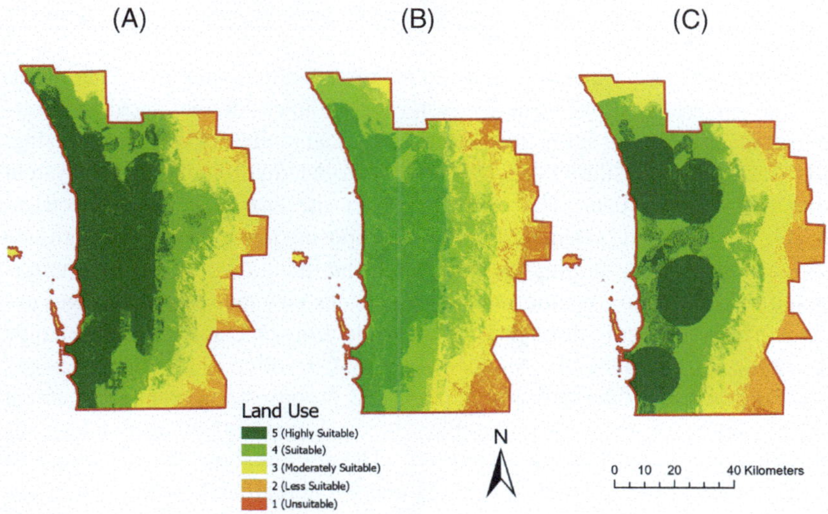

FIGURE 2.6 Suitability map for (A) decentralized, (B) semicentralized, and (C) centralized green hydrogen production plants in Perth, Western Australia.

To address this challenge, surplus solar energy can be utilized for hydrogen production, providing an efficient energy storage solution that enhances grid stability and security. According to the Australian Renewable Energy Agency (Australian Renewable Energy Agency, 2024), 80% of Australia's electricity is expected to come from renewable sources by 2050. Given this trajectory, developing effective energy storage solutions, including hydrogen, will be essential for ensuring a stable and resilient power grid while facilitating the broader transition to sustainable energy systems.

2.8 Key techno-economic and environmental considerations for residential application

Green hydrogen is an emerging technology still in its early stages, which poses challenges to its widespread adoption, particularly from an economic perspective. One major hurdle is the high cost of essential components such as electrolyzers, fuel cells, and cryogenic storage tanks used in the hydrogen lifecycle (production, storage, and utilization). Additionally, electrolyzers currently have a low capacity factor, meaning their operational efficiency is heavily influenced by fluctuations in variable renewable energy supply. When renewable generation is lower than electricity demand, electrolyzer operation is reduced, whereas excess renewable energy allows for extended operation. However, large-scale production of these devices can drive down costs, making green hydrogen more competitive. Studies have shown that increasing the production scale of polymer electrolyte fuel cells (PEFCs) and SOFCs, including their stacks and peripheral systems, significantly reduces manufacturing expenses (James & DeSantis, 2015; Okubo et al., 2023).

Several studies have also conducted techno-economic analyses of both on-grid and off-grid hydrogen energy systems for the residential sector (Yunez-Cano et al., 2016). The study presents an analytical model for sizing, analyzing, and assessing the feasibility of a hybrid photovoltaic/hydrogen (PV/H_2) system using real weather data. The study evaluates the energy balance, electrical variables, and overall system performance, proposing improvements to optimize efficiency by utilizing surplus energy. Using solar radiation data from a monitoring station in Mexico City, the study demonstrates how the system manages energy storage and supply through PV panels, electrolyzers, and proton exchange membrane (PEM) fuel cells. Another study by Cozzolino et al. (2016) designs a hybrid wind/solar/fuel cell power plant and proposes a power management strategy, relying primarily on renewable sources. A case study of artificial islands between Tunisia and Italy was analyzed. The results show the feasibility of a self-sustaining system with integrated energy storage and backup solutions, achieving a cost of power of 0.522 €/kWh (Türkay & Telli, 2011). The study focuses on a techno-economic analysis of a renewable energy and hydrogen storage system to meet the energy demand of a pilot region. This study evaluates the integration of solar and wind energy, both as stand-alone systems and alongside the conventional grid, using AN optimization model based on cost and resource availability. The pilot region in this case is the Electrics & Electronics Faculty at Istanbul Technical University. Mukherjee et al. (2017) study designs a renewable hydrogen-powered microgrid for backup power in Cornwall, Ontario, using the energy hub concept to optimize generation and meet community energy demands. This system incorporates solar PV, wind turbines, electrolyzers, hydrogen storage, fuel cells, and vehicle-to-grid services, with a required generation capacity of 2000 and 3000 kW backup fuel cell systems. While the system is technically feasible, the economic viability remains a challenge, emphasizing the need for additional incentives. These studies demonstrate the considerable potential of hydrogen energy systems in residential and community applications, highlighting their technical feasibility. However, the high costs associated with system components and the need for economic incentives remain significant barriers to widespread adoption. Further research and development, along with policy support, are crucial to overcoming these challenges and accelerating the transition to sustainable, hydrogen-powered energy solutions.

The successful implementation of community-based green hydrogen systems also involves addressing a range of economic, environmental, and social considerations. Key challenges include the establishment of effective organizational structures to support these systems, the adoption of specific technologies such as fuel cells, and raising public awareness to encourage acceptance and active participation in the development of hydrogen infrastructure. Overcoming these obstacles is essential not only for countries with well-established hydrogen sectors but also for those at the beginning stages of their energy transition journey.

These considerations directly influence regulatory processes, highlighting the need for a collaborative approach involving policymakers, industry stakeholders, and the public to ensure a smooth transition.

On the economic front, green hydrogen contributes to economic growth by creating a wide range of employment opportunities. Investment in production facilities, supply networks, and supporting infrastructure generates jobs not only during the construction phase but also through ongoing operations and maintenance. In addition, the development of the hydrogen sector stimulates other industries, such as the manufacturing of electrolyzers, renewable energy services, and specialized engineering fields, contributing to broader industrial growth (IEA, 2019). Additionally, the expansion of the hydrogen economy can stimulate related industries, such as electrolyzer manufacturing, renewable energy integration, and specialized engineering services, further boosting economic growth. However, the high investment costs required for establishing green hydrogen infrastructure and the currently high production costs remain significant barriers to large-scale deployment. These factors can limit the ability of many regions to adopt green hydrogen on a wide scale, especially in areas with limited renewable resources.

The large-scale use of green hydrogen contributes to achieving climate targets while also improving energy security by minimizing reliance on imported fossil fuels. In addition, establishing a domestic hydrogen sector encourages innovation, builds specialized knowledge, and creates economic opportunities for local communities, helping to ensure a fair and inclusive transition to sustainable energy.

2.9 Policy and urban planning implications

Transitioning to a hydrogen-based economy presents numerous benefits, including revitalizing economic activities, improving environmental conditions for urban dwellers, and securing a sustainable future for cities. Furthermore, it enhances energy security and mitigates the negative impacts of fossil fuel consumption on the environment (Tao et al., 2022).

While policymakers have introduced some initiatives to support the hydrogen sector, more targeted policy measures are required at every stage, from technological development to market entry and large-scale commercialization to facilitate its integration into the energy landscape (Ahmad et al., 2023).

Governments worldwide have recognized the role of hydrogen in future energy systems. By 2017 approximately 22 EU nations had enacted policies supporting hydrogen integration, with a focus on transportation (Yvonne Ruf et al., 2018). More recently, countries such as Norway, Germany, France, Italy, Chile, and Morocco have taken significant steps by formulating and publishing national hydrogen strategies (Borbujo et al., 2021; Martin Robinius et al., 2018). Despite these efforts, residential applications require tailored policy interventions to address sector-specific challenges.

To accelerate hydrogen adoption in the residential sector and buildings, policymakers must implement financial incentives, streamline regulatory approvals, and establish market-driven mechanisms such as contracts for difference (CfDs) to enhance investment security (Simon Schreck et al., 2023). Additionally, adapting existing gas infrastructure for hydrogen distribution can minimize costs and expedite deployment. Clear regulatory guidelines for pipeline conversion, safety protocols, and long-term storage solutions will be essential in ensuring a reliable hydrogen supply chain.

2.9.1 Policy considerations for hydrogen integration in buildings

The successful adoption of hydrogen in the residential sector requires a comprehensive policy framework addressing infrastructure, economic feasibility, and regulatory alignment. Key policy measures include:

- **Investment in infrastructure:** Expanding hydrogen storage and distribution networks, including retrofitting existing gas pipelines.
- **Financial incentives:** Grants, subsidies, and tax incentives to reduce the cost of electrolyzers and hydrogen-compatible appliances.
- **Regulatory streamlining:** Simplifying permitting processes and establishing clear safety and quality standards for hydrogen-based residential energy systems.
- **Market development mechanisms:** Introducing CfDs, hydrogen purchase obligations, and carbon pricing to support long-term hydrogen market viability.
- **Integration with renewables:** Coordinating hydrogen production with solar PV and wind energy to optimize efficiency and minimize grid instability.
- **Public–private partnerships:** Encouraging collaborations between governments, industry stakeholders, and research institutions to drive technological innovation and commercialization.
- **Safety and monitoring regulations:** Hydrogen's high flammability requires stringent safety standards, including advanced leak detection, ventilation, and continuous monitoring. Standardized safety codes and certification frameworks are essential for risk mitigation.
- **Technology investment:** Decentralized hydrogen integration in the building sector faces affordability challenges. Policy support through research funding, subsidies, and pilot programs is essential to advance cost-effective appliances, storage, and distribution systems for widespread adoption.

By implementing these policy measures, hydrogen can play a pivotal role in decarbonizing the residential sector, enhancing urban energy resilience, and contributing to a cleaner, more sustainable future.

2.10 Conclusion

Decarbonizing the **building and residential sectors** is crucial for advancing a cleaner energy future. While the increasing adoption of **solar PV systems** in homes presents opportunities, it also brings challenges, particularly in managing **excess electricity generation**. The **Duck Curve phenomenon**, where midday solar generation exceeds demand but evening demand spikes when solar output declines, disrupts grid stability, increases reliance on backup power sources, and raises operational costs. Without effective utilization strategies, surplus solar energy remains underutilized, reducing overall system efficiency.

Green hydrogen presents a complementary solution by converting excess solar power into a storable and versatile energy carrier. However, its effectiveness in residential applications depends on addressing **technological, economic, and infrastructure limitations**. This chapter emphasizes the **potential of hydrogen beyond electricity generation**, particularly in **hot water systems**, which could provide a more feasible pathway for residential hydrogen integration. Hydrogen-based water heating offers **efficiency benefits, reduces grid dependency, and aligns with existing gas infrastructure**, making it a viable option for decarbonizing homes.

The case study introduced in this chapter outlines a **strategic site selection framework** for green hydrogen production in **Western Australia**, identifying optimal locations for **decentralized, semicentralized, and centralized hydrogen plants**. Additionally, it explores the **role of surplus residential solar energy** in hydrogen production, offering a pathway to enhance **grid stability, energy security, and sectoral decarbonization**. While large-scale hydrogen applications remain critical, expanding research into **small-scale residential applications**, such as **hydrogen-powered water heaters, FCs for backup power, and hybrid heating solutions**, could accelerate adoption and feasibility.

Future research should focus on **techno-economic assessments, safe hydrogen storage** in residential settings, and **regulatory frameworks** to ensure hydrogen integration is both practical and secure. Advancements in **leak detection systems, storage safety protocols, and cost reductions in electrolysis technology** will be pivotal in scaling residential hydrogen applications.

Ultimately, achieving a **sustainable and resilient energy system** requires a shift from conventional energy models toward **holistic urban planning, policy-driven resource conservation, and long-term energy security measures**. By integrating **green hydrogen as a key enabler of decarbonization**, particularly in residential heating and storage, we can foster a **low-carbon future where renewable energy powers homes and communities efficiently and sustainably**.

References

Adams, Matthew, Richardson, S., Drinkwater, J., & Gamboa, C. (2019). *World Green Building Council. Bringing embodied carbon upfront: Coordinated action for the building and construction sector to tackle embodied carbon* https://www.architecturalrecord.com/ext/resources/Issues/2022/02-February/Bringing-Embodied-Carbon-Upfront.pdf.

Ahmad, F., Rafiq, K., Najam, T., Hussain, E., Sohail, M., Abid, M. Z., Mahmood, A., Javed, M. S., & Shah, S. S. A. (2023). Metal-organic frameworks for electrocatalytic water-splitting: Beyond the pyrolysis. *International Journal of Hydrogen Energy, 48*(90), 35075–35111. doi:10.1016/j.ijhydene.2023.05.247.

Al Busaidi, A. S., Kazem, H. A., & Khan, M. F. (2014). A review of optimum sizing techniques for off-grid hybrid PV-wind renewable energy systems. *International Journal of Students Research in Technology & Management, 2*, 93–102.

Ali Yildirim, M., Bartyzel, F., Vallati, A., Woźniak, M. K., & Ocłoń, P. ł. (2023). Efficient energy storage in residential buildings integrated with RESHeat system. *Applied Energy, 335*, 120752. doi:10.1016/j.apenergy.2023.120752.

Alsayegh, O., Albusairi, H., & Qasem, H. (2008). Examining the challenges of distributed energy resources interconnection with electric utility networks. In *Proceedings of the fourth International Conference on Energy Research & Development* Kuwait University, Fourth International Conference on Energy Research & Development (ICERD-4), Kuwait City, Kuwait.

Ammermann, H., Hoff, P., Atanasiu, M., Tisler, O., & Kaufmann, M. (2015). *Advancing europe's energy systems-stationary fuel cells in distributed generation* https://www.h2knowledgecentre.com/content/researchpaper1118.

Arapostathis, S., Pearson, P. J. G., & Foxon, T. J. (2014). UK natural gas system integration in the making, 1960-2010: Complexity, transitional uncertainties and uncertain transitions. *Environmental Innovation and Societal Transitions, 11*, 87–102. http://www.elsevier.com/wps/find/journaldescription.cws_home/724359/description#description. doi:10.1016/j.eist.2014.01.004.

Australian Energy Market Operator. (2019). Electricity statement of opportunities. *Analysis & Policy*. https://apo.org.au/node/254701.

Bahadori, A., & Nwaoha, C. (2013). A review on solar energy utilisation in Australia. *Renewable and Sustainable Energy Reviews, 18*, 1–5. doi:10.1016/j.rser.2012.10.003.

Baruah, A., Basu, M., & Amuley, D. (2021). Modeling of an autonomous hybrid renewable energy system for electrification of a township: A case study for Sikkim, India. *Renewable and Sustainable Energy Reviews, 135*, 110158. doi:10.1016/j.rser.2020.110158.

Barzola-Monteses, J., & Espinoza-Andaluz, M. (2019). Performance analysis of hybrid solar/H2/battery renewable energy system for residential electrification. *Energy Procedia*, 158. http://www.sciencedirect.com/science/journal/18766102. doi:10.1016/j.egypro.2019.01.024.

Basnet, S., Deschinkel, K., Le Moyne, L., & Péra, M. C. (2023). A review on recent standalone and grid integrated hybrid renewable energy systems: System optimization and energy management strategies. *Renewable Energy Focus, 46*, 103–125. doi:10.1016/j.ref.2023.06.001.

Bindoff, N. L. (2019). *Changing ocean, marine ecosystems, and dependent communities*. IPCC.

Blonsky, M., Nagarajan, A., Ghosh, S., McKenna, K., Veda, S., & Kroposki, B. (2019). Potential impacts of transportation and building electrification on the grid: A review of electrification projections and their effects on grid infrastructure, operation, and planning. *Current Sustainable/Renewable Energy Reports, 6*(4), 169–176. doi:10.1007/s40518-019-00140-5.

Borbujo, I. C., Pereirinha, P. G., Vega, M. G., Del Valle, J. A., & Anton, J. C. A. (2021). Heavy duty transport descarbonization: Legislation and Standards for hydrogen and battery electric buses and heavy-duty trucks. *IEEE Vehicle Power and Propulsion Conference, VPPC 2021* http://ieeexplore.ieee.org/xpl/mostRecentIssue.jsp?punumber=9698824. doi:10.1109/VPPC53923.2021.9699373.

Australian Renewable Energy Agency. (2024). Solar PV R&D. https://arena.gov.au/renewable-energy/solar-pv-rd/ (this is the citation to the best of my knowledge). Clean

Energy Council. Clean Energy Australia 2024. https://cleanenergycouncil.org.au/getmedia/0cb12425-37ab-479e-9a4b-529622cc9c02/clean-energy-australia-2024.pdf.

Clean Energy Council, (2024). Clean Energy Australia. Clean Energy Council.

Clean Energy Regulator, Australian Government, (2024). Small-scale installation postcode data. Clean Energy Regulator, Australian Government.

Cozzolino, R., Tribioli, L., & Bella, G. (2016). Power management of a hybrid renewable system for artificial islands: A case study. *Energy, 106*, 774–789. doi:10.1016/j.energy.2015.12.118.

Dawood, F., Anda, M., & Shafiullah, G. M. (2020). Hydrogen production for energy: An overview. *International Journal of Hydrogen Energy, 45*(7), 3847–3869. http://www.journals.elsevier.com/international-journal-of-hydrogen-energy/. doi:10.1016/j.ijhydene.2019.12.059.

Economics, F. (2022). *Cost of switching from gas to electric appliances in the home [Report prepared for the Gas Appliance Manufacturers Association of Australia]* https://gamaa.asn.au/wp-content/uploads/2022/07/Frontier-Economics-Report-GAMAA.pdf.

Ecosys, (2014). Flexibility options in electricity systems., Ecosys.

Elmallah, S., Brockway, A. M., & Callaway, D. (2022). Can distribution grid infrastructure accommodate residential electrification and electric vehicle adoption in Northern California? *Environmental Research: Infrastructure and Sustainability, 2*(4). https://iopscience.iop.org/journal/2634-4505. doi:10.1088/2634-4505/ac949c.

Emissions Gap Report 2024. No more hot air … please! With a massive gap between rhetoric and reality, countries draft new climate commitments. (2024), United Nations Environment Programme.

Fernández, L. (2024). Global wind power market - statistics & facts. Statista.

Geschwindt, S., (2024, August 29). Sweden's been stealthily using hydrogen to forge green steel. Now it's ready to industrialise. TNW Letter, https://thenextweb.com/news/sweden-hydrogen-green-steel-vattenfall-ssab-lkab-hybrit.

Global Energy Review. CO2 emissions in 2021. 2022, Global Energy Review.

Go, J., Byun, J., Orehounig, K., & Heo, Y. (2023). Battery-H2 storage system for self-sufficiency in residential buildings under different electric heating system scenarios. *Applied Energy, 337*, 120742. doi:10.1016/j.apenergy.2023.120742.

Gordon, J. A., Balta-Ozkan, N., & Nabavi, S. A. (2023). Divergent consumer preferences and visions for cooking and heating technologies in the United Kingdom: Make our homes clean, safe, warm and smart!. *Energy Research and Social Science, 104*. http://www.journals.elsevier.com/energy-research-and-social-science/. doi:10.1016/j.erss.2023.103204.

Griffiths, S., Sovacool, B. K., Kim, J., Bazilian, M., & Uratani, J. M. (2021). Industrial decarbonization via hydrogen: A critical and systematic review of developments, socio-technical systems and policy options. *Energy Research & Social Science, 80*, 102208. doi:10.1016/j.erss.2021.102208.

Groppi, D., Garcia, D. A., Lo Basso, G., Cumo, F., & De Santoli, L. (2018). Analysing economic and environmental sustainability related to the use of battery and hydrogen energy storages for increasing the energy independence of small islands. *Energy Conversion and Management, 177*, 64–76. doi:10.1016/j.enconman.2018.09.063.

Hybrit. (n.d.). Fossil free steel. a joint opportunity! Retrieved February 2025, https://www.hybritdevelopment.se/en/.

International Energy Agency, The future of hydrogen. (2019), International Energy Agency.

James, B. D., & DeSantis, D. A. (2015). *Manufacturing cost and installed price analysis of stationary fuel cell systems.Strategic Analysis*. Arlington, VA, USA: Inc.

Kaleem, A., Zaman, A., & Rajakaruna, S. (2024). Hydrogen at home: The current and future landscape of green hydrogen in residential settings. *Sustainable Energy Technologies and Assessments, 72*, 104058. doi:10.1016/j.seta.2024.104058.

Khan, T., Yu, M., & Waseem, M. (2022). Review on recent optimization strategies for hybrid renewable energy system with hydrogen technologies: State of the art, trends and future directions. *International Journal of Hydrogen Energy, 47*(60), 25155–25201. http://www.journals.elsevier.com/international-journal-of-hydrogen-energy/. doi:10.1016/j.ijhydene.2022.05.263.

Maestre, V. M., Ortiz, A., & Ortiz, I. (2022). The role of hydrogen-based power systems in the energy transition of the residential sector. *Journal of Chemical Technology and Biotechnology, 97*(3), 561–574. http://onlinelibrary.wiley.com/journal/10.1002/(ISSN)1097-4660. doi:10.1002/jctb.6938.

Martin Robinius, J. L., Grube, T., Reuß, M., Stenzel, P., & Konstantinos Syranidis, P. K. A. D. S. (2018). *Comparative analysis of infrastructures: Hydrogen fueling and electric charging of vehicles*. Forschungszentrum Jülich.

Melvin, J. (2018). The split incentives energy efficiency problem: Evidence of underinvestment by landlords. *Energy Policy, 115*, 342–352. doi:10.1016/j.enpol.2017.11.069.

Michelini, E., Ferretti, D., Miccoli, L., & Parisi, F. (2023). Autoclaved aerated concrete masonry for energy efficient buildings: State of the art and future developments. *Construction and Building Materials, 402*, 132996. doi:10.1016/j.conbuildmat.2023.132996.

Michelmersh Brick Holdings PLC. (n.d.). Michelmersh to produce the world's first 100% hydrogen fired clay bricks [Press release]. https://www.mbhplc.co.uk/wp-content/uploads/Michelmersh-Produces-Worlds-First-Hydrogen-Fired-Clay-Bricks.pdf.

Mohammed, Y. S., Mustafa, M. W., & Bashir, N. (2014). Hybrid renewable energy systems for off-grid electric power: Review of substantial issues. *Renewable and Sustainable Energy Reviews, 35*, 527–539. doi:10.1016/j.rser.2014.04.022.

Mukherjee, U., Maroufmashat, A., Ranisau, J., Barbouti, M., Trainor, A., Juthani, N., El-Shayeb, H., & Fowler, M. (2017). Techno-economic, environmental, and safety assessment of hydrogen powered community microgrids; case study in Canada. *International Journal of Hydrogen Energy, 42*(20), 14333–14349. doi:10.1016/j.ijhydene.2017.03.083.

Okubo, T., Shimizu, T., Hasegawa, K., Kikuchi, Y., Manzhos, S., & Ihara, M. (2023). Factors affecting the techno-economic and environmental performance of on-grid distributed hydrogen energy storage systems with solar panels. *Energy, 269*, 126736. doi:10.1016/j.energy.2023.126736.

Quarton, C. J., & Samsatli, S. (2018). Power-to-gas for injection into the gas grid: What can we learn from real-life projects, economic assessments and systems modelling? *Renewable and Sustainable Energy Reviews, 98*, 302–316. https://www.journals.elsevier.com/renewable-and-sustainable-energy-reviews. doi:10.1016/j.rser.2018.09.007.

Rana, M. d. M., Uddin, M., Sarkar, M. d. R., Meraj, S. T., Shafiullah, G. M., Muyeen, S. M., Islam, M. d. A., & Jamal, T. (2023). Applications of energy storage systems in power grids with and without renewable energy integration — A comprehensive review. *Journal of Energy Storage, 68*, 107811. doi:10.1016/j.est.2023.107811.

Roberts, C., & Geels, F. W. (2019). Conditions and intervention strategies for the deliberate acceleration of socio-technical transitions: Lessons from a comparative multi-level analysis of two historical case studies in Dutch and Danish heating. *Technology Analysis and Strategic Management, 31*(9), 1081–1103. http://www.tandf.co.uk/journals/titles/09537325.asp. doi:10.1080/09537325.2019.1584286.

Sandoval, N., Harris, C., Reyna, J. L., Fontanini, A. D., Liu, L., Stenger, K., White, P. R., & Landis, A. E. (2025). Achieving equitable widespread residential building electrification – examining barriers, strategies, and opportunities using Los Angeles as a case study. *Applied Energy, 384*, 125498. doi:10.1016/j.apenergy.2025.125498.

Schiro, F., Stoppato, A., & Benato, A. (2020). Modelling and analyzing the impact of hydrogen enriched natural gas on domestic gas boilers in a decarbonization perspective. *Carbon Resources Conversion, 3*, 122–129. www.keaipublishing.com/en/journals/carbon-resources-conversion/. doi:10.1016/j.crcon.2020.08.001.

Schöne, N., Khairallah, J., & Heinz, B. (2022). Model-based techno-economic evaluation of power-to-hydrogen-to-power for the electrification of isolated African off-grid communities. *Energy for Sustainable Development, 70*, 592–608. http://www.elsevier.com. doi:10.1016/j.esd.2022.08.020.

Scita, R., Raimondi, P. P., Noussan, M., (2020). Green hydrogen: The holy grail of decarbonisation? An analysis of the technical and geopolitical implications of the future hydrogen economy. https://www.ssrn.com/index.cfm/en/. doi:10.2139/ssrn.3709789.

Scrivener, K. L., John, V. M., & Gartner, E. M. (2018). Eco-efficient cements: Potential economically viable solutions for a low-CO_2 cement-based materials industry. *Cement and Concrete Research, 114*, 2–26. http://www.sciencedirect.com/science/journal/00088846. doi:10.1016/j.cemconres.2018.03.015.

Shafiullah, G. M., Oo, A. M. T., Shawkat Ali, A. B. M., & Wolfs, P. (2013). Potential challenges of integrating large-scale wind energy into the power grid–A review. *Renewable and Sustainable Energy Reviews, 20*, 306–321. doi:10.1016/j.rser.2012.11.057.

Simon Schreck, M. S., Żelisko, W., Prates, R., (2023). Policies for a sustainable hydrogen economy in Europe. Germanwatch e.V. https://www.germanwatch.org/en/node/89555.

Sovacool, B. K. (2016). How long will it take? Conceptualizing the temporal dynamics of energy transitions. *Energy Research and Social Science, 13*, 202–215. http://www.journals.elsevier.com/energy-research-and-social-science/. doi:10.1016/j.erss.2015.12.020.

Statista. (n.d.). Global cumulative installed solar PV capacity from 2010 to 2024 [Graph]. Statista. Retrieved February 2025, from https://www.statista.com/statistics/280220/global-cumulative-installed-solar-pv-capacity/.

Stegra. (2025). Green steel. Retrieved February 2025, from https://stegra.com/green-steel.

Tao, X., Zhao, Y., Wang, S., Li, C., & Li, R. (2022). Recent advances and perspectives for solar-driven water splitting using particulate photocatalysts. *Chemical Society Reviews, 51*(9), 3561–3608. doi:10.1039/d1cs01182k.

Türkay, B. E., & Telli, A. Y. (2011). Economic analysis of standalone and grid connected hybrid energy systems. *Renewable Energy, 36*(7), 1931–1943. doi:10.1016/j.renene.2010.12.007.

Synergy. (2021, October). Everything you need to know about the Duck Curve. Retrieved February 2025, from https://www.synergy.net.au/Blog/2021/10/Everything-you-need-to-know-about-the-Duck-Curve.

UNEP, Building materials and the climate: Constructing a new future. UN Environment Programme (2023).

Venegas, R. M., Acevedo, J., & Treml, E. A. (2023). Three decades of ocean warming impacts on marine ecosystems: A review and perspective. *Deep-Sea Research Part II: Topical Studies in Oceanography, 212*. http://www.journals.elsevier.com/deep-sea-research-part-ii-topical-studies-in-oceanography/. doi:10.1016/j.dsr2.2023.105318.

Viñuales, J. E. (2016). The Paris agreement on climate change: Less is more. *German Yearbook of International Law, 59*, 11–45. www.gyil.org/. doi:10.3790/978-3-428-55357-0.

Wu, Y., & Zhong, L. (2025). Decarbonizing urban residential communities with green hydrogen systems. *Nature Cities, 2*(1), 81–92. doi:10.1038/s44284-024-00178-7.

Xie, Y., Ueda, Y., & Sugiyama, M. (2021). Greedy energy management strategy and sizing method for a stand-alone microgrid with hydrogen storage. *Journal of Energy Storage, 44*, 103406. doi:10.1016/j.est.2021.103406.

Yunez-Cano, A., González-Huerta, R. d. G., Tufiño-Velázquez, M., Barbosa, R., & Escobar, B. (2016). Solar-hydrogen hybrid system integrated to a sustainable house in Mexico. *International Journal of Hydrogen Energy, 41*(43), 19539–19545. doi:10.1016/j.ijhydene.2016.06.203.

Yvonne Ruf, S. L., Pfister, J., Droege, C., Kaufmann, M., & Heieck, F. (2018). *Fuel cells and hydrogen for green energy in European cities and regions*. Roland Berger.

International Energy Agency, IEA (2021), *Global Hydrogen Review 2021*, IEA, Paris https://www.iea.org/reports/global-hydrogen-review-2021, Licence: CC BY 4.0

Chapter 3

Decarbonizing iron and steel: Fundamentals, hydrogen reduction, and technology pathways

Mohammed Sahab[a], Anisa Emrani[b], Jamil Abdelmajid[a] and Asmae Berrada[c]

[a] *Higher School of Technology of Fez (EST Fez), Sidi Mohamed Ben Abdellah University (USMBA), Fez, Morocco,* [b] *ENSAS/Laboratory of Materials, Processes, Environment and Quality (LMPEQ), Cadi Ayyad University (UCA), Safi, Morocco,* [c] *LERMA Lab, College of Engineering and Architecture, International University of Rabat, Sala Al Jadida, Morocco*

3.1 Introduction

The iron and steel industry is central to modern economies and infrastructures, yet it remains one of the hardest industrial sectors to decarbonize. Depending on the accounting boundary and year, iron and steel are responsible for roughly 5%–9% of global energy-related CO_2 emissions, reflecting the high temperatures and reducing conditions achieved with coal-derived carbon in blast furnace–basic oxygen furnace (BF–BOF) routes (Guo et al., 2025). Persistently rising steel demand in emerging economies adds urgency to the search for credible transition pathways that can reconcile growth with deep emissions cuts. In this context, hydrogen-based direct reduction combined with electric arc furnace (EAF) steelmaking has emerged as one of the most promising near-commercial options for "near-zero" primary steel—particularly when powered by low-carbon electricity (Kim et al., 2022).

Multiple decarbonization strategies are actively explored: (1) maximizing secondary steel via scrap-EAF, which already exhibits low direct emissions but is limited by scrap quality and availability; (2) deploying natural-gas–based DRI as an intermediate step; (3) retrofitting carbon capture on BF–BOF assets; and (4) transitioning to hydrogen direct reduction (H-DRI/DRI–EAF) supplied by clean electricity (Guo et al., 2025). Comparative technoeconomic and system studies increasingly converge on H-DRI/EAF as a mature high-abatement route. Indeed,

Decarbonizing through Sustainable Hydrogen and its Derivatives.
DOI: https://doi.org/10.1016/B978-0-443-27632-3.00003-7

it is capable of ~90%–95% CO_2 reduction relative to coal-based baselines when the electricity and hydrogen are low-carbon. Moreover, studies provided that power price, iron-ore quality, and hydrogen supply chains for this route are favorable (Gurgel et al., 2025).

Within the H-DRI/EAF route, two technology pillars dominate the energy and cost balance: electrolysis for H_2 production and shaft-furnace reduction of high-grade pellets to direct-reduced iron (DRI) or hot-briquetted iron (HBI), followed by melting and refining in an EAF (Wang et al., 2024). Because electrolysis is energy-intensive, the choice of electrolyzer (low-temperature devices like proton exchange membrane (PEM)/alkaline vs. high-temperature solid-oxide electrolysis [SOE]) critically shapes electricity demand and integration opportunities. Low-temperature systems are commercially established and modular, while SOE offers thermodynamic advantages at elevated temperature and the prospect of higher electrical efficiency by leveraging process heat or steam—potentially attractive when high-grade waste heat is available near the reduction shaft or off-gas handling system (Lahrichi et al., 2024). Recent reviews and prototypes underscore rapid progress in SOE performance and durability, even as scale-up and dynamic operation remain active R&D fronts.

For countries with strong renewable resources, the H-DRI/EAF pathway is not only an industrial decarbonization tool but also a power-sector integrator. By scheduling electrolysis and the EAF around variable renewable energy profiles and market signals, a steel complex can act as a controllable load, indirectly supporting grid balancing and the integration of wind and solar. This "sector coupling" logic motivates regionalized assessments rather than global averages: grid carbon intensity and wholesale price distributions materially shift the economics and life-cycle emissions of hydrogen steelmaking. Consequently, replication studies that preserve core assumptions while localizing the electricity system are essential to judge feasibility (Shahabuddin et al., 2023).

Current literature on H-DRI/EAF converges on several robust conclusions. First, abatement potential is high when hydrogen and electricity are low-carbon; published estimates commonly exceed 90% CO_2 reduction versus coal-based routes, with residual emissions stemming from lime calcination, electrodes, and ancillary fuels—targets for further mitigation via material substitutions or carbon capture on nonprocess streams. Second, energy use and levelized cost are dominated by electrolysis, making electrolyzer efficiency, capital cost, and utilization central to feasibility. Third, ore quality (DR-grade pellets) and scrap availability are real constraints that can shift costs, yields, and electricity intensity. Finally, operational flexibility—enabled by hydrogen and HBI buffers—can partially hedge electricity price volatility and lower effective costs in markets with significant intraday spread (Shahabuddin et al., 2023).

This chapter provides a roadmap to decarbonizing iron and steel. It first reviews conventional production routes—blast furnace and direct reduction for ironmaking, basic oxygen converters and EAFs for steelmaking, and downstream casting—then evaluates hydrogen-enabled pathways, including

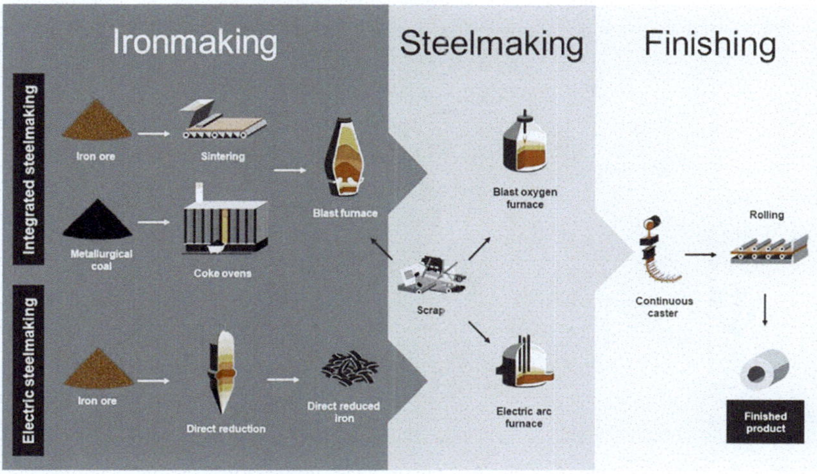

FIGURE 3.1 Steel making processes.

supply options (from reforming and gasification to water electrolysis) and their energy-carbon trade-offs. Thermodynamic and kinetic fundamentals of iron-oxide reduction are presented to explain rate-controlling regimes and the effects of temperature, gas composition, porosity, and particle size. An energy-and-climate mapping identifies emissions hotspots across the value chain, before synthesizing practical levers—raw-material choices, DRI–EAF configurations, scrap use, electrification, CCUS, product strategies, and circularity. The chapter concludes with the benefits case and a structured basis for technology, policy, and investment decisions.

3.2 Conventional iron and steel production routes

%Global iron and steel production is dominated by two distinct process routes, which together account for virtually all output worldwide (99.7%) the two are summarized in Fig. 3.1. Both routes share the same steelmaking and finishing stages but diverge fundamentally in their iron-ore reduction method, energy sources, and equipment. The "integrated" route, involves the reduction of iron-ore with concurrent carburization to produce liquid hot metal (pig iron) in the blast furnace. This hot metal is subsequently converted to steel in a BOF. The steel is then shaped by continuous casting into semifinished products (slabs, blooms, and billets) and, after reheating, hot-rolled into finished products (e.g., wire rod, sheet, sections, beams).

The second route consists of direct reduction of iron ore followed by melting and refining in an electric furnace, typically an EAF. In this pathway, iron is first reduced in the solid state, then melted in the electric furnace; downstream processing mirrors that of the integrated route—continuous casting and hot rolling.

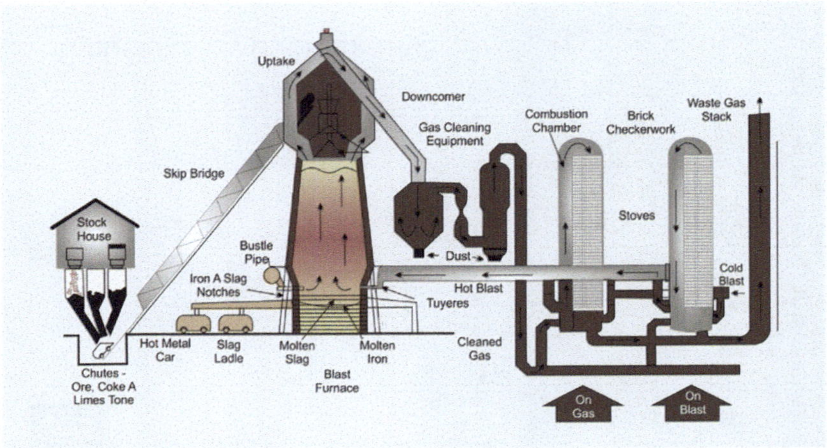

FIGURE 3.2 Hot-metal production facility centered on a blast furnace.

In terms of relative shares, the two routes are far from equivalent in scale. According to the World Steel Association the the blast furnace–BOF route accounted for 70.8% of global steel production in 2021, while the direct-reduction-based route accounted for 28.9% (World Steel, 2026)%. As indicated in Fig. 3.1, both routes commence with ironmaking (ore reduction), followed by steelmaking and finishing operations. The subsequent sections detail each stage.

3.3 Industrial iron ore reduction processes

3.3.1 The blast furnace

The more widely used of the two routes is the integrated route with over 1200 units in operation worldwide, concentrated primarily in China, which alone accounts for more than 50% of global pig iron output (World Steel, 2026). In this case, iron-ore reduction—carried out in a blast furnace—produces hot metal (pig iron). Fig. 3.2 schematically depicts a complete hot-metal production facility centered on a blast furnace. On the left of the figure is the "raw materials" section, supplying ore—typically Fe_2O_3 in pellet form—together with coke and lime. The quality and preparation of these raw materials—particularly coke strength and pellet reducibility—have been shown to significantly influence furnace productivity and energy efficiency (Rahmatmand et al., 2023). In the center is the heating system, generally fired with natural gas, which delivers hot blast to the furnace; modern installations enrich the hot blast with oxygen and inject pulverized coal or natural gas as auxiliary fuels to reduce coke consumption and lower operating costs (Zhang et al., 2022). On the right we find the blast-furnace proper. Pellets are generally produced from concentrated iron ore using one of two shaping technologies: disk pelletizer or a drum pelletizer (Fig. 3.3). The

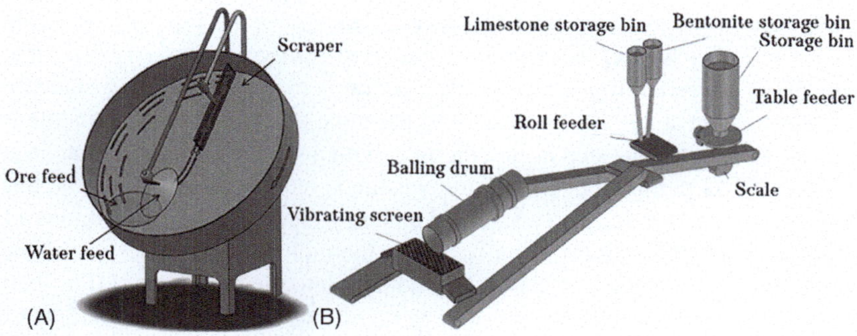

FIGURE 3.3 Disc-cutter excavator and drum-cutter excavator.

former consists of an inclined, rotating disk; the latter is an inclined cylinder (6°–10°), known as a balling drum. Both technologies have been continuously optimized over recent decades, with disk pelletizers generally favored for their superior size control and lower recycle rates, while drum pelletizers are preferred for higher throughput capacities in large-scale operations (Battle el al., 2014). In both cases, the moistened concentrate is introduced into the pelletizer with binders and, in some instances, additional water. As the feed circulates within the drum or on the disk, it agglomerates and forms balls (pellets). Pellet diameter and porosity are critical parameters governing reduction kinetics and gas diffusion during blast furnace and direct reduction operations (Cavaliere et al., 2024) that is why they are controlled either directly at the outlet of the disk pelletizer or downstream in the case of the drum pelletizer (Moraes et al., 2018).

Lime is introduced as a powder, either neat or premixed with the ore; it acts as a flux, capturing impurities during slag formation. The basicity ratio of the slag is a critical operational parameter, as it governs slag viscosity, sulfur partition, and overall furnace stability (Chenguang, 2024). The hot blast (950°C–1350°C), which may be oxygen-enriched, ignites incandescent coke (≈1500°C) at the tuyere level, generating carbon monoxide (CO). Oxygen enrichment of the blast, typically in the range of 3%–6%, is widely practiced in modern blast furnaces to increase productivity and reduce coke consumption (Nie et al., 2021). As it rises through the furnace, the CO reduces iron oxides prior to their melting (Chapter 1 - A brief description of the blast-furnace process, 1979). Carbon monoxide is produced through a sequence of reactions— Eq. (i) (CO formation where the standard enthalpy change at 298 K is −111 kJ per mole of carbon) and (Eq. ii) (coke gasification; where the calculated standard enthalpy change at 298 K is 172 kJ per mole of CO_2)—in the presence of CO_2 and excess carbon (Chapter 1 - A brief description of the blast-furnace process, 1979). The gases liberated in this zone reach temperatures of approximately 2000°C–2200°C (Philibert et al., 1997):

$$C + 1/2O_2 \rightarrow CO \quad \text{(i)}$$

$$C + CO_2 \rightarrow 2CO \quad \text{(ii)}$$

Reduction of iron ore by CO proceeds as shown in Eq. (iii), (iv) and (v) (Chapter 1 - A brief description of the blast-furnace process, 1979). First, hematite is converted to magnetite this reaction is exothermic, with a standard enthalpy change at 298 K of minus 48 kJ per mole of CO. Second, magnetite is reduced to wüstite, this step is endothermic, with a standard enthalpy change at 298 K of 50 kJ per mole of CO, Third, wüstite is reduced to metallic iron, this final step is exothermic, with a standard enthalpy change at 298 K of minus 11 kJ per mole of CO. Carburization occurs as the molten iron trickles over coke lumps, during which carbon dissolves gradually into the liquid phase (Philibert et al., 1997). The product of this stage is liquid hot metal at roughly 1500°C, containing about 2%–6.7% carbon:

$$3Fe_2O_3 + CO \rightarrow 2Fe_3O_4 + CO_2 \tag{iii}$$

$$1.2Fe_3O_4 + CO \rightarrow 3.8Fe_{0.947}O + CO_2 \tag{iv}$$

$$Fe_{0.947}O + CO \rightarrow 0.947Fe + CO_2 \tag{v}$$

A representative blast furnace operates at a hot metal production rate of approximately 170 t/h (Liu et al., 2018). Per ton of hot metal, typical consumptions are ~1600 kg of iron ore, 320 kg of coke at 90% carbon (for carburization), 160 kg of coal at 75%–85% carbon (for CO generation), and 1000 Nm^3 of air (Philibert et al., 1997).These input intensities have declined significantly over the past three decades through process optimization. For example, increasing the O_2 concentration from 21% to 25% has been shown to reduce the coke rate by approximately 90 kg per ton of hot metal (Rahmatmand et al., 2023). From an environmental standpoint, the blast furnace steps result in approximately 1476 kg CO_2-eq per ton of hot metal, about 70% of which is attributable to coal use (Fan & Friedmann, 2021).

3.4 Direct reduction processes

The second iron production route is direct reduction, a process that uses either coal or natural gas (Ghosh & Chatterjee, 2008; Ling et al., 2024). This route has grown steadily over the past two decades and reached over 120 million tonnes in 2022. This growth is due to the expansion of natural-gas-based shaft furnace installations in the Middle East, India, and North Africa (World Steel, 2026). For gas-based operation, the reactors employed are predominantly shaft furnaces, while fluidized-bed and retort reactors are used in specific applications.For coal-based processes, the furnaces used are predominantly rotary hearth furnaces, as well as rotary kilns or multiple-hearth furnaces (Ghosh & Chatterjee, 2008; Ling et al., 2024). The finished product of direct reduction is called DRI or sponge iron. It contains about 0.1% carbon for coal-based direct reduction and about 2% when gas is used (Ghosh & Chatterjee, 2008). DRI is characterized by its degree of reduction (R) defined in Eq. (3.1) where $(nO)_R$ is the total molar amount of oxygen atoms removed from the ore during reduction, and $(nO)_T$ is the total molar amount of oxygen atoms initially bound to iron in the ore before reduction. The metallization degree (M) is defined as in Eq. (3.2), where Fe^0 is the metallic iron content of the reduced sample, and Fe^{total} is the total iron content in the ore before reduction. In DRI, the amount of gangue (impurities) is significant, as it will influence the subsequent stages of steel production (Astier, 2005):

$$Reduction\ degree(R, pct) = \frac{(nO)_R}{(nO)_T} \times 100 \tag{3.1}$$

$$Metallization\ Degree\ (M, pct) = \frac{Fe^0}{Fe^{total}} \times 100 \tag{3.2}$$

The principal advantages of direct reduction over blast-furnace reduction are twofold. First, the absence of reliance on coke production because here the product is not heavily carburized and reduction proceeds via gases, This is significant given the declining availability and rising cost of coking coal globally, as well as the environmental tax associated with coke (Fan & Friedmann, 2021). Second, the smaller scale of the installations, which entails a lower capital expenditure and a reduced site-level environmental impact. However, certain disadvantages should be noted: (1) lower productivity per unit reactor volume; (2) the final product has a low-carbon content, which can create difficulties during conversion to steel (Chapter 1 - A brief description of the blast-furnace process, 1979). A further notable drawback is that, owing to the porosity and high specific surface area of DRI, reoxidation can occur if it is heated to around 200°C, or even at ambient temperature in the presence of moisture (Chapter 1 - A brief description of the blast-furnace process, 1979). The environmental impact of these methods is 1048 kg CO_2-eq per ton of metal when coal is used and 522 kg CO_2-eq per ton of metal when natural gas is used (Fan & Friedmann, 2021).

3.5 Coal-based direct reduction process

The primary reactor used to produce DRI with coal is the rotary kiln. Rotary kilns used in direct reduction typically measure 60–125 m in length and 4–6 m in diameter, operating at rotation speeds of 0.3–0.5 rpm to ensure adequate mixing and heat transfer between the solid burden and reducing gases. Fig. 3.4 presents a simplified schematic of the process. The system comprises a hopper for feed supply, a rotary kiln, an off-gas-treatment system, a water-cooled rotary cooler, a filter, and a magnetic separation system.

The feed materials (iron ore, coal, and limestone) are introduced into the rotary kiln, where coal serves as both the fuel and reducing agent for the reduction of iron oxides. Limestone is added primarily to absorb sulfur released during coal combustion; this prevents sulfur pickup by the DRI and maintains product quality. Once the ore has been reduced, the DRI is cooled in a rotary cooler. A magnetic separation is then performed to remove residual coal before the DRI is

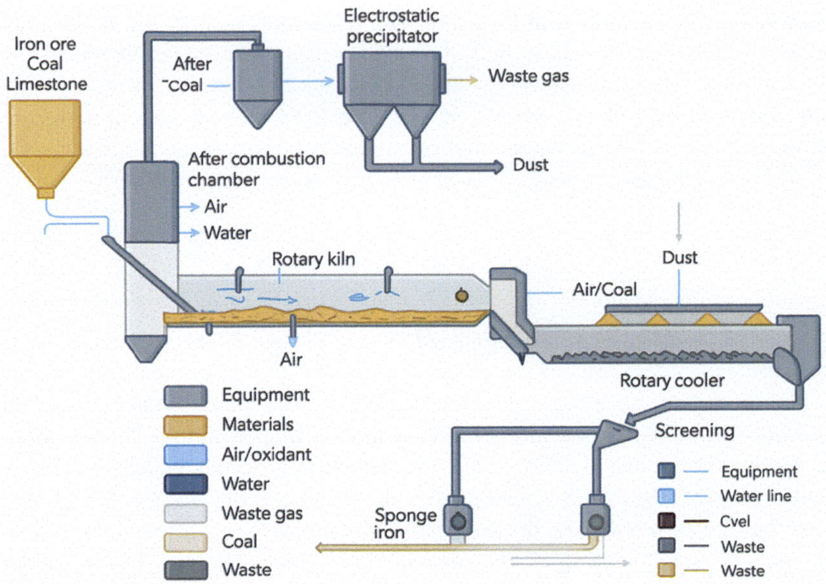

FIGURE 3.4 Simplified diagram of the DRI production process using a rotary kiln.

transported or briquetted. Accretion formation in rotary kilns, driven by high coal ash content and excess dolomite, produces two types of deposits: agglomerated DRI by-products and sintered DRI particles (Shankar Rao & Gurugubelli, 2021). The target temperature for the solid burden (ore and coal) is 950°C–1050°C in order to avoid a large temperature difference with the gas, which would lead to the formation of accretions. A notable alternative to the rotary kiln for coal-based direct reduction is the rotary hearth furnace. Fig. 3.5 presents the Fastmet process developed by Midrex and Kobe, which has a production capacity of 450 $kt\cdot yr^{-1}$. Unlike the rotary kiln, this rotary hearth furnace processes a single layer of composite pellets which contains both iron ore and coal. This enables faster reduction times of typically 6–12 minutes at temperatures of 1250°C–1350°C (Adav et al., 2023).

The feedstock for this process consists of fine-grained concentrated iron ore mixed with pulverized coal and binders. The precise sizing of the iron ore concentrate—typically below 100 μm—is critical to ensure intimate contact between iron oxide particles and coal. The mixture is pelletized, dried, and charged into the furnace in a single layer. The single-layer charging configuration distinguishes the rotary hearth furnace from shaft-based processes and is key to achieving the rapid, uniform heating required for short residence times (Heidari et al., 2021). The burden is then heated to approximately 1250°C–1400°C by combustion of gas, hydrocarbons, or pulverized coal, achieving 90%–95% metallization in 6–12 min (Ghosh & Chatterjee, 2008). Reduction is affected by carbon from the coal within the pellets, in direct contact with the iron oxides, and also by gaseous species (CO and H_2) generated by combustion of the fuel

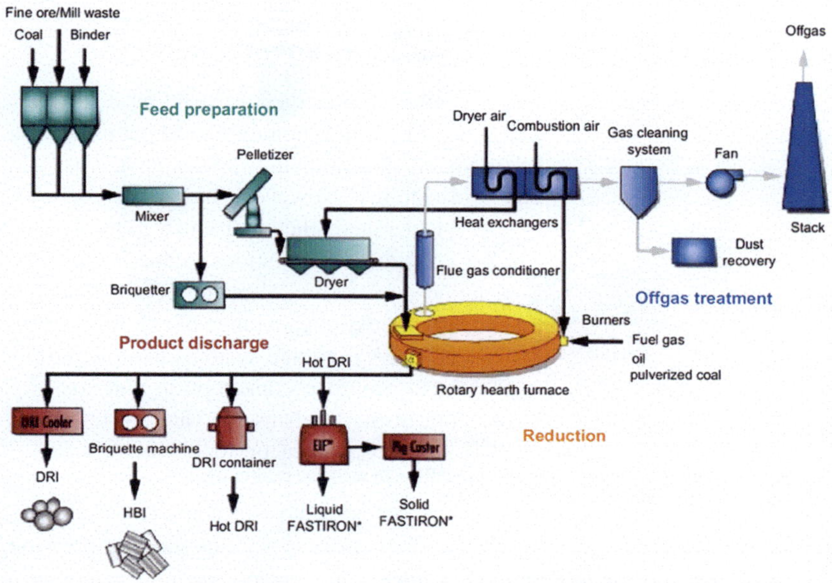

FIGURE 3.5 Descriptive process diagram of the Fastmet coal-based direct-reduction process (Midrex and Kobe Steel).

supplying the heat. The carbon content can be controlled between 1% and 6%, yielding a carbon-bearing prereduced feed with a high degree of metallization, well suited to steelmaking (Ghosh & Chatterjee, 2008).

3.6 Natural-gas–based direct reduction process

The gases generally used in gas-based direct reduction are dihydrogen (H_2) and carbon monoxide (CO). These can be produced via steam reforming of methane, whose principal reaction is given below as Eq. (vi) whith a standard enthalpy change at 298 K of 206 kJ per mole of methane. (Xu & Froment, 1989). Steam methane reforming is a strongly endothermic process that requires operating temperatures of 800°C–1000°C and is typically carried out over nickel-based catalysts.

$$CH_4 + H_2O \rightarrow CO + 3H_2 \qquad \text{(vi)}$$

As noted previously, several reactor types are employed for gas-based direct reduction; however, the most widely deployed is the Midrex process, which is based on a shaft furnace. The dominance of the Midrex process reflects decades of continuous technological refinement, with the latest generation of shaft furnaces. Indeed, this route accounts for 80% of shaft-furnace DRI production and 60% of global DRI production (2021) (Midrex Technologies, Inc., 2025). Fig. 3.6 presents the process. Its production capacity ranges from 300 to 2000 kt • yr^{-1}, depending on unit size (Astier, 2005). The flowsheet comprises a heat

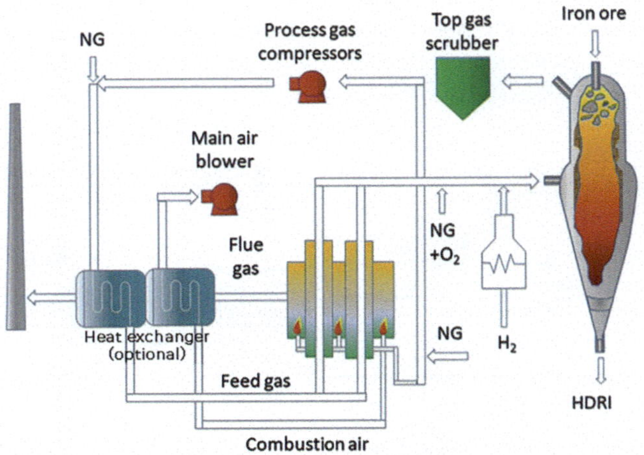

FIGURE 3.6 Midrex process for producing DRI with methane reforming.

recuperator, a natural-gas reformer, a shaft furnace, various gas compression, scrubbing, and cooling systems, and a stack.

Natural gas is preheated and then reformed upstream of the shaft furnace (in the reformer) to produce carbon monoxide and hydrogen. The H_2–CO mixture is then fed directly to the shaft furnace for reduction. Reduction occurs in the solid state between the ascending gases and the descending iron ore. The reaction pathway follows Eq. (iii), (iv) and (v) for carbon monoxide and Eq. (vii), (viii) and (ix) for hydrogen (Longbottom & Kolbeinsen, 2008). The reduction of hematite to magnetite is slightly exothermic, with an enthalpy change of -1.0 kJ mol^{-1} of Fe. The subsequent reduction of magnetite to wüstite is endothermic and requires 24.9 kJ mol^{-1} of Fe; therefore, it represents the most energy-intensive stage. In contrast, the final reduction of wüstite to metallic iron is exothermic and releases 25.4 kJ mol^{-1} of Fe. Thus, the heat absorbed during the intermediate step is nearly compensated by the heat released in the final stage.

$$3Fe_2O_3 + H_2 \rightarrow 2Fe_3O_4 + H_2O \quad \text{(vii)}$$

$$Fe_3O_4 + H_2 \rightarrow 3FeO + H_2O \quad \text{(viii)}$$

$$FeO + H_2 \rightarrow Fe + H_2O \quad \text{(ix)}$$

After reduction, the off-gas contains typically 40%–45% CO, 30%–35% H_2, and 15%–20% CO_2, It exits at the furnace top, is scrubbed, and is used as a fuel for methane reforming. Beyond the reduction zone, the DRI enters a conical cooling section which maintains an inert or slightly reducing atmosphere to prevent reoxidation of the freshly reduced DRI. At the end of which it is discharged for transfer to an EAF. Moreover, when high-carbon DRI is required, methane can be injected into this final zone, where it is cracked in the presence of reactive metallic DRI. In general, DRI metallization is around 93%–94%% (Ghosh & Chatterjee, 2008).

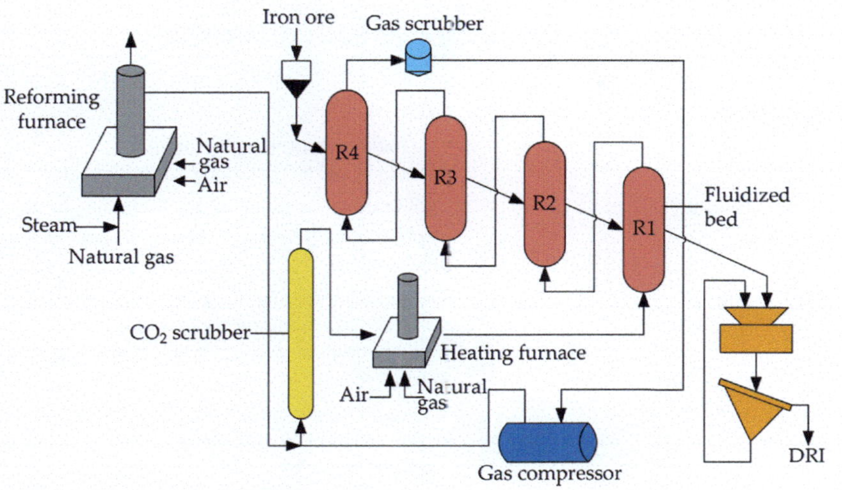

FIGURE 3.7 Finmet process of DRI production with reforming.

3.7 Fluidized-bed direct reduction: The FINMET process

Another reactor mentioned earlier is the fluidized-bed type (Fig. 3.7), which has a capacity of 550 kt • yr^{-1} (Astier, 2005). The system comprises, first, a steam reformer, followed by four fluidized-bed reactors (R1 to R4), various gas-treatment units (scrubber, compressors, etc.), and a hopper system that transfers the feed from one reactor to the next (Astier, 2005).

In this process, solids and gases flow in opposite directions across the reactor cascade. Indeed, the feedstock—lumps of concentrated iron ore smaller than 12 mm—is introduced into the first reactor, R1, whereas the reducing-gas mixture enters the last reactor, R4. This countercurrent configuration is thermodynamically advantageous as it maximizes the driving force for reduction by ensuring that the most reduced iron contacts the freshest, most reactive gas (Zhu, Hu, & Dou, 2024). The gas mixture, whose composition is controlled according to the desired final carbon content, is at approximately 820°C at R4 and about 550°C at R1 (Ghosh & Chatterjee, 2008). The progressive temperature decrease from R4 to R1 drives the sequential reduction reactions while simultaneously preheating and drying the incoming ore burden without requiring external energy input. DRI is recovered at the outlet of R4 as pieces metallized to about 93% and is subsequently briquetted for transport and storage (Astier, 2005).

3.8 Conversion to steel

Here, two technologies are presented: the basic oxygen converter and the EAF. The former is used to refine hot metal through a sequence of dedicated metallurgical units targeting specific impurities or thermal adjustments, as summarized in Table 3.1 . The EAF is employed for the same purpose but accepts

TABLE 3.1 Refining steps for hot metal and their reactors (Steinfeld et al., 1993).

Metallurgical unit	Role
Torpedo ladle	Hot metal transport; temperature homogenization; preliminary desulfurization
AOD Converter	Deep Decarburization, desiliconization and dephosphorization for stainless and special steel grades
Steel ladle	Deoxidation; settling of inclusions; chemical and thermal adjustment (composition and temperature); desulfurization by metal–slag stirring
RH degasser	Vacuum Decarburization (for ultra-low-carbon steel); dehydrogenation; deoxidation for high-purity grades

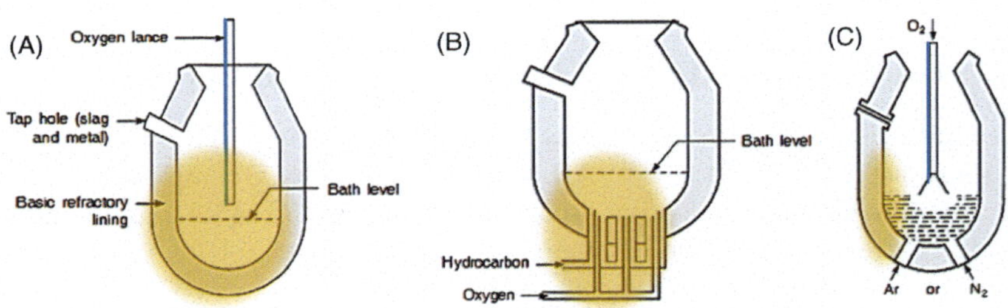

FIGURE 3.8 Different types of oxygen-blown converters: (A) bottom-blown, (B) top-blown, and (C) combined-blown.

a wider range of iron-bearing feedstocks (scrap steel, DRI, briquettes, and sometimes hot metal).

3.9 Basic oxygen converter

Among the hot-metal refining operations, the principal one, called conversion, is carried out in a basic oxygen converter. This metallurgical unit one of the highest-productivity reactors in any industrial process with typical tap-to-tap times of 30–40 minutes. Its main objective is to oxidize the impurities present in the iron, agglomerate them with the slag already present, and oxidize the carbon so as to markedly reduce its concentration. Scrap may be added at this stage as a coolant and supplementary iron source, with scrap rates typically ranging from 15% to 25% of the total metallic charge (Philibert et al., 1997).

There are several types of oxygen converters, shown in Fig. 3.8, differing in how oxygen is blown into the hot steel, either from above or from below, and how the blowing tuyeres are cooled, either by water or by hydrocarbon. There is also a final, so-called mixed (combined) converter that incorporates both approaches. Moreover, the latter enables slag–hot-metal mixing, which promotes effective agglomeration of oxidized particles into the slag (Philibert et al., 1997).

3.10 Electric arc furnace

Once DRI has been produced, the next step is its melting and refining, which is carried out in part in an EAF. The EAF may operate on alternating current (AC) or direct current (DC) (Seetharaman et al., 2013). The choice between AC and DC technology involves trade-offs in capital cost, electrode consumption, power supply infrastructure, and furnace size. the DC furnaces is generally favored for larger heat sizes due to their lower electrode consumption and more stable arc behavior. This technology accounts for roughly 24% of total steel production and emits, 0.23–0.46 tCO_2-eq per ton of steel, depending on the electricity source, efficiencies, and the origin of the steel (scrap, DRI, etc.) (Fan & Friedmann, 2021). When the charge consists solely of scrap, furnace consumption is estimated at 670 kWh per ton of liquid steel. When DRI replaces scrap as the primary charge, energy consumption increases by roughly 100–150 kWh per ton due to the additional energy required to melt and reduce residual iron oxides (Vogl et al., 2018). Fig. 3.9 shows two schematic EAF layouts: the AC furnace above and the DC furnace below. The main advantages of these furnaces over basic oxygen converters are their relatively small size, lower environmental impact, greater flexibility in the grades produced, and more precise control. Furthermore, the EAF is able to reach full production capacity within minutes while blast furnace needs weeks to reline and restart. This gives steelmakers the operational agility in responding to market demand fluctuations when using EAF (Astier, 2005).

Steelmaking in the EAF proceeds in three phases: arc initiation, melting, and refining. The total duration of these three phases typically ranges from 40 to 60 minutes in modern high-productivity EAF operations. At the outset, during arc initiation, the arc power is not set to its maximum in order to limit radiative heating of the roof. In this step, the electrodes descend and create their wells in the cold charge. The aim is to lower the electrodes rapidly so as to move quickly to the melting phase and minimize arc radiation on the roof. During melting, the scrap is liquefied by direct radiation and/or by thermal transfer from the molten pool to the remaining solid scrap, with electrode power nearly at its maximum. Early in this phase, the arc is surrounded by scrap and therefore poses no risk to the furnace shell. However, as melting progresses, the shell becomes increasingly exposed and thus subject to wear; accordingly, arc power is reduced. Finally, the refining phase serves to raise the bath temperature for tapping and to perform the necessary refining operations; at this point, arc power is managed to limit refractory wear (Fours à arc : Dossier complet, 2025).

The remainder of the refining and the adjustment of composition are then performed in a steelmaking ladle during secondary metallurgy (Astier, 2005; Seetharaman et al., 2013). In a DC furnace, the arc is directed from the electrode

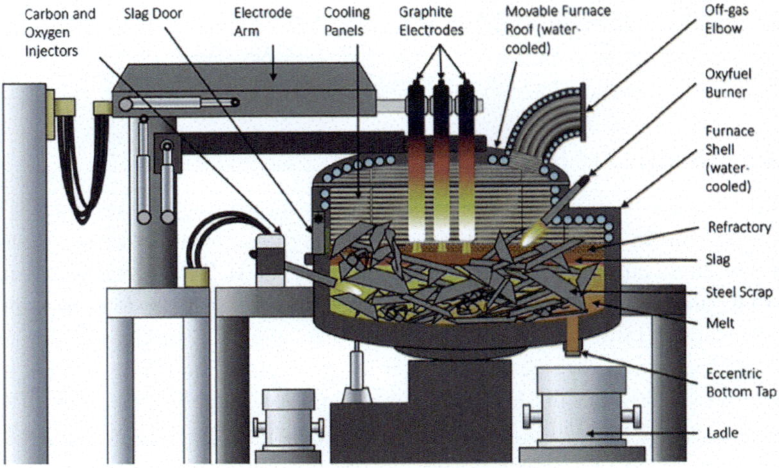

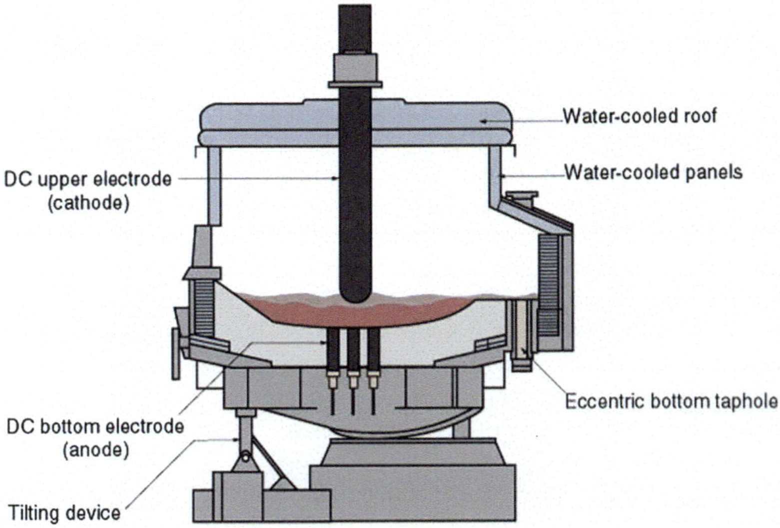

FIGURE 3.9 Diagrams of electric arc furnaces—AC (top) and DC (bottom).

tip toward the charge at an angle of roughly 30°–45° toward the walls, rather than vertically as in AC furnaces, where the arc is emitted from the electrode tip toward the charge. This allows faster, more uniform melting and reduces thermal losses. This feature adds to other advantages of the DC furnace, such as reduced furnace-induced electrical disturbances, lower noise, and reduced electrode consumption. However, there are drawbacks: electrode cost rises with larger diameters, and furnace shell maintenance is also slightly more expensive (Ayed et al., 2005).

3.11 Steel refining prior to casting

Secondary-metallurgy operations are carried out in a ladle or a ladle furnace and consist of a series of targeted refining steps. The sequence and combination of these operations vary according to the steel grade being produced. Indeed, commodity grades may require only basic deoxidation and temperature adjustment, while high-performance grades for automotive or energy applications demand the full suite of secondary metallurgy treatments. These operations consist of: deoxidation by adding deoxidizing elements such as silicon, aluminum, and others; desulfurization by inert-gas stirring in the presence of a lime-rich slag; decarburization by oxygen injection applied only for low-carbon steels grades; denitriding and dehydrogenation by vacuum degassing; and, finally, grade adjustment by adding alloying elements via wire feeding from coils. Precise temperature control at this stage is critical, as deviations of as little as 10°C from the target casting temperature can lead to casting defects. Generally, superheating to casting temperature is achieved by aluminothermic heating for ladle treatments and by an electric arc for ladle furnaces (Philibert et al., 1997).

3.12 Continuous casting

This process follows immediately after secondary-metallurgy operations so as to minimize the additional heat input required for shaping. The thermal integration between secondary metallurgy and continuous castingis a critical factor in reducing energy consumption and improving plant productivity (Pourfathi & Tavakoli, 2023) Fig. 3.10 shows the principal elements of continuous casting. First, a ladle (in which the secondary-metallurgy steps are performed) transfers the liquid steel to a preheated tundish which allows inclusion flotation and temperature homogenization before the steel enters the mold. The steel is then teemed into a water-cooled copper mold to initiate solidification. At the mold exit, a system is installed to control the strand withdrawal speed. In the configuration schematized in the following, the steel strand is bent while still malleable, before being shaped and cut into a semifinished product. The products of continuous casting are slabs, blooms, or billets, which are sent to a reheating furnace and hot-rolled into flat products (sheet or plate) or long products (sections, beams, wire rod, bars) (Philibert et al., 1997).

3.13 Decarbonized reduction with hydrogen

To minimize environmental impacts, a reductant other than carbon monoxide is possible and has already been introduced, namely, hydrogen. The principal

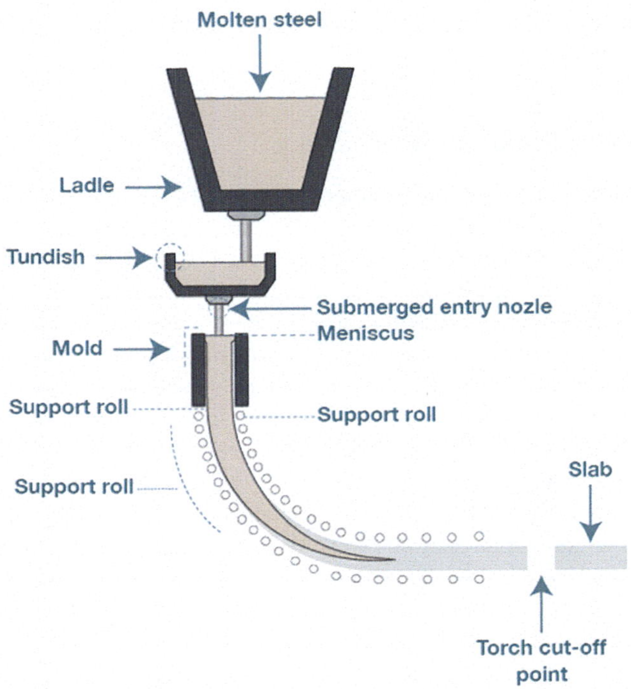

FIGURE 3.10 Main elements required for the continuous casting of steel.

advantage of this reductant is that its reaction with the ore emits no greenhouse gases, and that it can be produced by water electrolysis. However, the thermodynamic and kinetic behavior of hydrogen as a reductant differs significantly from that of carbon monoxide. Indeed, hydrogen reduction is endothermic above 810°C, which requires continuous external heat input to sustain the reaction. This will have important implications for shaft furnace design and energy integration.

As noted earlier, iron reduction emits a large quantity of greenhouse gases. These emissions are due, in part, to the reduction reaction of iron ore, which generally employs carbon-based reagents. Hydrogen enables iron reduction without greenhouse gas emissions and thus fits within the decarbonization of the reduction step. Indeed, the use of hydrogen in reduction can allow a reduction close to 90% in greenhouse gas emissions relative to conventional direct-reduction methods. However, the actual emissions reduction achieved depends critically on the carbon intensity of the electricity used to produce the hydrogen. Indeed, green hydrogen produced via renewable-powered electrolysis delivers near-zero lifecycle emissions, while hydrogen produced from fossil fuels offers only marginal improvement over conventional routes (Patisson & Mirgaux, 2020).

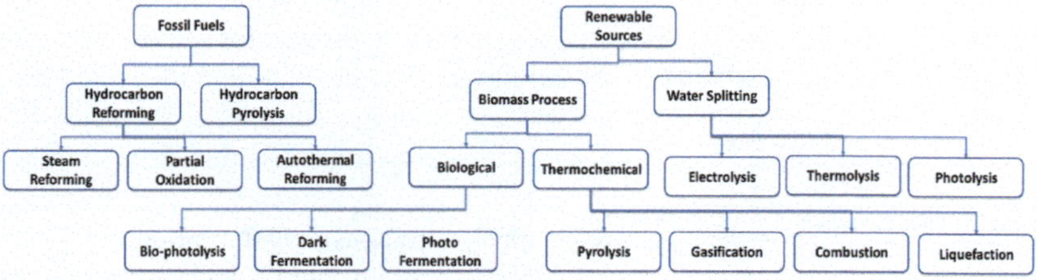

FIGURE 3.11 Different hydrogen-production technologies.

From a production standpoint, 54 kg of hydrogen are required to produce 1 ton of iron; producing 1 megaton of iron would therefore require a hydrogen-production unit with a capacity of 70,000 $Nm^3 \cdot h^{-1}$ (Ivy, 2004). A simple calculation gives a sense of what DRI production with hydrogen would represent. In 2022 125 Mt of DRI were produced according to the World Steel Association; producing this with hydrogen would require 6.75 Mt of the reductant. Yet electrolysis requires between 53.4 and 70.1 $kWh \cdot kg^{-1}$ of H_2 (Ivy, 2004). Taking the average value, 416 TWh of electricity would be needed to supply productioni.e., 1.3%% of global electricity generation in 2025'. It thus appears difficult to convert all DRI production to hydrogen. Moreover, these calculations do not account for production via the integrated route, which represents the larger share of output; producing all steel this way would bring the requirement close to one-quarter of global electricity generation. However, if one adopts a more moderate stance in terms of production, hydrogen can have a role as an alternative to the carbon-based reducing gas (i.e., carbon monoxide, CO).

3.14 Hydrogen production

In response to the high greenhouse gas emissions associated with the use of fossil fuels in industry, hydrogen is attracting many sectors (automotive, aviation, metallurgy, etc.). However, it is currently produced using technologies that emit large quantities of greenhouse gases. Indeed, the International Energy Agency (IEA) reports that of the 90 Mt of hydrogen produced in 2020 79% came from hydrogen-production facilities based on fossil energies: 59% from natural gas and 19% from coal; the remaining 21% is a by-product of hydrocarbon refining; all of this is complemented by other methods (IEA, 2021). To decarbonize production, the most promising route is water electrolysis, which today represents 0.03% of global production, with roughly 30 kt of hydrogen produced in 2020 (IEA, 2021). As for hydrogen use, 72 Mt are used directly in the petroleum industry and for ammonia production, and 18 Mt are destined for methanol production and for the direct reduction of iron ore (IEA, 2021).

Fig. 3.11 presents all current hydrogen-production pathways (Nikolaidis & Poullikkas, 2017). Two main families can be distinguished: those based on fossil fuels and those based on renewable sources. In the fossil route, two

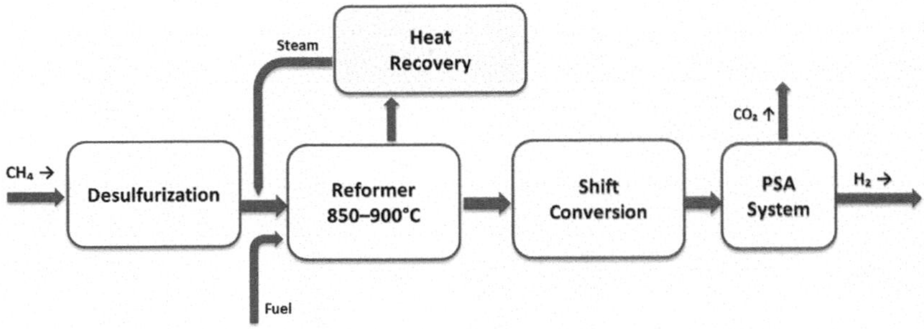

FIGURE 3.12 Block diagram of the steps in steam-methane reforming.

major groups exist: pyrolysis of hydrocarbons and their reforming. The latter can be implemented by three techniques: reforming or steam reforming, partial oxidation (gasification), and autothermal reforming. For the renewable route, biomass or water can be used. For biomass, two pathways are possible: via a biological process or via thermochemical routes. The first includes several techniques, namely biophotolysis, dark fermentation, and photofermentation. The second comprises four techniques: pyrolysis, gasification, combustion, and liquefaction (Nikolaidis & Poullikkas, 2017). From Fig. 3.11, it is apparent that a very large number of techniques are available today. However, some are more mature and costlier than others: at present, the most widely used (>78%) are hydrocarbon steam reforming (natural gas) and coal gasification.

3.15 Steam reforming of natural gas

Steam reforming of natural gas consists of decomposing natural gas (primarily methane, CH_4) into carbon dioxide (CO_2) and hydrogen (H_2). Fig. 3.12 presents the main steps of this technique.

The desulfurization step is a critical safeguard for long-term reformer operability and economics. Indeed, sulfur-bearing compounds are present in the feed to the reforming system and can lead to catalyst poisoning inside the reformer. To remove them, an initial desulfurization step is required, as shown by Nikolaidis & Poullikkas (2017)). This purification is carried out by the irreversible combination of hydrogen sulfide and zinc oxide. Initially, the sulfur compounds may be present in several forms: hydrogen sulfide (H_2S), mercaptans (R-HS), and carbonyl sulfide (COS). A preliminary catalytic hydrogenation step is needed to convert the sulfur compounds to hydrogen sulfide before zinc oxide adsorption can proceed (Giroudière & Le Gall, 2012). This two-stage desulfurization approach is now standard practice in industrial steam reforming plants as it achieves outlet sulfur concentrations below 0.05 ppm which fits the tolerance of commercial reforming catalysts. After purification, the natural gas is sent to the reformer, where it is combined with pressurized hot steam. The reformer outlet gas reaches temperatures on on the order of 800°C–950°C, although the process is activated at about 700°C–800°C under a pressure ranging from 3 to 28 atm, depending on the reformer type and downstream applications (Ogden, 2002; Rostrup-Nielsen et al., 2025). Lower pressures favor thermodynamic conversion of methane but require larger equipment, while higher pressures reduce vessel size and favor downstream compression for hydrogen storage or pipeline transport. The equation for methane reforming is given in Eq. (x), the reaction is strongly endothermic, with a standard enthalpy change of +206 kJ mol^{-1} CH_4 at 298 K(Ogden, 2002):

$$CH_4 + H_2O \rightarrow CO + 3H_2 \tag{x}$$

a nickel-based catalyst (IEA, 2021). Because the reforming reaction is highly endothermic, a substantial amount of thermal energy (63.3 kJ per mole of H_2) is required to activate the reaction (Nikolaidis & Poullikkas, 2017). This energy is supplied by the combustion of natural gas, which accounts for 30%–35% of the process's total gas consumption (Muradov, 2001).Process efficiency is improved through heat integration to minimize fuel consumption. Indeed, heat recovery is used to preheat the feed to approximately 450°C–650°C (Rostrup-Nielsen et al., 2003). Finally, to avoid coke formation in the reformer, the steam-to-carbon ratio in the reactor must be greater than 3; a lower value leads to reduced process efficiency as well as poorer product-gas quality (Ogden, 2002). Maintaining an adequate steam-to-carbon ratio also helps preserve catalyst activity and extend catalyst lifetime by limiting carbon deposition on the catalyst surface. Once reforming is complete, a water–gas shift (WGS) conversion of carbon monoxide follows. This reaction, given by Eq. (xi), greatly increases the amount of hydrogen produced (Huang & Dincer, 2014):

$$CO + H_2O \rightarrow CO_2 + H_2 \tag{xi}$$

The final step is purification by pressure swing adsorption (PSA) to achieve very high purities of 98–99.999 mol% H_2. The feed gas. which is cooled to approximately 20°C–40°C, contains about 70%–80% H_2 together with CO_2, CH_4, CO, and trace amounts of N_2%%%%%%% (Sircar & Golden, 2000). In practical terms, the pressurized inlet gas passes through an adsorber on which impurities are captured; a portion of the gas is withdrawn directly and constitutes the first stream of pure hydrogen. Hydrogen is recovered as the product gas because of its relatively low adsorption affinity compared with the other gas components. Commercial PSA units commonly employ adsorbents such as activated carbon, zeolites, and silica gel to selectively remove impurities. This step is halted before the adsorber becomes saturated with the inlet gas. The adsorber is then brought to a lower pressure by sending part of the clean gas to a neighboring adsorber. This step is repeated two additional times. The adsorber is subsequently depressurized to the maximum in countercurrent flow, the outlet gases being off-gases containing a portion of the desorbed impurities. A purge then follows using a stream of hydrogen from a neighboring adsorber. At this point, the remaining impurities are desorbed and extracted via an exhaust line.

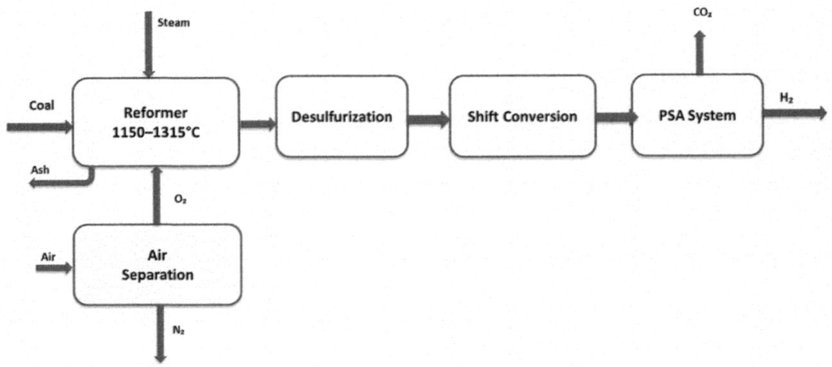

FIGURE 3.13 Block diagram of the main steps in coal gasification.

The adsorber is next brought back up to pressure to the initial fill level using a neighboring adsorber. Once re-pressurized, the adsorber is ready for a new cycle (Sircar & Golden, 2000). Although PSA can achieve extremely high hydrogen purities, a compromise often exists between hydrogen purity and recovery. The off-gas stream, which contains desorbed impurities together with a fraction of hydrogen, is commonly reused as a fuel source within the plant to improve the overall energy efficiency of the process.

Owing to its technological maturity, large production capacity, and relatively low production cost, steam methane reforming remains the dominant industrial route for hydrogen production worldwide. In total, this process requires 50.9 kWh per kg H_2 (Sircar & Golden, 2000). The actual efficiency depends on several factors, including reformer operating conditions, heat-integration performance, and the effectiveness of downstream purification units.The production capacity of a hydrogen unit is approximately 20,000 kg/h, with a conversion efficiency of 74%–85% (Nikolaidis & Poullikkas, 2017; Giroudière & Le Gall, 2012, 2025).

3.16 Coal gasification

Coal gasification, or partial oxidation of coal, consists of oxidizing coal in the presence of water and oxygen (O_2).This exothermic–endothermic balance can be tuned by adjusting the O_2/steam ratio to favor H_2 production while minimizing CO_2 formation Fig. 3.13 presents the main steps of coal gasification. Units similar to those in methane steam reforming can be identified: desulfurization, CO conversion (WGS), and purification (PSA). One also notes the presence of a reformer and an air-separation system (Nikolaidis & Poullikkas, 2017). The air-separation unit (ASU) alone accounts for up to 15%–20% of the plant's auxiliary power consumption, driving research into oxygen-transport membranes and chemical-looping concepts. In these systems, coal is introduced into the reactor with oxygen at high temperature and/or with steam, the use of supercritical water as a gasification medium is also emerging as an alternative to acheive higher carbon conversion and reduced tar formation (Huang & Dincer, 2014). This process is thermodynamically favored at elevated pressures and is often catalyzed by alkali compounds (e.g., K_2CO_3), which enhance gasification kinetics and promote tar-reforming reactions.

In practical terms, the oxygen and steam present in the reactor oxidize the coal, yielding a gas mixture containing water vapor (H_2O), carbon monoxide (CO), carbon dioxide (CO_2), and hydrogen. The H_2/CO ratio, typically around 0.5–1.0, can be adjusted via the water-gas shift (WGS) stage to meet downstream requirements for methanol synthesis or fuel cells. Undesirable compounds such as tar, phenols, or ash may also be present, advanced plasma-based tar cracking and catalytic filters are being investigated to reduce these contaminants to parts-per-million levels (Huang & Dincer, 2014). Coal gasification is a complex process carried out at very high temperatures, from 1250°C to 1600°C. It proceeds via several chemical reactions; the principal ones being given in Eqs. (xii–xv) (Huang & Dincer, 2014; Seyitoglu et al., 2017). The thermal efficiency of this process ranges from 60% to 75% when calculated on a higher heating value (HHV) basis (Holladay et al., 2009). This efficiency can be improved by co-feeding biomass or municipal solid waste, which enhances carbon utilization and reduces greenhouse gas emissions per unit of energy output. Moreover, the coal-to-gas conversion yield varies between 80% and more than 95%, depending on the reactor type (moving-bed, fluidized-bed, and entrained-flow) (Gräbner, 2014). In terms of production, the largest coal-to-hydrogen unit in Mongolia produces 20,455 kg·h^{-1} of hydrogen (Gräbner, 2014)This scale, however, is being challenged by smaller modular gasifiers coupled with carbon capture, which offer lower transportation costs and greater geographical flexibility for hydrogen distribution (Sánchez-Ramírez, Juarez-García, & Huerta-Rosas, 2026).:

$$C + 1/2O_2 \rightarrow CO \tag{xii}$$

$$C + H_2O \rightarrow CO + H_2 \tag{xiii}$$

$$H_2 + 1/2O_2 \rightarrow H_2O \tag{xiv}$$

$$C + CO_2 + H_2O \rightarrow CO + H_2 \tag{xv}$$

3.17 Hydrocarbon steam cracking

Hydrocarbons are chemical compounds obtained from petroleum refining. These feedstocks vary widely in composition depending on the crude oil source, with heavier fractions (e.g., naphtha, gas oil) requiring different cracking severities than lighter ethane or propane feeds. These petrochemical crudes consist solely of carbon and hydrogen atoms and have the general formula C_nH_m. The ratio n/m influences product distribution. Indeed, the higher hydrogen-to-carbon ratios increases ethylene yields and reduces propylene and butadiene formation. Once extracted, petrochemical crudes are fractionated, yielding natural gas

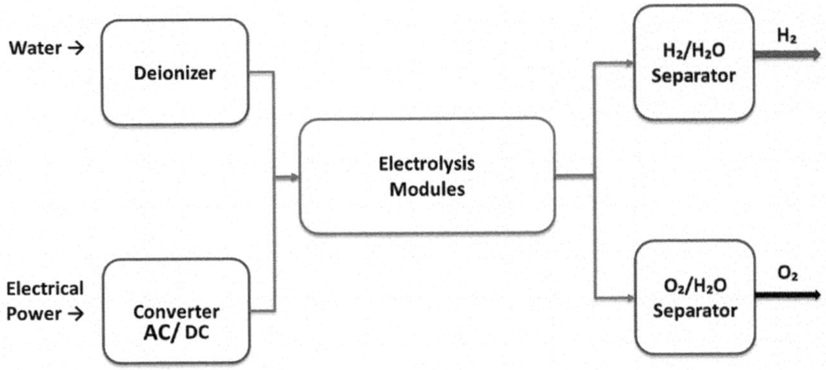

FIGURE 3.14 Block diagram of the main components of an electrolyzer for hydrogen production.

and petroleum fractions composed mainly of saturated hydrocarbons, paraffins, and naphthalenes. These compounds exhibit low chemical reactivity and do not allow for a great diversity of chemical products (Raimbault & Lefebvre, 1995). This low reactivity stems from the strong and stable C–C and C–H sigma bonds, which require high energy inputs to break, making steam cracking one of the most energy-intensive processes in the petrochemical industry. It is through steam cracking that these materials are transformed into unsaturated hydrocarbons, which offer numerous reaction possibilities and are therefore widely used in organic synthesis (Raimbault & Lefebvre, 1995).

Steam cracking of hydrocarbons favors C–C bond rupture over C–H cleavage due to lower bond energy (345 vs. 413 $kJ \cdot mol^{-1}$). the output is primarily olefins with minor H_2 production. Steam enhances selectivity and reduces coke via partial pressure reduction and active gasification of carbon deposits (Raimbault & Lefebvre, 1995).

3.18 Water electrolysis

Water electrolysis accounts for a very small share of hydrogen production (0.03% in 2020), more than 40% of which is produced in Europe (IEA, 2021). By 2050 the share of electrolytic production is expected to reach 50%–60%, depending on the scenario (IEA, 2021). Today, three technologies are used to perform electrolysis (Holladay et al., 2009; Nikolaidis & Poullikkas, 2017):

- Alkaline electrolysis
- Membrane electrolysis (PEM)
- Solid-oxide electrolysis (solid-oxide electrolysis cells [SOEC])

Overall, as shown in Fig. 3.14, electrolysis systems comprise a water deionization unit, a power source, an electrolyzer, and gas separators (H_2/H_2O or O_2/H_2O). The deionization unit is critical for preventing membrane degradation and catalyst poisoning, particularly in PEM electrolyzers where impurities such as metal ions can drastically reduce performance and lifetime. These technologies offer good efficiencies (about 60%–70%) but are highly energy-intensive: 53.4–70.1 kWh per kg of H_2 produced in total. This energy consumption corresponds to a lower heating value (LHV) efficiency range of 56%–74%, with the higher values achieved by SOEC which operates at elevated temperatures (Haryanto et al., 2005). The most widespread technology is alkaline electrolysis, representing a production capacity of 380,000 kg H_2 per year, i.e., 4,228,000 m^3 per year (Nikolaidis & Poullikkas, 2017). The dominance of alkaline electrolysis is due to its long operational track record (over 100 years) and lower capital costs (approximately 500–1000 €/kW). The use of inexpensive nickel-based electrodes is also an advantage. However, its dynamic response is slower than that of PEM systems and it operates at lower current densities.

3.19 Alkaline electrolysis

Alkaline electrolysis is the most widely used of the three technologies. Typically, the cathode may be made of steel, stainless steel, steel with a nickel coating, or nickel. Nickel-based cathodes are preferred due to their high electrocatalytic activity for the hydrogen evolution reaction (HER) and excellent stability in concentrated alkaline media, while research continues on Raney nickel and nickel–molybdenum alloys to further reduce overpotentials and improve performance. For the anode, common materials are nickel oxide, nickel spinel, and cobalt supported on titanium (Santos et al., 2013). Anode materials are critical for the oxygen evolution reaction (OER); cobalt and nickel spinels are investigated as low-cost alternatives to precious metal oxides like IrO_2 or RuO_2 used in PEM systems. The electrolyte is generally an aqueous solution containing 25–30 wt% potassium hydroxide (KOH) to improve electrical conductivity; sodium hydroxide (NaOH) or sodium chloride (NaCl) may also be encountered, the choice of KOH is preferred over NaOH due to its higher ionic mobility and conductivity, while NaCl is rarely used in practice because chloride ions can lead to chlorine evolution at the anode, competing with oxygen formation and causing corrosion (Holladay et al., 2009; Santos et al., 2013). During operation, the electrolyte typically runs at high current densities of 100–300 mA• cm^{-2}. Higher current densities improve hydrogen productivity but increase ohmic losses and gas bubble formation at the electrode surface. This may reduce effective electrode area and increase overpotentials(Turner et al., 2008). In practice, the two electrodes are immersed in the electrolyte and separated by a membrane or diaphragm. A DC is applied, driving electrons from the negative pole of the power supply toward the cathode. Consequently, hydrogen cations (H^+) migrate to the cathode and hydroxide anions (OH^-) to the anode. In alkaline media, the actual charge carriers are OH^- ions migrating from cathode to anode through the diaphragm, while H^+ concentration is extremely low due to the high pH; the overall ionic transport is governed by diffusion and migration under the applied electric field. Once at the cathode, the hydrogen cations are reduced there and recombine to form dihydrogen (Santos et al., 2013). The reactions governing water decomposition are given in Eqs. (xvi–xviii) (Nikolaidis & Poullikkas, 2017):

$$\text{Anode}: 4OH^- \rightarrow O_2 + 2H_2O + 4e^- \qquad \text{(xvi)}$$

$$\text{Cathode}: 2H_2O + 2e^- \rightarrow H_2 + 2OH^- \qquad \text{(xvii)}$$

$$\text{Complete}: \ H_2O \rightarrow H_2 + \frac{1}{2}O_2 \qquad \text{(xviii)}$$

3.20 Proton exchange membrane electrolysis

The second technology used is PEM electrolysis. PEM electrolysis has gained significant attention in recent years due to its high current density capability and fast dynamic response. Also, it has the ability to operate at high differential pressures which makes it suitable for coupling with intermittent renewable energy sources such as wind and solar. The electrolyzer consists of two electrodes and a membrane.The membrane is typically made of a perfluorosulfonic acid polymer such as Nafion. It acts as a solid electrolyte that conducts protons while providing electrical insulation and gas separation. Its most influencing parameters is the thickness (typically 50–200 μm) and water content. For operation, high-purity deionized water with a minimum resistivity of 1 $M\Omega \bullet cm$ is required. This is essential because metalic impurities can poison the membrane and catalyst layers which leads to irreversible performance degradation and shortened stack lifetime. In this case filtration and ion-exchange systems are employed upstream to maintain water quality (Holladay et al., 2009; Turner et al., 2008). First, deionized water is fed to the anode side, and the applied current produces hydrogen cations (H^+) and oxygen. The protons then pass through the membrane under the electric field, whereas the oxygen remains on the anode side and is vented at atmospheric pressure. Dihydrogen is formed at the cathode by electrons supplied by the power source (Turner et al., 2008). At the anode, the OER occurs under highly acidic conditions which requires the use of noble metal catalysts such as iridium oxide (IrO_2) or ruthenium oxide (RuO_2). This causes a major cost barrier for widespread commercialization as those catalysts are expensive and scarce. At the cathode, the HER occurs on platinum-based catalysts, which are also costly but essential to achieve acceptable kinetics. For this reaction to proceed, a high current density (>1600 $mA \bullet cm^{-2}$) is necessary; as this density increases, overall efficiency decreases, but the quantity of hydrogen produced increases (Turner et al., 2008). The advantages of this technology are: rapid start-up and shut-down times (seconds to minutes), excellent load-following capability (0%–200% of nominal load), high hydrogen purity (>99.99%), a large pressure difference can be maintained between anode and cathode, thereby avoiding high oxygen pressures; and the use of hazardous products such as potassium hydroxide is avoided (Santos et al., 2013). The equations governing the electrolysis are given in Eqs. (xix–xxi) (Holladay et al., 2009):

$$\text{Anode}: 2H_2O \rightarrow O_2 + 4H^+ + 4e^- \quad \text{(xix)}$$

$$\text{Cathode}: 4H^+ + 4e^- \rightarrow 2H_2 \quad \text{(xx)}$$

$$\text{Complete}: H_2O \rightarrow H_2 + 1/2O_2 \quad \text{(xxi)}$$

3.21 Solid-oxide electrolysis cell

The last technology is the SOEC, notable for its solid electrolyte with electrodes on either side. In practice, steam is introduced at the porous cathode and diffuses to the electrolyte/cathode interface. When current is applied, water is dissociated into dihydrogen and oxygen anions. The anions traverse the electrolyte to the anode, where they are oxidized to oxygen and then diffuse through the porous anode (Ni et al., 2008). At the anode (oxygen electrode), materials such as lanthanum strontium manganite (LSM) or lanthanum strontium cobalt ferrite (LSCF) are commonly used, and the OER at these high temperatures exhibits significantly lower overpotentials compared to low-temperature electrolyzers. In this technology, the electrodes are commonly referred to as the hydrogen electrode or the oxygen electrode, depending on the gas produced at each (Nechache & Hody, 2021). For proper cell operation, the electrolyte must be chemically stable while exhibiting good ionic and electronic conduction; it must also be gas-tight to prevent recombination of hydrogen with oxygen and be sufficiently thin to avoid ohmic overpotentials. The electrolyte is typically yttria-stabilized zirconia (YSZ) or scandia-stabilized zirconia (ScSZ), which exhibit excellent ionic conductivity above 700°C. The electrodes, for their part, must be chemically stable in reducing or oxidizing environments, porous, and possess good electronic conductivity. Finally, their thermal expansion coefficients should be similar to avoid damage from mechanical stresses (Ni et al., 2008). %

This technology concludes the overview of current hydrogen-production methods. We have seen that the most widespread is methane steam reforming, which accounts for 59% of global production, followed by coal gasification at 19%; finally, 21% of production is obtained as by-products of hydrocarbon cracking. However, steam reforming and gasification emit large quantities of greenhouse gases: 11.7 and 20 kgCO_2-eq per kg H_2, respectively (IEA, 2021). As explained earlier, once produced, hydrogen is used in several sectors, including the direct reduction of iron ore (IEA, 2021). It is therefore necessary to decarbonize hydrogen production to improve the impact of hydrogen-based iron reduction. One of the routes presented for this purpose is electrolysis: in France, this technique reduces emissions to 2.77 kgCO_2-eq per kg H_2, although when a more carbon-intensive power mix is used, this technology remains as emissive

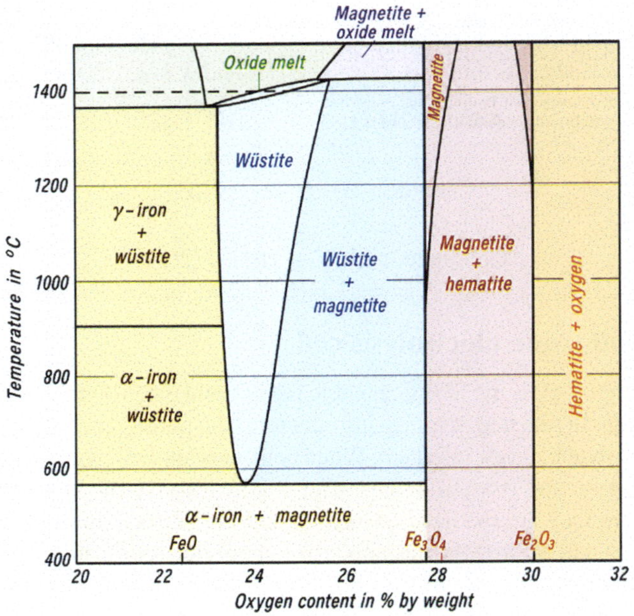

FIGURE 3.15 Phase diagram of the iron–oxygen system for temperatures from 400°C to over 1400°C and oxygen contents from 20% to 32%.

as fossil routes (19.8 kgCO_2-eq per kg H_2 with the European mix) (Nechache & Hody, 2021). It should be kept in mind that electrolysis consumes large amounts of electricity (53.4–70.1 kWh per kg H_2); consequently, the country in which it is implemented must have a low-carbon energy mix for it to be advantageous.

As noted earlier, reduction of iron ore is one of hydrogen's applications. This reaction is known and used industrially. The reduction reaction will be presented in detail in the next section, covering thermodynamic aspects, the reduction mechanism, and the parameters influencing reaction kinetics.

3.22 Thermodynamic aspects of hydrogen-based iron-ore reduction

In essence, iron oxide is reduced to metallic iron by hydrogen, forming water; the overall reaction is given in Eq. (xxii) (Longbottom & Kolbeinsen, 2008). The reaction has a standard enthalpy change of +49.4 kJ mol^{-1} Fe at 298 K, which indicates an endothermic process.:

$$3Fe_2O_3 + H_2 \rightarrow 2Fe_3O_4 + H_2O \quad \text{(xxii)}$$

The global reaction is therefore relatively endothermic and, industrially speaking, to convert 1 ton of iron ore into about 700 kg of iron, it is theoretically necessary to supply 420 Nm^3 of hydrogen, i.e., about 38 kg of H_2. Fig. 3.15 shows the phase diagram of the iron–oxygen system, correlating temperature and oxygen percentage with each existing phase. One observes hematite (Fe_2O_3), magnetite (Fe_3O_4), and wüstite ($Fe_{)1-y)}O$) appearing in sequence as the oxygen content decreases. It is also apparent that for oxygen contents below 27.6% and temperatures below roughly 570°C, wüstite does not form. These data indicate that the reduction pathway depends on temperature and that several reactions proceed in sequence to achieve complete reduction [44]. Eqs. (xxiii and xxiv) describe the reduction for temperatures below 570°C, while Eqs. (xxv–xxvii) apply for temperatures above 570°C (Longbottom & Kolbeinsen, 2008; von Bogdandy & Engell, 2013):

$$3Fe_2O_3 + H_2 \rightarrow 2Fe_3O_4 + H_2 \tag{xxiii}$$

$$Fe_3O_4 + H_2 \rightarrow 3FeO + H_2O \tag{xxiv}$$

$$3Fe_2O_3 + H_2 \rightarrow 2Fe_3O_4 + H_2O \tag{xxv}$$

$$Fe_3O_4 + H_2 \rightarrow 3FeO + H_2O \tag{xxvi}$$

$$FeO + H_2 \rightarrow Fe + H_2O \tag{xxvii}$$

For this reaction, the Gibbs free energy is given in Eq. (3.3). This expression shows that these reactions are influenced by temperature and by the partial pressures of water and hydrogen. The reaction may therefore fail to proceed if the water content is too high relative to the proportion of hydrogen:

$$\Delta G_n^0(T) = \Delta G_n^0(T) + RT\ln\left(\frac{p_{H_2O}}{p_{H_2}}\right) \tag{3.3}$$

Fig. 3.16 presents the Chaudron diagram for the reduction of iron ore under hydrogen (Patisson & Mirgaux, 2020). The solid lines represent equilibria between two solid phases. For example, the horizontal line around 900°C corresponds to the transition from α-Fe to γ-Fe. Another example: at 40% water in the gas mixture and at 1000°C, the system passes from γ-Fe to wüstite of type $Fe_{0.94}O$. This diagram shows that, with increasing temperature, iron formation can occur even with high proportions of water in the reaction atmosphere (up to 40% at above 1000°C). Conversely, if the water proportion is too high, even at very high temperatures, complete reduction of iron will not be possible. Moreover, at lower temperatures, the maximum allowable water fraction is smaller. For instance at temperatures below 570°C—i.e., in the case where wüstite formation is avoided—the maximum water content is about 20%.

3.23 Physicochemical aspects of hydrogen-based reduction

3.24 Gas–solid reaction

3.24.1 Macroscopic aspects

To begin with, it is important to note that pellets will hereafter be treated as spheres. In reality, pellets have varied shapes, although they are predominantly

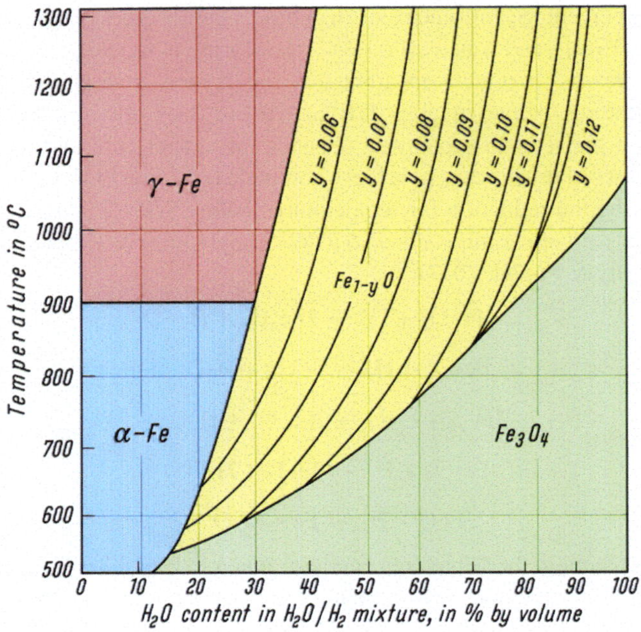

FIGURE 3.16 Chaudron diagram for the reduction of iron ore under hydrogen.

spheroidal. This geometric simplification is standard in the literature but it introduces a degree of error in surface area and diffusion path length calculations. This degree or error becomes more significant for irregularly shaped or cracked pellets. Moreover, hydrogen reduction of iron ore is a reaction in which the particle does not change size during the reaction (Levenspiel, 1999). In this case, two models are generally used: the progressive-conversion model (PCM) and the shrinking-core model (SCM) (Levenspiel, 1999). In the PCM, the gas diffuses into the particle and reacts throughout the reaction time. The local kinetics may differ, but the particle is converted continuously. In the SCM, the reaction starts at the external surface of the particle and then progresses toward the center, leaving a layer of transformed material called "ash." The diameter of the unreacted core shrinks as conversion proceeds (Levenspiel, 1999). Generally, because the iron layer is dense, homogeneous diffusion through the ore is not possible. In addition, since the reducing gas is in close contact with the ore, reduction begins at the outer surface. Logically, therefore, the SCM is preferred to describe iron-ore reduction (Heidari et al., 2021). However, Spreitzer and Schenk (2019) noted that the applicability of the SCM deteriorates at high temperatures where sintering modifies the iron layer morphology and creates preferential diffusion pathways. So they suggest that the model selection should account for the operating temperature range and the specific reduction stage being modeled.

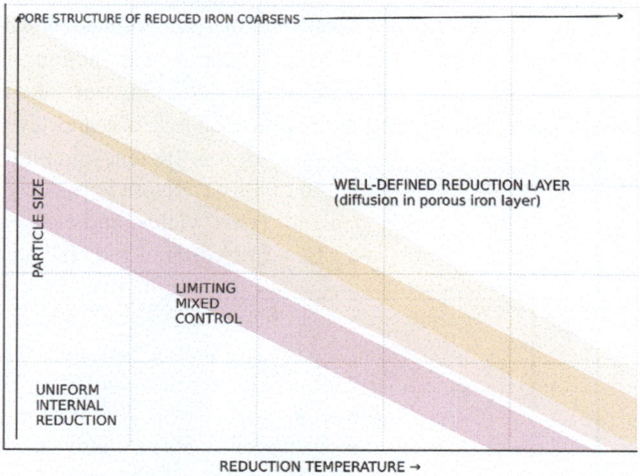

FIGURE 3.17 Different controlling regimes of reduction as a function of particle size and reduction temperature.

In this model, several phenomena can be rate-limiting depending on the conditions: diffusion of the gas from the bulk atmosphere to the reaction surface, diffusion of the gas through the transformed portion of the particle, and the chemical reaction at the core surface (Levenspiel, 1999).

More specifically, Turkdogan and Vinters (1971) studied in depth the rate-limiting phenomena in the hydrogen reduction of iron ore. They identified three controlling processes, which may occur simultaneously:

- Uniform internal reduction: Occurs when small, porous particles are reduced at low temperature such that diffusion does not hinder reduction. In this case, the reaction time is independent of particle radius. This regime is rarely encountered in industrial pellets, which are typically 8–16 mm in diameter and exhibit significant internal diffusion resistance even at moderate temperatures, but it serves as a useful theoretical reference for interpreting fine powder reduction experiments
- Mixed-control regime: This regime applies at the beginning of the reduction for larger pellets and at elevated temperature. A thin but rapidly permeable iron layer forms, moreover porosity and surface reaction are the rate-limiting factors with reaction time proportional to pellet radius. This regime can occasionally produce reduced zones ahead of the main wüstite-to-iron front.
- Diffusion through the formed iron: As the iron layer thickens and reduction progresses, diffusion through the pores of the iron becomes rate-limiting. In this case, the reaction time is proportional to the square of the pellet radius.

These different limiting regimes are presented in Fig. 3.17 as a function of particle size and temperature. As explained, the uniform internal reduction

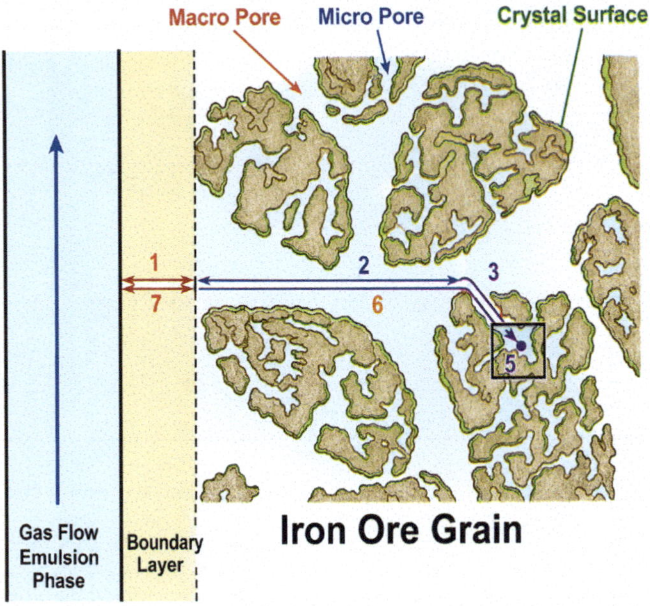

FIGURE 3.18 Physicochemical step in the reduction of iron-ore grains under hydrogen.

regime corresponds to low temperatures and small particle sizes; diffusion through the formed iron corresponds to larger particles and higher temperatures; and the mixed-control regime occupies the transition between the two. One may also note that the mixed-control regime applies to small particles at high temperature and to relatively large particles at low temperature (Turkdogan & Vinters, 1971).

3.25 Microscopic aspects

From a microscopic standpoint, the reduction of an iron-oxide grain proceeds via seven physicochemical steps, referred to as the reaction sequence. These seven steps are common to most gas-solid reactions that involves porous solids. The overall reduction rate is determined by the slowest among them either the chemical reaction at the active site or the diffusion of gases through the pore structure, depending on the temperature and the grain size. Fig. 3.18 schematically presents this sequence (von Bogdandy & Engell, 2013). First, hydrogen is transferred from the gas stream to the ore surface (1). This external mass-transfer step is influenced by the gas velocity, turbulence, and the boundary layer thickness around the grain; under typical industrial conditions with high gas flow rates, this step is rarely rate-limiting unless the gas velocity is very low or the grain is exceptionally large. The gas then diffuses through the macropores (2) and then the micropores (3) of the sample to the active site, where the chemical reaction

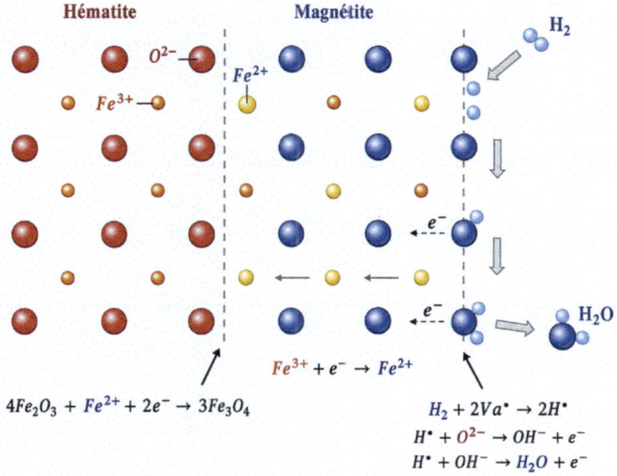

FIGURE 3.19 Schematic of the transformation of hematite to magnetite.

occurs (4). Macropores (diameters > 50 nm) facilitate rapid gas transport via molecular diffusion, while micropores (diameters < 2 nm) offer higher surface area for reaction but impose greater diffusional resistance. Water vapor then diffuses outward through the micropores (5) and then through the macropores (6) before being transferred back to the gas stream (7) The outward diffusion of water vapor must compete with the inward diffusion of hydrogen, and the accumulation of water vapor within the pores can lead to a local decrease in the reducing potential (von Bogdandy & Engell, 2013). Each step in this sequence is detailed in the following.

Among these steps, the fourth (i.e., the chemical reaction at the active site) is the most critical, as it governs the actual transformation of iron oxide into metallic iron. The reaction occurring at active sites combines adsorption of hydrogen, solid-state transport of ions and electrons, formation of H_2O, and its desorption. It is shown in Figs. 3.19 and 3.20 for the reduction of magnetite and of wüstite, respectively. In practice, if conditions are met (activation energy, sufficient hydrogen availability, etc.), hydrogen in contact with the surface is adsorbed on two vacant interstitial sites (2V_a$^+$) (bonding to surface atoms constituting the ore) according to Eq. (xxviii). Its interaction with surface oxygen leads to water formation and the release of two electrons according to Eqs. (xxix and xxx) (Ma et al., 2022). This water-formation step is exothermic and releases electrons that are injected into the oxide lattice, increasing the electronic conductivity of the oxide and facilitating subsequent charge-transfer processes . These electrons affect the reduction of Fe^{3+} to Fe^{2+} (Eq. xxxi) and thereby enable the production of magnetite according to Eq. (xxxii) (Feilmayr et al., 2004; Kim et al., 2021). Moreover, a gradient in chemical potential is created between the outer surface of magnetite and the hematite–magnetite front, driving

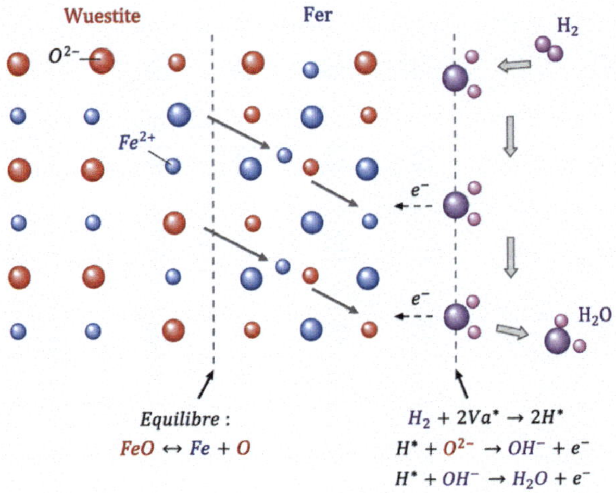

FIGURE 3.20 Schematic of the transformation of wüstite (FeO) to iron.

the migration of electrons and Fe^{2+} ions toward the hematite and producing additional magnetite. This chemical potential gradient serves as the driving force for solid-state diffusion, with Fe^{2+} ions migrating outward through the magnetite layer toward the hematite core while electrons move in the opposite direction, a process that is rate-limited by cation diffusion and is strongly temperature-dependent (activation energy approximately 100–150 $kJ \cdot mol^{-1}$) . The same process underlies the transformation of magnetite to wüstite via Eq. (xxxiii) (Feilmayr et al., 2004; Kim et al., 2021). During the reduction of wüstite to iron, the process is different. Once an iron layer has formed, the reaction proceeds thanks to the gradient in oxygen activity, which drives the migration of oxygen atoms from the Fe/FeO interface—where the equilibrium FeO ↔ Fe + O prevails—toward the surface, where steam is formed. This is because the iron layer formed on wüstite has no surface oxygen, so the reducing gas cannot react directly with it (Kim et al., 2021). Thus, the reduction of wüstite is controlled not by the chemical reaction at the surface but rather by the solid-state transport of oxygen through the product layer. This stage is sensitive to temperature and to the morphology (the porosity and thickness) of the iron layer formed. It is important to note that hydrogen adsorption and the reduction steps are successive processes that are virtually impossible to dissociate in the course of the reaction (von Bogdandy & Engell, 2013):

$$H_2 + 2Va^+ \rightarrow 2H^+ \qquad \text{(xxviii)}$$

$$H^+ + O^{2-} \rightarrow OH^- + e^- \qquad \text{(xxix)}$$

$$H^+ + OH^- \rightarrow H_2O + e^- \qquad \text{(xxx)}$$

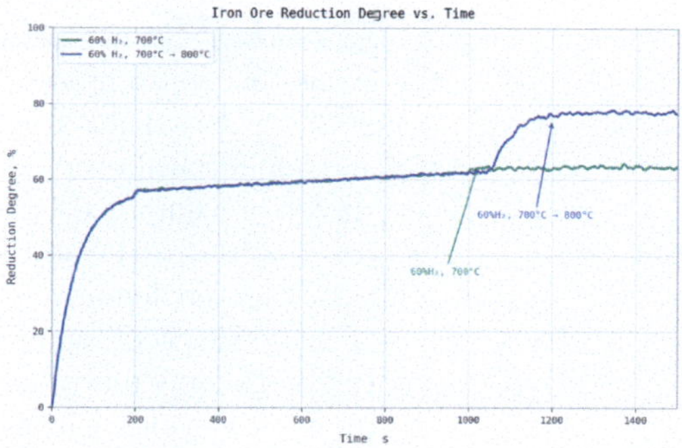

FIGURE 3.21 Evolution of the reduction rates of hematite powder in 60%/40% H_2/N_2 between 700°C and 800°C.

$$Fe^{3+} + e^- \rightarrow Fe^{2+} \quad \text{(xxxi)}$$

$$4Fe_2O_3 + Fe^{2+} + 2e^- \rightarrow 3Fe_3O_4 \quad \text{(xxxii)}$$

$$Fe_3O_4 + Fe^{2+} + 2e^- \rightarrow 4FeO \quad \text{(xxxiii)}$$

3.26 Parameter influencing the kinetics of hydrogen-based reduction

Hydrogen-based reduction is a complex chemical reaction involving several parameters. Among these, some exert a greater influence on the reduction kinetics and will be discussed here.

3.27 Impact of temperature on reduction performance

The kinetic constant of reduction follows the Arrhenius law described in Eq. (3.4); it is therefore strongly linked to temperature. Moreover, as shown in Eq. (xxii), the reduction of iron by hydrogen is overall endothermic, so the greater the heat input, the more the reduction is favored:

$$k = Ae^{-\frac{Ea}{RT}} \quad (3.4)$$

Numerous studies have experimentally investigated the impact of temperature. Fruehan et al. (2005) reduced 100 mg of predried ore powder; the atmosphere was a 60%/40% mixture of hydrogen and nitrogen at a flow rate of 2 L• min^{-1}. Fig. 3.21 shows the evolution of the reduction degree for two specimens. The first is reduced at 700°C, and the second is first reduced at 700°C and then raised to 800°C after 1000 s of reduction. The measurements clearly show an increase in the reduction rate with increasing temperature (Fruehan et al., 2005). The use of fine ore powder rather than pellets in this study

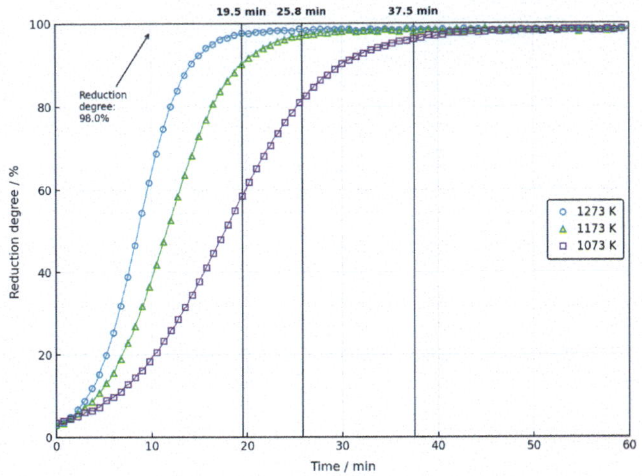

FIGURE 3.22 Evolution of the reduction rates of ore pellets under hydrogen over time at 800°C, 900°C, and 1000°C.

artificially suppresses intraparticle diffusion resistance, so the observed kinetic improvement with temperature reflects the intrinsic reaction rate rather than the relevant behavior of industrial pellets. To observe a wider temperature range, Wagner et al. (2008) performed thermogravimetric analyses of pure hematite powder under 100% hydrogen at different temperatures. Their results show that the highest temperature (900°C) enables the fastest reduction and that the lower the temperature, the longer the time required to achieve complete reduction (Wagner, 2008). In practice, steam accumulation at lower gas flow rates can partially neutralize the temperature-driven acceleration. This effect becomes increasingly significant as reactor scale increases. Since the feedstock used in the steel industry consists of spherical iron-ore pellets, some authors have carried out reductions on such specimens. For example, Zuo et al. (2015) reduced iron-ore pellets under different H_2:CO ratios at various temperatures. Their results demonstrate the same trends as mentioned previously.

Indeed, Fig. 3.22 shows the time evolution of pellet reduction under pure hydrogen at 800°C, 900°C, and 1000°C: one can see that reductions at 800°C and 900°C reach 98% after 37.5 and 25.8 min, respectively. At 1000°C, the reduction is faster still, since in 19.5 min the specimen is reduced to 98%—a gain of about 20 min for a 200°C difference. Zuo et al. (2015) also showed that diffusion coefficients increase with temperature. This is consistent with the Arrhenius dependence of gas-phase diffusion and explains why the kinetic advantage of higher temperatures is amplified in pellets relative to powders. Indeed, as temperature rises, both the intrinsic reaction rate and the gas transport through the porous structure accelerate simultaneously which lead to the improvement of the overall reduction rate.

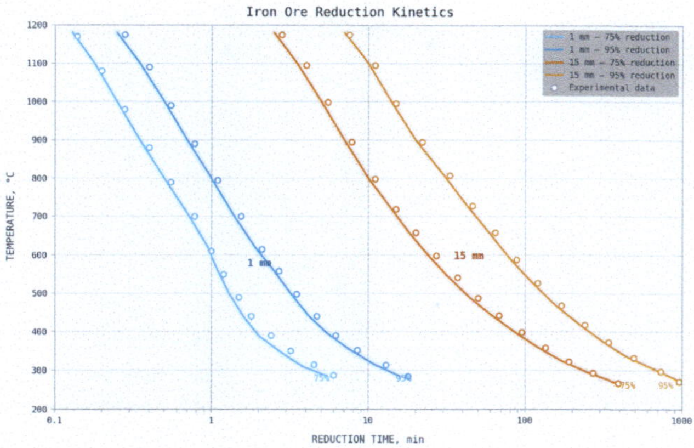

FIGURE 3.23 Effect of temperature on the reduction time under 0.96 atm H_2 for hematite ore particles 1 or 15 mm in diameter.

However, some authors have observed a slowdown in reduction at temperatures around 650°C. For example, as shown in Fig. 3.23, Turkdogan and Vinters (1971) observed a reduction rate slowdown for small ores (1 mm) between 500°C and 800°C. This slowdown was not observed for larger ores (15 mm). This size-dependent behavior is counterintuitive. Generally, smaller particles are expected to reduce faster due to their shorter diffusion path lengths, so the governing mechanism is not diffusion-limited but rather related to microstructural transformations at the particle surface that are more pronounced at smaller scales.

3.28 Impact of reducing gas composition on reduction performance

First, since direct reduction is classically carried out with carbon monoxide, it is necessary to compare its performance with that of hydrogen. This is what Bonalde et al. (2005) did by reducing ore pellets at 850°C under flows of different reducing gases, i.e., hydrogen, carbon monoxide, and Midrex gas (55.7% H_2, 34.0% CO, 6.3% CO_2, and 4.0% CH_4). The results, shown in Fig. 3.24, are clear. One sees a distinct advantage for reduction under hydrogen, followed by Midrex gas, and finally carbon monoxide. Complete reduction is reached in about 15 min for hydrogen, whereas the other two reach roughly 80% and 50% reduction for Midrex gas and carbon monoxide, respectively (Bonalde et al., 2005).

This trend was also shown by Zuo et al., 2015 proving that hydrogen's viscosity is much lower than that of carbon monoxide and its molecular size is smaller. Therefore, H_2 molecules make more effective contact with the reducible surface, which improves reduction kinetics (Zuo et al., 2015). It is worth noting, however, that these experiments were conducted on a single pellet type at a fixed temperature, which limits the generalizability of the results to industrial

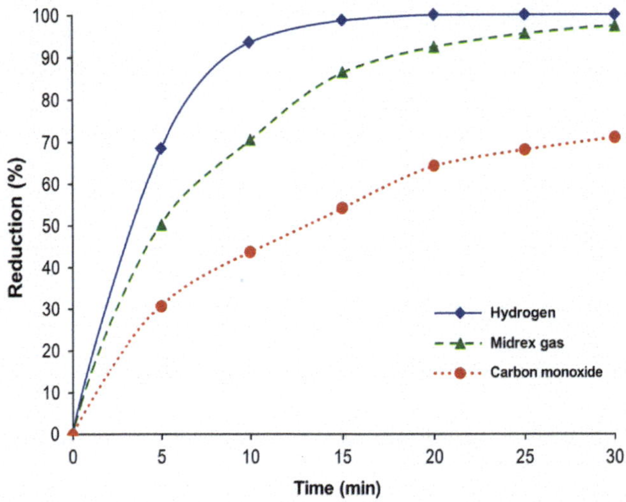

FIGURE 3.24 Reduction rate of an iron-ore pellet versus time under different atmospheres.

conditions, where pellet porosity, bed geometry, and gas flow distribution vary significantly

Since reduction under hydrogen produces water, it is important to discuss the impact of steam on the reduction. First,(von Bogdandy and Engell, 2013) shows that if the amount of water is too high, complete reduction is not possible even at very high temperatures. Indeed, thermogravimetric analyses of hematite reduction in atmospheres composed of dihydrogen, steam, and nitrogen at 400°C showed that the presence of water in the atmosphere does not significantly affect the first step of the reduction (hematite to magnetite), but that the subsequent reduction is markedly slowed. This slowdown depends on the amount of water, and even with little water, complete reduction was not observed Longbottom & Kolbeinsen, 2008). This thermodynamic limitation is often overlooked in kinetic studies conducted under pure hydrogen with continuous gas renewal, which artificially suppress steam partial pressure and therefore overestimate achievable metallization under industrial conditions. Another effect of water is the reoxidation of iron. Indeed, Brinkman et al. (2021) observed that iron oxide reduced by hydrogen can be reoxidized in the presence of water. The reoxidation process is favored at temperatures between 100°C and 500°C which makes atmosphere control during post-reduction a critical but often underestimated operational challenge.

3.29 Impacts of the iron and steel industry on climate and energy use

Global steel demand is projected to increase by more than one-third by 2050 with the majority of growth concentrated in emerging and rapidly industrializing

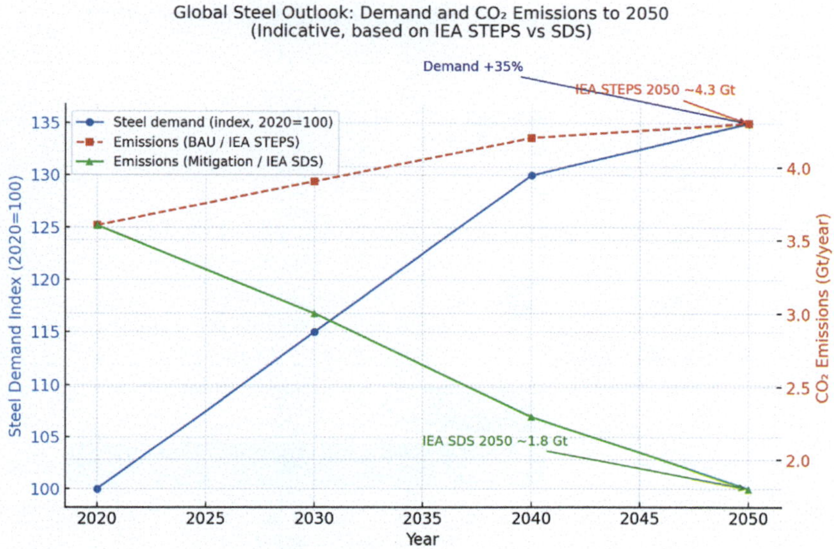

FIGURE 3.25 Global steel outlook: Demand and CO2 emissions to 2025.

economies. The industry has already demonstrated its exposure to systemic shocks. In 2020 the COVID-19 pandemic delivered a sharp disruption to global markets, triggering a ~5% contraction in crude steel production. While severe, this downturn was short-lived: by 2021 the sector rebounded, illustrating its close coupling with global GDP and construction cycles. Such volatility highlights the structural vulnerability of steel to both economic crises and policy-driven transformations in energy-and-climate regulation (Kim et al., 2022).

Looking forward, baseline scenarios from the IEA suggest that the sector will resume a robust upward trajectory once short-term shocks are absorbed. However, without the rapid deployment of mitigation measures and breakthrough technologies, greenhouse gas emissions from iron and steel are set to rise substantially. As illustrated in Fig. 3.25, by 2050 emissions could be ~2.7 Gt CO_2 per year higher than today despite the global climate agenda. This outcome would position steel as one of the largest remaining industrial emitters, complicating efforts to meet mid-century net-zero targets (Kim et al., 2022).

Such a scenario is especially concerning because the industry already accounts for ~7%–9% of global energy-related CO_2 emissions, and demand is locked in by the capital-intensive and long-lived nature of steelmaking assets. Once a blast furnace or integrated plant is built, it typically operates for 30–40 years, making early decisions on decarbonization technologies critical. Without decisive action, today's investment choices could lock in several decades of high emissions, further straining the global carbon budget (Kim et al., 2022).

A useful starting point for assessing the industry's climate impacts is to map the energy and carbon-intensive unit operations. The iron and steel sector emits

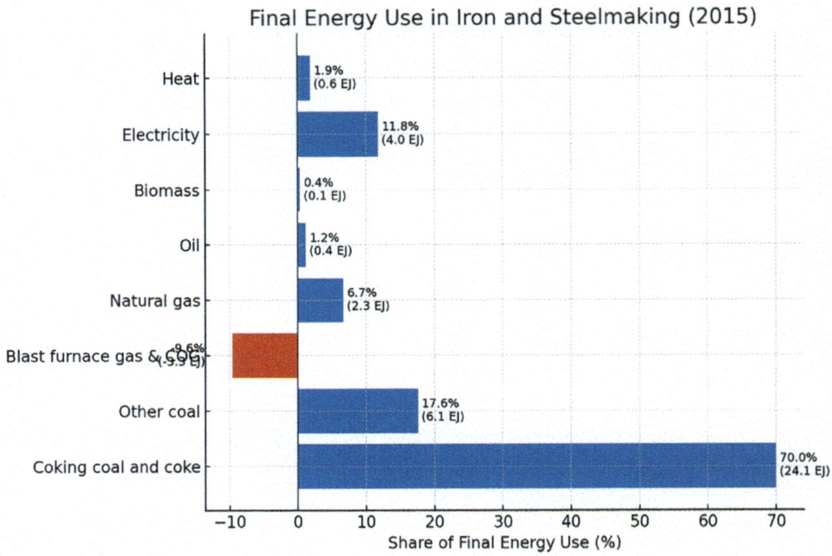

FIGURE 3.26 Contribution of major energy sources across iron- and steelmaking steps.

GHGs from (1) raw materials and process chemistry, (2) fuel combustion across ovens, boilers, stoves, furnaces, and ancillary equipment from sintering through to finishing, and (3) indirect electricity use (e.g., in EAFs) (Hasanbeigi et al., 2014). Fig. 3.26 summarizes the contribution of major energy sources across iron- and steelmaking steps. The figure makes it easier to see how coking coal and coke dominate (70%), followed by other coal (17.6%) and electricity (11.8%). Recovered energy offsets about –9.6% of the total.

In a typical integrated BF/BOF plant, as shown in Fig. 3.27, CO_2 arises predominantly from coal use, with additional emissions from fluxes. Of the roughly 1.8 t CO_2 per ton of rolled coil, about 1.7 t CO_2 is attributable to coal, and the remaining ~0.1 t CO_2 to lime use (Hasanbeigi et al., 2014).

3.30 Overview of other existing decarbonization technologies and practices

The iron and steel sector offers a wide range of decarbonization options. Fig. 3.28 groups them into five classes: (1) raw materials, (2) iron and steelmaking processes, (3) steel products and use, (4) waste and recycling, and (5) breakthrough innovations.

On the feedstock side, the industry's reliance on coal, coke, and natural-gas–based hydrogen for DRI makes it highly carbon-intensive, but transitions to low-carbon hydrogen, solid recovered fuels (SRF), and bioenergy present significant mitigation opportunities. Several hydrogen-based projects, such as ULCOS (Abdul Quader et al., 2016), (Henrekson, 2026), SuSteel (center), Salcos-Macor (Salzgitter, 2025), ArcelorMittal MIDREX (Midrex Technologies, Inc., 2025),

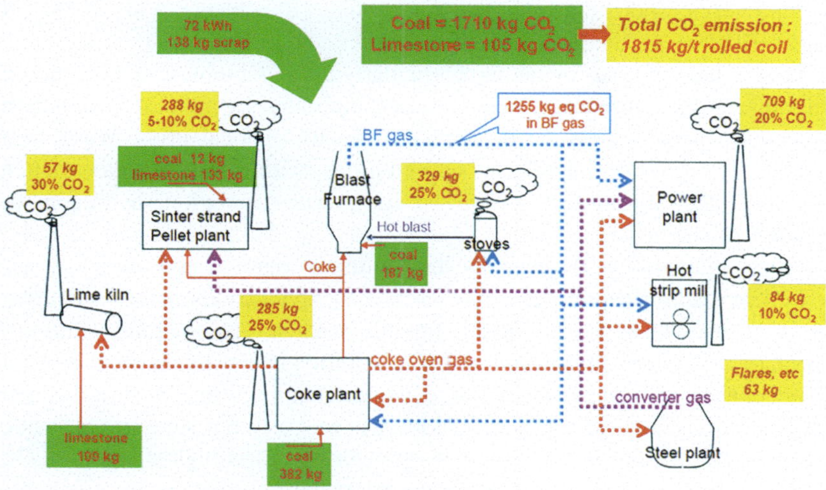

FIGURE 3.27 CO_2 emissions from a typical steel mill.

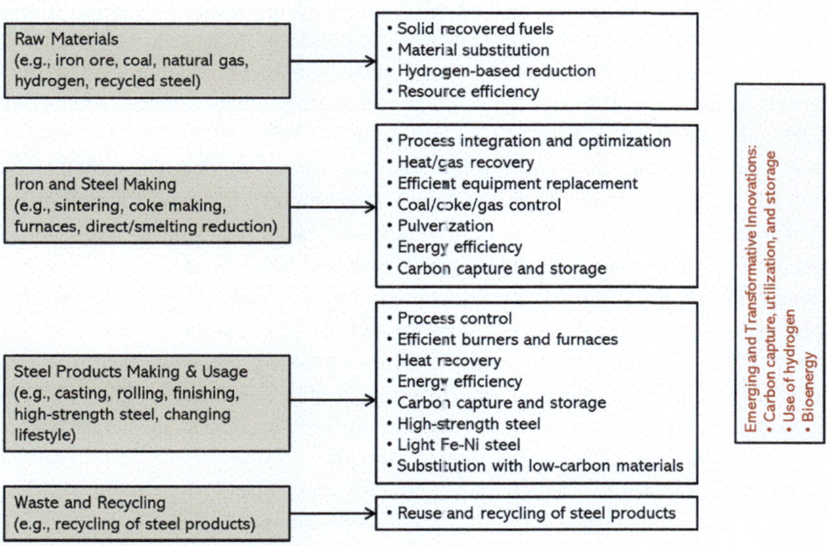

FIGURE 3.28 Decarbonization options in the steel industry.

and flash ironmaking in the U.S. (Sohn, 2014), demonstrate the momentum of H_2 deployment, with modeling showing up to a 91% reduction in direct CO_2 emissions compared to natural gas, provided the hydrogen is green or blue.

In iron and steelmaking, decarbonization levers include energy efficiency, renewables, waste-heat recovery, and process integration. Efficiency gains remain substantial, with studies indicating nearly half of purchased energy is lost, and country-specific assessments (e.g., Switzerland, Germany, China, India) showing varying potentials and payback profiles (Bhadbhade et al., 2019; Na et al., 2020).

Renewables and fuel switching, particularly through biomass and biochar in sintering and blast furnaces, can reduce CO_2 emissions by up to 50%, although outcomes depend heavily on the carbon intensity of local grids. Waste-heat recovery also provides significant opportunities, with coke-oven gas, molten slag, and other by-products harnessed for power generation, hydrogen, or chemical production.

Process integration frameworks that optimize material and energy flows, as well as carbon capture and storage (CCS) applications across sintering, coking, and downstream processes, provide scalable cross-cutting solutions, especially as higher CO_2 prices improve feasibility. Downstream, steel product manufacturing and usage stages also demand attention: efficient furnaces and burners, process control, CCS, and waste-heat recovery remain relevant, while lightweighting strategies such as high-strength and ultra-high-strength steels, HISTAR® beams, and design changes in sectors like automotive offer life-cycle GHG reductions of up to 40%.

On the recycling front, increasing secondary steel production through EAFs and scrap recovery is critical, with EAF routes offering 62%–90% lower emissions compared to primary ore-based steel, depending on grid carbon intensity. Globally, more than 22 billion tons of steel have been recycled since 1900 avoiding vast amounts of iron ore and coal use, and projections suggest secondary steel could account for nearly 40% of output by 2050. Finally, transformative innovations hold the key to long-term decarbonization. A global scan has identified 86 promising technologies, ranging from mature but under-deployed to experimental. Low-carbon hydrogen stands out, with potential to abate more than 3 Gt CO_2 annually at competitive costs, and is being advanced by firms such as SSAB, POSCO, ArcelorMittal, voestalpine, and Salzgitter. Hydrogen direct reduction, especially when coupled with CCS or carbon capture and utilization (CCU), is a leading pathway, and rapidly falling renewable energy costs are accelerating its viability. At the frontier, molten oxide electrolysis (MOE) promises to eliminate process CO_2, achieving zero-carbon steel production if powered by clean electricity. Together, these pathways underscore a multifaceted and evolving transition in which hydrogen, efficiency, circularity, and innovation converge to transform one of the most emissions-intensive global industries.

3.31 Conclusion

Integrating hydrogen into steelmaking is arguably the most promising route to decarbonize the sector and to align with net-zero targets by 2050. This chapter assembled, in one arc, the technical foundations of iron and steelmaking and the pathways to deep decarbonization. We began by mapping the two dominant production routes—integrated BF–BOF and DRI–EAF—and the downstream chain of conversion (BOF/EAF), secondary metallurgy, and continuous casting, clarifying where energy is used, where carbon enters, and how process chemistry (reduction and carburization) drives emissions. We detailed blast-furnace and contrasted it with direct reduction in coal- and gas-based units, including product properties and operational trade-offs. These fundamentals set the stage for a rigorous treatment of hydrogen-enabled ironmaking and its integration with EAF steelmaking, framed by hydrogen supply pathways with their distinct efficiency and integration attributes. At the reaction level, we presented the thermodynamics and kinetics/mechanisms that explain why temperature, particle size/porosity, gas composition, and steam content so strongly govern reduction rates and final iron quality. We also linked these fundamentals to practice by noting EAF requirements when charging DRI—the role of carbon in completing metallization, supplying chemical heat via oxygen–carbon reactions, and sustaining foamy slag for thermal efficiency and arc stability. Together, these sections ground the reader in how process physics and chemistry shape both emissions and abatement levers.

Building on that base, the chapter developed an energy-and-climate mapping of the value chain, identifying hotspots from sintering/coking and stoves through ironmaking to power-intensive EAF steps, and situating the sector in global scenarios where, absent major shifts, steel remains a leading source of industrial CO_2. From there, we synthesized a portfolio of decarbonization options to provide a structured basis for technology selection and investment sequencing across diverse power systems and ore/scrap endowments.

References

Abdul Quader, M., Ahmed, S., Dawal, S. Z., & Nukman, Y. (2016). Present needs, recent progress and future trends of energy-efficient ultra-low carbon dioxide (CO2) steelmaking (ULCOS) program. *Renewable and Sustainable Energy Reviews, 55*, 537–549. https://www.journals.elsevier.com/renewable-and-sustainable-energy-reviews. doi:10.1016/j.rser.2015.10.101.

Adav, S., Srishilan, C., & Shukla, A. K. (2023). Thermodynamic Model of MIDREX Ironmaking Process Using FactSage^{TM} and Macro Facility. *Metall Mater Trans B, 54*, 3508–3525. https://doi.org/10.1007/s11663-023-02928-9.

Astier, J. (2005). Réduction directe. In Ferrous metals: Primary metal production (Article M7580). *Techniques de l'Ingénieur*. https://doi.org/10.51257/a-v3-m7580.

Ayed, P., Ouvradou, C., & Astier, J. (2005). Aciérie électrique. *Élaboration et recyclage des métaux*. doi:10.51257/a-v3-m7700.

Battle, T., Srivastava, U., Kopfle, J., Hunter, R., & McClelland, J. (2014). Chapter 1. 2—The Direct Reduction of Iron. In S. Seetharaman (Ed.), Treatise on Process Metallurgy, Elsevier (pp. 89–176). https://doi.org/10.1016/B978-0-08-096988-6.00016-X.

Bhadbhade, N., Zuberi, M. J. S., & Patel, M. K. (2019). A bottom-up analysis of energy efficiency improvement and CO2 emission reduction potentials for the Swiss metals sector. *Energy, 181*, 173–186. www.elsevier.com/inca/publications/store/4/8/3/. doi:10.1016/j.energy.2019.05.172.

Bonalde, A., Henriquez, A., & Manrique, M. (2005). Kinetic analysis of the iron oxide reduction using hydrogen-carbon monoxide mixtures as reducing agent. *ISIJ International, 45*(9), 1255–1260. doi:10.2355/isijinternational.45.1255.

Brinkman, L., Bulfin, B., & Steinfeld, A. (2021). Thermochemical hydrogen storage via the reversible reduction and oxidation of metal oxides. *Energy and Fuels, 35*(22), 18756–18767. https://pubs.acs.org/journal/enfuem. doi:10.1021/acs.energyfuels.1c02615.

Cavaliere, P., Sadeghi, B., Dijon, L., Laska, A., & Koszelow, D. (2024). Three-dimensional characterization of porosity in iron ore pellets: A comprehensive study. *Minerals Engineering, 213*, 108746. https://doi.org/10.1016/j.mineng.2024.108746.

Chenguang, B. (2024). Basicity and Acidity of Slag. In The ECPH Encyclopedia of Mining and Metallurgy (pp. 128–129). Singapore: Springer. https://doi.org/10.1007/978-981-99-2086-0_1148.

de Moraes, S. L., de Lima, J. R. B., & Ribeiro, T. R. (2018). *Iron ore pelletizing process: An overview*. IntechOpen. doi:10.5772/intechopen.73164.

Fan, Z., & Friedmann, S. J. (2021). Low-carbon production of iron and steel: Technology options, economic assessment, and policy. *Joule, 5*(4), 829–862. https://www.journals.elsevier.com/joule. doi:10.1016/j.joule.2021.02.018.

Feilmayr, C., Thurnhofer, A., Winter, F., Mali, H., & Schenk, J. (2004). Reduction behavior of hematite to magnetite under fluidized bed conditions. *ISIJ International, 44*(7), 1125–1133. https://www.jstage.jst.go.jp/browse/isijinternational. doi:10.2355/isijinternational.44.1125.

center, Metallurgical Competence Center. (n.d.). K1-MET – Metallurgical Competence Center. Retrieved September 6, 2025,https:https://www.k1-met.com/en/non_comet/susteel.

Levenspiel, O. (1999). *Chemical reaction engineering* (3rd ed.). Wiley.

Fours à arc : Dossier complet, Techniques de l'Ingénieur. (2025), https://www.techniques-ingenieur.fr/base-documentaire/energies-th4/electrothermie-industrielle-42270210/fours-a-arc-d5920/.

Fruehan, R. J., Li, Y., Brabie, L., & Kim, E. J. (2005). Final stage of reduction of iron ores by hydrogen. *Scandinavian Journal of Metallurgy, 34*(3), 205–212. doi:10.1111/j.1600-0692.2005.00722.

Ghosh, A., & Chatterjee, A. (2008). *Iron making and steelmaking: Theory and practice*. PHI Learning https://books.google.co.ma/books?id=7_GcmB4i_dsC.

Giroudière, F., & Le Gall, A. (2012). Production des gaz de synthèse par vaporeformage. In Opérations unitaires. Génie de la réaction chimique. *Techniques de l'Ingénieur*. https://doi.org/10.51257/a-v3-j5480.

Gräbner, M. (2014). *Industrial coal gasification technologies covering baseline and high-ash coal*. Wiley. doi:10.1002/9783527336913.

Guo, X., Huang, L., Miao, H., Mi, L., & Han, Z. (2025). Exploring carbon reduction pathways in the steel industry from the perspective of emerging technologies for achieving carbon neutrality. *Journal of Environmental Management, 385*. https://www.sciencedirect.com/science/journal/03014797. doi:10.1016/j.jenvman.2025.125768.

Gurgel, A., Benavides, K., Morris, J., Mignone, B. K., Chapman, B., Kheshgi, H., Herzog, H., & Paltsev, S. (2025). Role of advanced steelmaking technologies in global climate change mitigation scenarios. *Journal of Cleaner Production, 513*, 145726. doi:10.1016/j.jclepro.2025.145726.

Haryanto, A., Fernando, S., Murali, N., & Adhikari, S. (2005). Current status of hydrogen production techniques by steam reforming of ethanol: A review. *Energy and Fuels, 19*(5), 2098–2106. doi:10.1021/ef0500538.

Hasanbeigi, A., Arens, M., & Price, L. (2014). Alternative emerging ironmaking technologies for energy-efficiency and carbon dioxide emissions reduction: A technical review. *Renewable and Sustainable Energy Reviews, 33*, 645–658. doi:10.1016/j.rser.2014.02.031.

Heidari, A., Niknahad, N., Iljana, M., & Fabritius, T. (2021). A review on the kinetics of iron ore reduction by hydrogen. *Materials, 14*(24). https://www.mdpi.com/1996-1944/14/24/7540/pdf. doi:10.3390/ma14247540.

Henrekson, M. (2026). In *HYBRIT: A hubristic hydrogen-based steel: 49* (pp. 187–201). Springer Nature. https://doi.org/10.1007/978-3-032-15512-2_10.

Holladay, J. D., Hu, J., King, D. L., & Wang, Y. (2009). An overview of hydrogen production technologies. *Catalysis Today, 139*(4), 244–260. doi:10.1016/j.cattod.2008.08.039.

Huang, J., & Dincer, I. (2014). Parametric analysis and assessment of a coal gasification plant for hydrogen production. *International Journal of Hydrogen Energy, 39*(7), 3294–3303. doi:10.1016/j.ijhydene.2013.12.054.

Ivy, J., Summary of electrolytic hydrogen production: Milestone completion report. (2004). doi:10.2172/15009552.

IEA. (2021). *Global Hydrogen Review 2021. Analysis* https://www.iea.org/reports/global-hydrogen-review-2021.

Kim, J., Sovacool, B. K., Bazilian, M., Griffiths, S., Lee, J., Yang, M., & Lee, J. (2022). Decarbonizing the iron and steel industry: A systematic review of sociotechnical systems, technological innovations, and policy options. *Energy Research & Social Science, 89*, 102565. doi:10.1016/j.erss.2022.102565.

Kim, S.-H., Zhang, X., Ma, Y., Filho, I. R. S., Schweinar, K., Angenendt, K., Vogel, D., Stephenson, L. T., El-Zoka, A. A., Mianroodi, J. R., Rchwerder, M., Gault, B., & Raabe, D. (2021). Influence of microstructure and atomic-scale chemistry on the direct reduction of iron ore with hydrogen at 700°C. *Acta Materialia, 212*, 116933. doi:10.1016/j.actamat.2021.116933.

Lahrichi, A., El Issmaeli, Y., Kalanur, S. S., & Pollet, B. G. (2024). Advancements, strategies, and prospects of solid oxide electrolysis cells (SOECs): Towards enhanced performance and large-scale sustainable hydrogen production. *Journal of Energy Chemistry, 94*, 688–715. https://www.sciencedirect.com/science/journal/20954956. doi:10.1016/j.jechem.2024.03.020.

Longbottom, R. J., & Kolbeinsen, L. (2008). Iron ore reduction with CO and H_2 gas mixtures – Thermodynamic and kinetic modelling. University of Wollongong. https://hdl.handle.net/10779/uow.27698079.v1.

Chapter 1 - A brief description of the blast-furnace process., W.K. Lu, ed., McMaster University. The Iron Blast Furnace (pp. 1–15). (1979). doi: 10.1016/B978-0-08-023218-8.50007-8.

Ling, J., Yang, H., Tian, G., Cheng, J., Wang, X., & Yu, X. (2024). Direct reduction of iron to facilitate net zero emissions in the steel industry: A review of research progress at different scales. *Journal of Cleaner Production, 441*, 140933. https://doi.org/10.1016/j.jclepro.2024.140933.

Liu, L., Jiang, Z., Zhang, X., Lu, Y., He, J., Wang, J., & Zhang, X. (2018). Effects of top gas recycling on in-furnace status, productivity, and energy consumption of oxygen blast furnace. *Energy, 163*, 144–150. https://doi.org/10.1016/j.energy.2018.08.114.

Ma, Y., Souza Filho, I. R., Bai, Y., Schenk, J., Patisson, F., Beck, A., van Bokhoven, J. A., Willinger, M. G., Li, K., Xie, D., Ponge, D., Zaefferer, S., Gault, B., Mianroodi, J. R., & Raabe, D. (2022). Hierarchical nature of hydrogen-based direct reduction of iron oxides. *Scripta Materialia, 213*, 114571. doi:10.1016/j.scriptamat.2022.114571.

Midrex Technologies, Inc., World DRI production reaches 119.2 Mt in 2021 Midrex publishes world direct reduction statistics. (2025), Midrex Technologies, Inc. https://www.midrex.com/company-news/world-dri-production-reaches-119-2-mt-in-2021-midrex-publishes-world-direct-reduction-statistics/.

Muradov, N. (2001). Hydrogen via methane decomposition: An application for decarbonization of fossil fuels. *International Journal of Hydrogen Energy, 26*(11), 1165–1175. doi:10.1016/S0360-3199(01)00073-8.

Na, H., Du, T., Sun, W., Sun, J., & He, J. (2020). Evaluation and improvement of energy utilization efficiency in typical iron and steel smelting route based on input-use-end model. *Energy Technology, 8*(3). http://onlinelibrary.wiley.com/journal/10.1002/(ISSN)2194-4296. doi:10.1002/ente.201901230.

Nechache, A., & Hody, S. (2021). Alternative and innovative solid oxide electrolysis cell materials: A short review. *Renewable and Sustainable Energy Reviews, 149*, 111322. doi:10.1016/j.rser.2021.111322.

Ni, M., Leung, M. K. H., & Leung, D. Y. C. (2008). Technological development of hydrogen production by solid oxide electrolyzer cell (SOEC). *International Journal of Hydrogen Energy, 33*(9), 2337–2354. doi:10.1016/j.ijhydene.2008.02.048.

Nie, H., Li, Z., Kuang, S., Yan, L., Zhong, W., Yu, A., … Xu, H. (2021). Numerical investigation of oxygen-enriched operations in blast furnace ironmaking. *Fuel, 296*, 120662. https://doi.org/10.1016/j.fuel.2021.120662.

Nikolaidis, P., & Poullikkas, A. (2017). A comparative overview of hydrogen production processes. *Renewable and Sustainable Energy Reviews, 67*, 597–611. https://www.journals.elsevier.com/renewable-and-sustainable-energy-reviews. doi:10.1016/j.rser.2016.09.044.

Ogden, D. J. M., (2002) Review of small stationary reformers for hydrogen production, Princeton University.

Patisson, F., & Mirgaux, O. (2020). Hydrogen ironmaking: How it works. *Metals, 10*(7), 1–15. https://www.mdpi.com/2075-4701/10/7/922/pdf. doi:10.3390/met10070922.

Philibert, J., Combrade, P., Bréchet, Y., & Vignes, A. (1997). *Métallurgie: Du minerai au matériau.* Masson https://books.google.co.ma/books?id=O6UGwQEACAAJ.

Salzgitter, SALCOS®. (2025). *Program for Low CO_2 Steel Production* https://salcos.salzgitter-ag.com/en/https://salcos.salzgitter-ag.com/en/salcos.html.

Rostrup-Nielsen, J. R., Sehested, J., & Noerskov, J. K. (2025). Hydrogen and synthesis gas by steam- and CO2 reforming. *ChemInform, 34.* https://onlinelibrary.wiley.com/doi/10.1002/chin.200317288?msockid=1c7f14dd70e4646117b307f5712365e1.

Pourfathi, A., & Tavakoli, R. (2023). Thermal optimization of secondary cooling systems in the continuous steel casting process. *International Journal of Thermal Sciences, 183*, 107860. https://doi.org/10.1016/j.ijthermalsci.2022.107860.

Rahmatmand, B., Tahmasebi, A., Lomas, H., Honeyands, T., Koshy, P., Hockings, K., & Jayasekara, A. (2023). A technical review on coke rate and quality in low-carbon blast furnace ironmaking. *Fuel, 336*, 127077. https://doi.org/10.1016/j.fuel.2022.127077.

Rahmatmand, B., Tahmasebi, A., Lomas, H., Honeyands, T., Koshy, P., Hockings, K., & Jayasekara, A. (2023). A technical review on coke rate and quality in low-carbon blast furnace ironmaking. *Fuel, 336*, 127077. https://doi.org/10.1016/j.fuel.2022.127077.

Raimbault, C., Lefebvre, G., et al. (1995). Vapocraquage des hydrocarbures. In Opérations unitaires. Génie de la réaction chimique. *Techniques de l'Ingénieur*. https://doi.org/10.51257/a-v2-j5460.

Rostrup-Nielsen, J. R., Sehested, J., & Noerskov, J. K. (2003). Hydrogen and synthesis gas by steam- and CO 2 reforming. *ChemInform, 34*(17). doi:10.1002/chin.200317288.

Sánchez-Ramírez, E., Juarez-García, M., Huerta-Rosas, B., et al. (2026). Waste Gasification for Hydrogen Production: Regional Challenges, Sustainability Pathways, and Future Research Directions. *Bioenerg. Res., 19*, 34. https://doi.org/10.1007/s12155-026-10971-y.

Santos, D. M. F., Sequeira, C. A. C., & Figueiredo, J. L. (2013). Hydrogen production by alkaline water electrolysis. *Quimica Nova, 36*(8), 1176–1193. http://www.scielo.br/pdf/qn/v36n8/v36n8a17.pdf. doi:10.1590/S0100-40422013000800017Portugal.

Seetharaman, S., McLean, A., Guthrie, R., & Sridhar, S. (2013). Treatise on process metallurgy. *Treatise on Process Metallurgy, 2*, 1–860. http://www.sciencedirect.com/science/book/9780080969848. doi:10.1016/C2010-0-67120-3.

Seyitoglu, S. S., Dincer, I., & Kilicarslan, A. (2017). Energy and exergy analyses of hydrogen production by coal gasification. *International Journal of Hydrogen Energy, 42*(4),

2592–2600. http://www.journals.elsevier.com/international-journal-of-hydrogen-energy/. doi:10.1016/j.ijhydene.2016.08.228.

Shahabuddin, M., Brooks, G., & Rhamdhani, M. A. (2023). Decarbonisation and hydrogen integration of steel industries: Recent development, challenges and technoeconomic analysis. *Journal of Cleaner Production, 395*, 136391. doi:10.1016/j.jclepro.2023.136391.

Shankar Rao, V., & Gurugubelli, S. N. (2021). A Case Study on Accretion Formation in Rotary Kiln of DRI Process. *Trans. Indian. Inst. Met., 74*, 45–49. https://doi.org/10.1007/s12666-020-02118-0.

Sircar, S., & Golden, T. C. (2000). Purification of hydrogen by pressure swing adsorption. *Separation Science and Technology, 35*(5), 667–687. doi:10.1081/SS-100100183.

Sohn, H. Y. (2014). From sulfide flash smelting to a novel flash ironmaking technology. *TMS Annual Meeting of Minerals, Metals and Materials Society*. United States https://www.tms.org/. doi:10.1002/9781118889657.ch4.

Spreitzer, D., & Schenk, J. (2019). Reduction of Iron Oxides with Hydrogen—A Review. *steel research int., 90*, 1900108. https://doi.org/10.1002/srin.201900108.

Steinfeld, A., Kuhn, P., & Karni, J. (1993). High-temperature solar thermochemistry: Production of iron and synthesis gas by Fe3O4-reduction with methane. *Energy, 18*(3), 239–249. doi:10.1016/0360-5442(93)90108-p.

Turkdogan, E. T., & Vinters, J. V. (1971). Gaseous reduction of iron oxides: Part I. Reduction of hematite in hydrogen. *Metallurgical Transactions, 2*(11), 3175–3188. doi:10.1007/bf02814970.

Turner, J., Sverdrup, G., Mann, M. K., Maness, P.-C., Kroposki, B., Ghirardi, M., Evans, R. J., & Blake, D. (2008). Renewable hydrogen production. *International Journal of Energy Research, 32*(5), 379–407. doi:10.1002/er.1372.

Vogl, V., Åhman, M., & Nilsson, L. J. (2018). Assessment of hydrogen direct reduction for fossil-free steelmaking. *Journal of Cleaner Production, 203*, 736–745. https://www.journals.elsevier.com/journal-of-cleaner-production. doi:10.1016/j.jclepro.2018.08.279.

von Bogdandy, L., & Engell, H. J. (2013). Scientific basis and technology. *The reduction of iron ores*. Springer https://books.google.co.ma/books?id=pV7_CAAAQBAJ.

Wang, L., Wang, G., & Li, S. (2024). Energy and exergy analyses of hydrogen direct reduction iron by the fluidized bed. *International Journal of Hydrogen Energy, 80*, 779–787. https://www.sciencedirect.com/science/journal/03603199. doi:10.1016/j.ijhydene.2024.07.130.

World Steel, Association. (2026). World Steel in Figures 2026. World Steel Association https://worldsteel.org/data/world-steel-in-figures/.

Xu, J., & Froment, G. F. (1989). Methane steam reforming, methanation and water-gas shift: I. Intrinsic kinetics. *AIChE Journal, 35*(1), 88–96. doi:10.1002/aic.690350109.

Zhang, C., Vladislav, L., Xu, R., Sergey, G., Jiao, K., Zhang, J., … Wang, G. (2022). Blast furnace hydrogen-rich metallurgy-research on efficiency injection of natural gas and pulverized coal. *Fuel, 311*, 122412. https://doi.org/10.1016/j.fuel.2021.122412.

Zhu, Gm., Hu, Mw., Dou, An., et al. (2024). Effect of temperature and reaction path interaction on fluidization reduction kinetics of iron ore powder. *J. Iron Steel Res. Int., 31*, 1840–1849. https://doi.org/10.1007/s42243-024-01182-wDownloadcitation.

Zuo, H-b., Wang, C., Dong, J-j., Jiao, K-x., & Xu, R-s. (2015). Reduction kinetics of iron oxide pellets with H2 and CO mixtures. *International Journal of Minerals, Metallurgy and Materials, 22*(7), 688–696. https://www.springer.com/materials/journal/12613. doi:10.1007/s12613-015-1123-x.

Wagner, D. (2008). *Étude expérimentale et modélisation de la réduction du minerai de fer par l'hydrogène (Doctoral dissertation, Institut National Polytechnique de Lorraine). HAL* https://hal.univ-lorraine.fr/tel-01753016/document.

Part II

Hydrogen derivatives

Chapter 4

Potential use of green ammonia as a fuel for transportation

Umar M. Jafar[a], Umar H. Nuhu[a], Mustafa M. Amin[a], Bashir Hashim[a], Rasheed M. Ibrahim[a] and Mohammad M. Hossain[a,b]

[a] *Department of Chemical Engineering, King Fahd University of Petroleum and Minerals, Dhahran, Saudi Arabia,* [b] *Interdisciplinary Research Center for Refining and Advanced Chemicals, King Fahd University of Petroleum and Minerals, Dhahran, Saudi Arabia*

4.1 Introduction

The continuous dependence on fossil fuels for energy has led to rising global temperatures and unprecedented climate change. Numerous studies have identified CO_2 and CH_4 as the key greenhouse gases driving this phenomenon (Hartfield et al., 2018). Fig. 4.1 illustrates the carbon dioxide emissions contributing to climate change. As a result of human actions, along with changes in land use, a substantial amount of CO_2 is released, increasing its atmospheric concentration (Oreskes, 2018). From the beginning of the industrial era to date, the concentration of atmospheric CO_2 has increased by more than 40%. Methane concentrations have also reached levels unprecedented in centuries, with a sharp rise through the 20th century significantly impacting the climate (Loulergue et al., 2008; Tiseo, 2025).

In response, several developed nations have initiated programs to shift from fossil fuel-based energy production to renewable sources, intending to lower greenhouse gas emissions and attain carbon neutrality. The shift toward renewable energy, which enables carbon-free power generation, has garnered substantial support from stakeholders (Gielen et al., 2019). However, the intermittent nature of renewable sources poses challenges in maintaining a stable electricity supply. As a result, energy storage solutions have become essential to ensuring a reliable and continuous power supply in modern energy systems (Kurien & Mittal, 2022).

Water electrolysis is a promising method for storing surplus hydrogen. Hydrogen as a highly sustainable and clean source of energy, having the potential to serve a crucial role in the decarbonization of energy and transportation sectors (Tornatore et al., 2022). However, its direct use as a fuel presents significant

Decarbonizing through Sustainable Hydrogen and its Derivatives.
DOI: https://doi.org/10.1016/B978-0-443-27632-3.00002-5

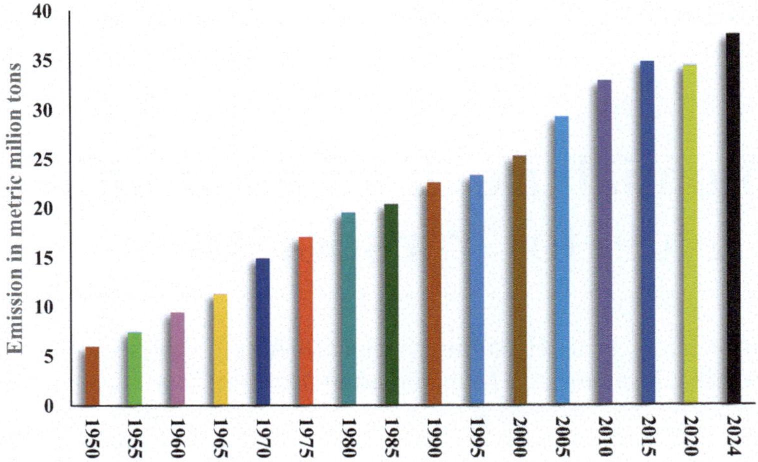

FIGURE 4.1 Global carbon emissions.

storage and safety challenges due to its properties, including wide flammability range (4%–75% concentration in air) (Kurien & Mittal, 2022).

To address these challenges, various hydrogen carriers have been proposed over time. Among them, NH_3 stands out as a viable alternative, offering several key advantages to hydrogen. These include lower energy storage costs, simpler production processes, higher volumetric energy density, improved handling and distribution efficiency, and greater scale-up feasibility.

In recent years, NH_3 has received significant attention as both a fuel and an effective hydrogen carrier. Due to its high hydrogen density, it serves as a practical means of transporting hydrogen. In fact, each 1 mole of NH_3 contains 1.5 moles of molecular hydrogen. Moreover, the challenges associated with hydrogen storage limit its use in vehicles. In contrast, liquid NH_3 can be safely stored at 9 bar pressure at 20°C or at atmospheric pressure at −33°C, making its storage and transportation considerably more manageable (Tornatore et al., 2022).

Hydrogen can be extracted from NH_3 through various processes. The major processes used commercially include thermal decomposition and catalytic cracking. However, NH_3 decomposition via thermal decomposition is a slow reaction that requires significant energy input. To enhance the efficiency of hydrogen production, catalytic cracking using metallic catalysts is commonly employed. Significant studies have concentrated on developing effective catalysts for NH_3 decomposition. Among them, ruthenium (Ru) is widely regarded as the most effective catalyst for this reaction at 400°C. However, its limited availability remains a challenge (Ashik et al., 2018). Other processes applied for NH_3 decomposition include electrolysis, electro-oxidation, etc.

NH_3 can also be utilized as a fuel in internal combustion engines (ICEs) and other combustion engine systems. The two main types of ICEs are spark-ignition (SI) and compression-ignition (CI) engines (Cardoso et al., 2021). SI engines, typically gasoline-based, ignite the air-fuel mixture using a spark plug, while CI engines, commonly diesel-based, rely on heat generated by mechanical compression to ignite the injected fuel (Yapicioglu & Dincer, 2019). The properties of NH_3 make it a viable fuel for ICEs, requiring minimal engine modifications to integrate with conventional vehicle fleets. It can also be used in dual-fuel systems alongside other fuels, offering a practical transition toward low-carbon combustion technologies (Cardoso et al., 2021). Despite this potential, real-world applications of NH_3-fueled ICEs remain limited. Therefore, further research and development are needed to optimize engine performance, increase efficiency, ensure safety, and comply with regulatory standards (El-Adawy et al., 2024).

Several studies have explored NH_3 production and its use as a fuel, covering topics such as NH_3-fueled engines, its application in generating power, the production and economic potential, and its use in fuel mixtures with hydrogen and methane. However, this chapter provides a comprehensive overview of the sustainable synthesis of NH_3, its potential applications in various engine types, and its viability as a fuel when mixed with other energy sources. Additionally, techno-economic considerations are examined. Specifically, this chapter evaluates the feasibility of NH_3 as a fuel and summarizes NH_3 production approaches with an emphasis on innovative green production methods. Furthermore, it explores the most effective strategies for integrating NH_3 as fuel and discusses the key advantages and challenges of using NH_3 in ICEs.

4.2 Ammonia as an energy carrier—Historical perspective

Carbon emissions associated with fossil fuels necessitate the need for advancement of renewable and alternative energy sources, and NH_3 has emerged as a promising candidate. NH_3 has been used as a fuel for ICEs since the early 1800s. Its use gained prominence during World War II when oil shortages led to an increase in alternative fuels. Despite this early adoption, research into NH_3-fueled CI and SI engines remained limited until the 2010s.

The feasibility of using NH_3 as a fuel was demonstrated during World War II by Norsk Hydro, which converted a truck to run on NH_3, and a Belgian public transportation company that powered a fleet of buses with the fuel. Fig. 4.2 illustrates the advancement of NH_3 vehicles since 1933 (Mounaïm-Rousselle et al., 2021). In the United States, a few studies explored NH_3 as a fuel source in the decades following the war. However, these investigations were largely limited to isolated vehicle tests and experimental applications (Tornatore et al., 2022).

The Haber-Bosch (H-B) process, introduced in the early 1900s, transformed NH_3 production by making it possible to synthesize NH_3 from atmospheric nitrogen. This innovation earned Fritz Haber the Nobel Prize in Chemistry in 1918 and Carl Bosch the same honor in 1931.

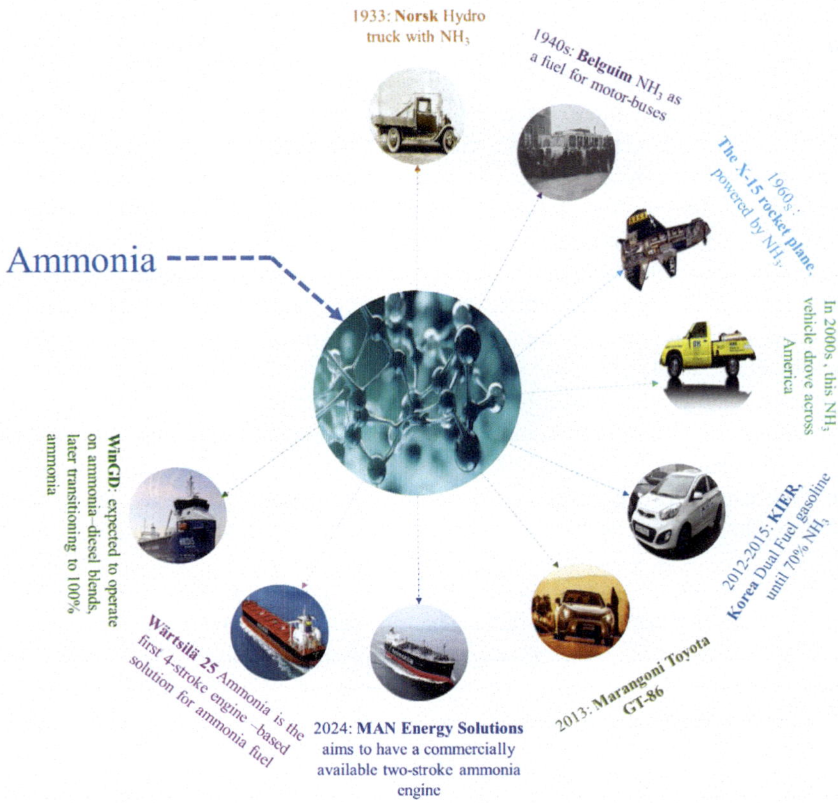

FIGURE 4.2 Advancements in NH_3 vehicles.

The significance of NH_3 production was first highlighted by Sir William Crookes in 1898. He warned that global wheat supplies would be insufficient to sustain a growing population, predicting that without a solution, population growth would be capped at 2 billion by around 1930. The development in the synthesis of NH_3 directly tackled this "wheat issue" by providing a means to produce synthetic fertilizers, thereby boosting agricultural yields and supporting continued population growth (Grasham et al., 2019).

4.3 Physicochemical properties of ammonia as fuel

NH_3 offers substantial benefits over pure hydrogen. It can be used directly in an internal combustion engine without emitting greenhouse gases. Some physicochemical properties of NH_3 as fuel compared to other fuels are listed in the following Table 4.1. The key properties determining the efficiency of NH_3 fuel include its lower heating value (LHV), research octane number (RON), and flame velocity.

TABLE 4.1 Comparison of combustion properties between ammonia and other fuels under ambient conditions (Ishaq & Crawford, 2024; Tornatore et al., 2022).

Property	Ammonia	Methane	Hydrogen	Gasoline	Diesel
Molar mass (g/mol)	17	16	2	114	167
Flammability limit (% vol)	15.8–28	4.4–17	4–76	1.4–7.6	1–6
Octane number	>130	130	>100	87-93	0
Stoichiometric air/fuel ratio (mass)	6	17.1	34.3	15	14.6
Autoignition temperature (°C)	651	537	530	440	225
Lower heating value (MJ/kg fuel)	18.8	50.1	120	42.5	45
Stoichiometric energy content (MJ/kg air)	3.13	2.92	3.5	2.83	3.08
Laminar flame speed (m/s)	0.07	0.38	3.51	0.42	–

4.3.1 Lower heating value

The LHV is a critical heating parameter for evaluating an engine power output and the efficiency of the fuel, as it represents the energy content per unit mass (MJ/kg) of a fuel or combustible material. Typically, NH_3 has a LHV of approximately 18.6 MJ/kg (Smith et al., 2020), though this value varies depending on conditions and the presence of hydrogen, nitrogen, or other substances in the fuel mixtures.

For instance, an 80% NH_3 and 20% hydrogen blend has an LHV of 21.5 MJ/kg, while an NH_3–H_2–N_2 mixture exhibits an LHV of 19.0 MJ/kg for a specified composition (Imhoff et al., 2021). These variations are crucial in determining the feasibility of NH_3 as a fuel, especially in applications where mass and volume are key considerations (Imhoff et al., 2021; Lhuillier et al., 2019). Table 4.1 compares the properties of NH_3 with conventional energy sources, including hydrogen. In terms of LHV, NH_3 has the lowest value among other fuels.

4.3.2 Cetane number and research octane number

Cetane number (CN) serves as a key indicator of diesel fuel ignition quality, measuring combustion speed and the compression required for ignition. Although NH_3 is not typically used as a diesel fuel and does not have a standard CN, it possesses an exceptionally high-octane number, close to 130. This makes NH_3 an excellent antiknocking agent for ICEs (Zheng et al., 2024). This, in turn, improves efficiency, reduces consumption of fuel, and enhances power output (Rehbein et al., 2019).

The octane number, which includes both the RON and motor octane number (MON), assesses the resistance of fuel to knocking during combustion, with larger values indicating high resistance (Singh et al., 2017). In contrast, the CN significantly influences cold-start performance, exhaust emissions, combustion behavior, and noise levels in diesel engines. The CN is determined by measuring its ignition delay in a cooperative fuel research (CFR) engine, using reference fuels such as n-hexadecane (n-$C_{11}H_{10}$), which has a CN of 0. These substances are blended to match the combustion characteristics of the test fuel.

The high antiknocking tendency of NH_3 can result in higher compression ratios (CR), improving engine efficiency. Therefore, the high RON value of NH_3 can be seen as an advantage for SI engines (Cornelius et al., 1996).

4.3.3 Flame velocity

Flame velocity refers to the speed at which the flame front moves through the premixed combustible gas. It is an essential factor in understanding combustion processes (Christensen, 2016). The flame velocity of NH_3 varies depending on factors like pressure, temperature, and mixture composition, which in turn affects engine performance and emission characteristics (Zitouni et al., 2022). For most hydrocarbons-air mixtures, the flame velocity is approximately 40 cm/s, increasing at higher temperatures and lower pressures (Pischinger et al., 2002).

The laminar burning velocity can be determined numerically by analyzing fundamental reaction mechanisms and molecular transport properties, using software like Chemkin or Cantera. Experimentally, techniques such as the heat flux method and stretched flame method are used to measure the laminar flame speed. Overall, the laminar flame speed is a key factor in understanding combustion processes (Christensen, 2016).

4.3.4 Emissions from exhaust of ammonia internal combustion engines

NH_3 fuel decreases greenhouse gas emissions but results in NO_x and unburned NH_3 emissions from the exhaust gases. Chemical kinetic principles have been instrumental in understanding combustion at a microscopic level, explaining the rate and progression of chemical reactions. After validation against combustion parameters and NO_x emission data, several efficient and concise kinetic models for NH_3 oxidation have been developed (Cai et al., 2023).

Compared to hydrocarbons, NH_3 has simpler chemistry. Hence, recent research has identified comparable NH_3 oxidation pathways. However, rate constants and branching ratios for elementary reaction steps vary depending on the scenarios of the application (Cai et al., 2023). Studies show that in lean NH_3 combustion, OH radicals and HNO primarily drive NO formation, whereas in rich combustion, H radicals promote the formation of NH_x radicals (Chai et al., 2021).

Studies on conventional fossil fuels indicate that NO_x emissions peak during NH_3 combustion under lean conditions, but with NH_3/hydrogen blends, these peaks occur at even leaner operating conditions. Additionally, NH_3-hydrogen combustion requires 35% more air than hydrocarbon combustion (Westlye et al., 2013).

Research on SI engines has shown that partially replacing NH_3 with hydrogen reduces zero NO_x emissions due to the nitrogen content in the fuel. Instead, the presence of hydrogen increases NO_x emissions.

Jin et al. (2023) investigated how varying ammonia-to-energy ratios (AER) impact combustion and emissions in NH_3-diesel dual-fuel engines. The study found that increasing AER reduced NO emissions compared to pure diesel engines, as amine denitrification promoted the formation of stable N_2 through reduction reactions. However, lower combustion temperatures led to a significant increase in unburned NH_3.

Wang et al. (2021) conducted simulations of premixed hydrogen-NH_3 combustion, identifying NO as the primary source of NO_x emissions. Two key mechanisms influenced NO production: (1) increased hydrogen content raised reactive O/H generation and NH_3 conversion, and (2) reduced NH_3 concentrations lowered NO-related reaction rates. Their study also found that NO_2 formed rapidly during the flame phase but was later converted back to NO in the postflame region through the reaction NO_2 and O to NO and O_2, ultimately limiting overall NO_2 emissions (Mounaïm-Rousselle et al., 2021; Westlye et al., 2013).

4.4 Sustainable methods of ammonia production

The production of NH_3 presently accounts for over 1.8% of global fossil fuel consumption and contributes to more than 1% of total CO_2 emissions. As presented in Fig. 4.3, the demand volume of 184 million tons per year (MT/year) in 2021 is projected to rise to 470 MT/year in 2050, a logical increase considering that fertilizer production alone constitutes over 80% of NH_3 demand. These projections, coupled with the imperative to carbon-free emissions, underscore the environmental importance of decarbonizing NH_3 manufacturing processes (Ashik et al., 2018). Presently, hydrogen for NH_3 synthesis is primarily obtained through steam methane reforming or coal gasification (Department, 2025; Ye et al., 2017).

To address the emissions challenge, three primary strategies are being explored: (1) continuing conventional production with the addition of carbon capture and storage (CCS), (2) generating hydrogen via water electrolysis powered by renewable energy sources, and (3) developing innovative synthesis methods, such as thermocatalytic and electrochemical approaches. The renewable energy

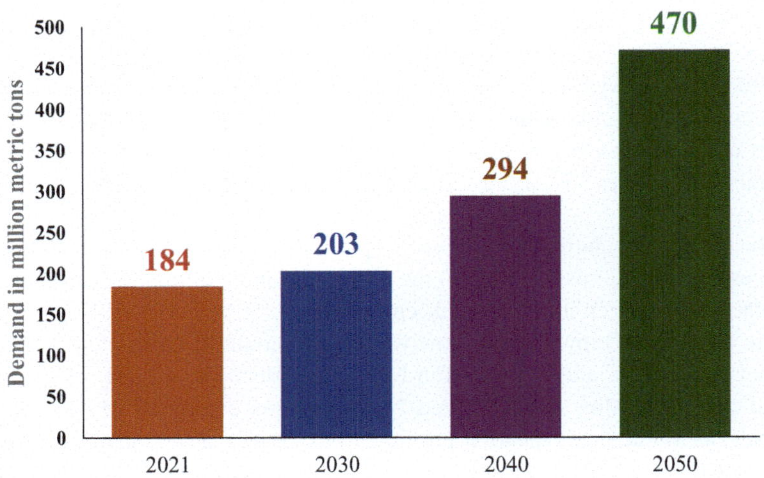

FIGURE 4.3 Global demand for ammonia.

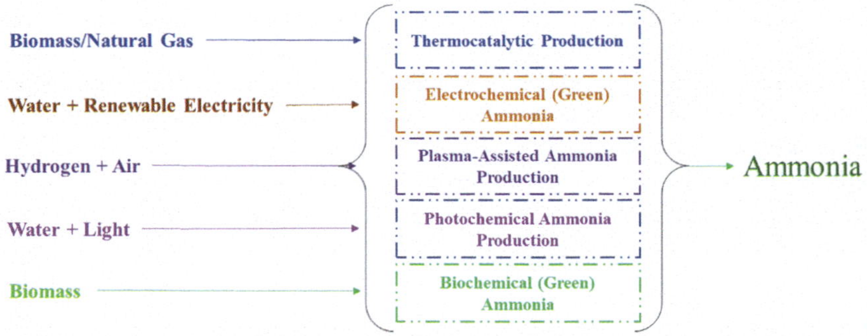

FIGURE 4.4 Methods of ammonia production.

sources proposed for this process include solar, wind, and tidal energy, as illustrated in Fig. 4.4 (Ishaq & Crawford, 2024).

Sustainable NH_3 production methods encompass a variety of techniques, including thermocatalytic, electrochemical, biochemical, photochemical, plasma-assisted, chemical looping, and hybrid approaches that combine two or more of these methods. Each technique offers distinct benefits and drawbacks. In the following sections, the primary focus will be on the first five methods.

4.4.1 Thermocatalytic method of production

Fritz Haber was the first to implement iron-based catalysts for commercial NH_3 production, utilizing the discovery made by Carl Bosch and Alwin Mittasch. This innovation laid the foundation for the thermocatalytic NH_3 synthesis method.

The exothermic nature of this reaction benefits from lower temperatures and higher pressures, which enhances NH_3 yield by facilitating the reaction between nitrogen and hydrogen. The H-B process remains the predominant technique for large-scale NH_3 production (Wang et al., 2019). In this process, hydrogen and nitrogen are combined in a 3:1 ratio over an iron-based catalyst at operating conditions ranging from 723 to 873 K and pressures between 10 and 25 MPa (Kitano et al., 2012). While the process achieves an NH_3 yield of approximately 15%–25% per cycle, the continuous recycling of unreacted gases ultimately results in an overall conversion efficiency of about 97% (Bañares-Alcántara et al., 2015). The thermocatalytic reaction underlying the H-B process is represented in Eq. (4.1) as follows:

$$N_2 + 3H_2 \leftrightarrow 2NH_3 \tag{4.1}$$

The NH_3 synthesis reaction faces significant kinetic challenges due to the high stability of the nitrogen molecule (N_2), which possesses a strong triple bond with a binding energy of 911 kJ/mol (Hagen et al., 2003). Breaking this bond on the catalyst surface requires substantial energy, necessitating high reaction temperatures. Additionally, the reaction is reversible, meaning high-pressure is applied to shift the equilibrium toward NH_3 formation, thereby enhancing the conversion of reactants and minimizing the decomposition of NH_3 back into nitrogen and hydrogen. Developing efficient heterogeneous catalysts capable of driving this reaction under milder conditions has been the focus of extensive research. Historically, the industrial H-B process relied on iron-based catalysts, which required high temperatures and pressures for effective NH_3 production. In 1971, researchers identified Ru supported on carbon as a more active catalyst than iron, particularly under lower temperature conditions (573–723 K) and moderate pressures (4–15 MPa) (Hagen et al., 2003). These Ru-based catalysts, often referred to as second-generation catalysts, offer multiple advantages, including improved energy efficiency, greater equipment reliability, and reduced operational costs (Izumi et al., 1996). In addition to catalyst development, Zhu et al. (1999) introduced a process involving reactors arranged in series, which enhanced NH_3 conversion, decreased gas recycling, and reduced energy consumption. As a result, the NH_3 industry continues to prioritize the development of catalysts capable of maintaining high activity at lower temperatures and pressures to optimize energy use and operational efficiency (Chonggen et al., 2011).

The H-B process produces hydrogen primarily through steam methane reforming, where methane reacts with water to yield hydrogen and carbon dioxide ($CH_4 + 2H_2O \rightarrow 4H_2 + CO_2$). Nitrogen is obtained via air separation, while other hydrogen sources include refinery gas, coal, naphtha, and heavy fuel oil, depending on regional availability (Nazemi et al., 2018). Globally, natural gas accounts for over 72% of NH_3 production, making NH_3 prices closely tied to natural gas costs. In China, coal serves as the predominant feedstock, resulting in higher energy consumption and greenhouse gas emissions compared to other countries (Bicer et al., 2017). Since the use of fossil fuels releases CO_2, recent research has focused on producing CO_2-free NH_3 by generating hydrogen through renewable electrolysis of water (Nanba et al., 2019).

4.4.2 Electrochemical method of production

Electrochemical NH_3 synthesis represents another versatile and widely applicable method for NH_3 production. Based on the type of electrolyte utilized, this process can be categorized into four distinct groups. These groups, classified according to the nature of the electrolyte membrane, include: (1) liquid electrolytes, which function near-ambient temperature; (2) molten-salt electrolytes, which operate at elevated temperatures ranging from 573 to 773 K; (3) composite electrolytes, which are composed of a blend of low-melting-point salts and solid electrolytes; and (4) solid electrolytes, which can function across a broad temperature spectrum, from near-ambient conditions to as high as 1073 K (Chen et al., 2017). Each of these electrolyte systems offers multiple configurations in terms of electrode materials, electrolyte types, and operational conditions (Amar et al., 2011). Among these, solid electrolyte membrane-based electrochemical methods are particularly advantageous due to their operational flexibility and the ease with which hydrogen can be separated from the resulting product. The electrochemical synthesis of NH_3 can either proceed through Eqs. (4.1) or (4.2) (Wang et al., 2019).

$$N_2 + 3H_2O \leftrightarrow 2NH_3 + 1.5O_2 \quad (4.2)$$

The electrochemical synthesis of NH_3 typically utilizes purified hydrogen and nitrogen, achieving high conversion rates but requiring elevated temperatures above 473 K. Hydrogen for this process can be sourced either from natural gas or through water electrolysis, similar to the hydrogen production step in the H-B process (Nazemi et al., 2018). Notably, when hydrogen is generated via water electrolysis using renewable energy, the resulting NH_3 production is CO_2-free. This method also eliminates the risk of catalyst poisoning caused by sulfur compounds and carbon monoxide (CO), contaminants commonly associated with hydrogen produced through natural gas steam reforming (Clarke et al., 2009).

In the electrochemical reaction, hydrogen is supplied at the anode, where it dissociates into protons at the interface between the electrode and the electrolyte. These protons traverse the electrolyte and subsequently participate in a cathodic reaction with nitrogen, combining with electrons to form NH_3. Alternatively, when water and nitrogen serve as reactants, water undergoes oxidation at the anode to produce oxygen, protons, and electrons. The protons and electrons then migrate toward the cathode, where they react with nitrogen to synthesize NH_3. This mechanism has driven increased research interest, particularly in efforts to enhance the efficiency and selectivity of NH_3 synthesis under moderate conditions.

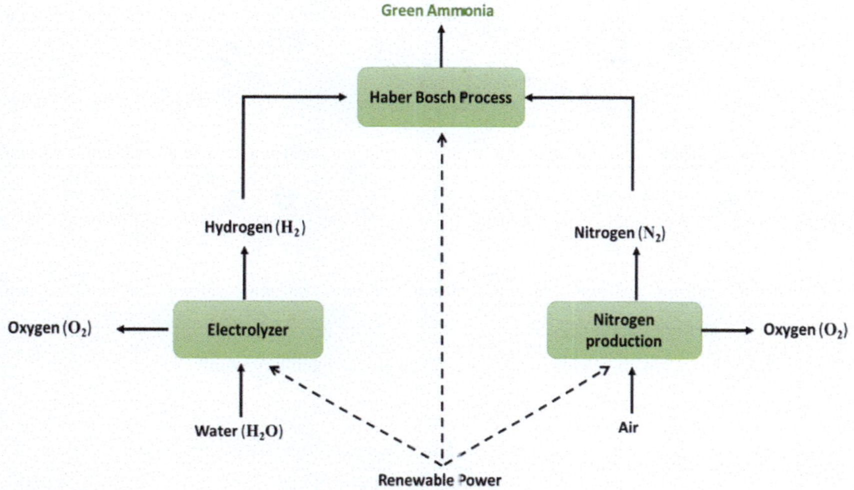

FIGURE 4.5 Combination of Haber-Bosch method and water electrolysis for green ammonia production.

The electrochemical method is generally considered as a more cost-effective alternative to the H-B process due to its potential to operate across a broader range of temperatures (Giddey et al., 2013), as illustrated in Fig. 4.5 (Tornatore et al., 2022). However, the performance of the electrochemical process is strongly influenced by the operating temperature. Recent reports suggest that the energy efficiency of this method can reach approximately 60%. Despite these promising developments, the technology remains under evaluation due to limited understanding of material degradation rates and concerns regarding the long-term durability of materials exposed to NH_3-rich environments.

4.4.3 Plasma-assisted method of production

Plasma-assisted NH_3 synthesis offers a promising alternative for producing NH_3 by facilitating the reaction between H_2 and N_2 through the generation of NH radicals. This method leverages plasma technology to enhance the synthesis process, providing a more efficient pathway for NH_3 production. Research in this field has focused on understanding and optimizing plasma-catalyst interactions under ambient conditions. A study by Li et al. (2023) revealed that the introduction of plasma was the sole factor responsible for increasing the rate of NH_3 synthesis. Andersen et al. (2023), along with Sun et al. (2022), applied the open-source software ZDPlasKin to conduct a detailed numerical analysis, identifying that plasma-assisted NH_3 synthesis is governed predominantly by surface reactions involving Langmuir–Hinshelwood (L-H) and Eley–Rideal (E-R) mechanisms. The core reaction pathways involved in this plasma-catalyzed process are depicted in Fig. 4.6 and were concisely summarized by Engelmann et al. (2021).

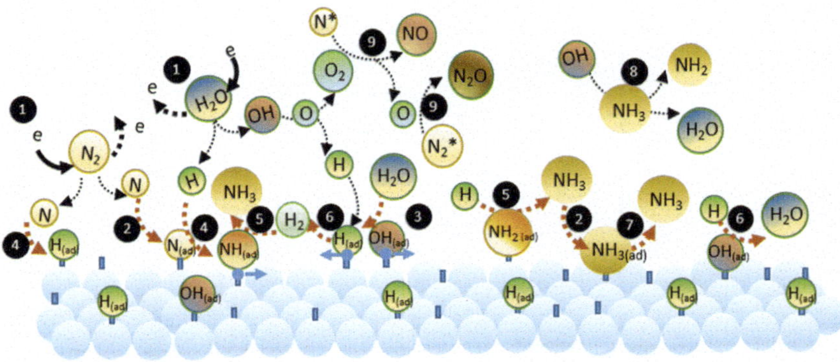

FIGURE 4.6 The main chemical reaction in plasma-assisted ammonia synthesis.

The synthesis process begins with the ionization, excitation, adsorption, and dissociation of N_2 and H_2 molecules due to electron collisions within the plasma environment. Once these molecules are dissociated into free radicals, they adsorb onto the catalyst surface, where the L-H reactions occur between adsorbed species. Simultaneously, atomic diffusion enables E-R reactions involving gas-phase species. NH_3 is subsequently formed through these reactions and is desorbed from the catalyst surface, completing the synthesis cycle. The interaction between plasma and catalyst is crucial, as it directly influences the efficiency and yield of NH_3 production.

When compared to the traditional H-B process, plasma-assisted synthesis presents a distinct advantage by breaking the strong nitrogen-nitrogen triple bond (N≡N) at ambient temperatures and pressures, unlike the H-B process, which requires high-temperature and high-pressure conditions. However, the energy requirements for plasma-based synthesis remain a significant challenge. The bond dissociation energy for the N≡N bond is 946 kJ/mol, substantially higher than the energy required to break the nitrogen-hydrogen bond (N–H) at 389 kJ/mol. This energy disparity affects the overall efficiency of the plasma-catalyzed process. As a result, ongoing research efforts focus on identifying catalysts that improve selectivity and reduce energy consumption in plasma-based NH_3 synthesis, as highlighted by Hong et al. (2022) and Ma et al. (2023).

An effective catalyst should exhibit strong adsorption properties for reactants while simultaneously facilitating the easy release of products, which presents a seemingly contradictory requirement. This challenge can, however, be addressed through the use of alloy catalysts, where the bond strength between N atoms and metal atoms on the catalyst surface is specifically tuned. A study conducted by Xie et al. (2019) demonstrated this principle by synthesizing a high-entropy alloy (HEA) composed of five commonly available metallic elements. The alloy achieved complete conversion (100%) at a temperature of 600°C. Notably, this HEA catalyst outperformed the noble metal Ru, which only reached a 73% conversion rate under similar conditions. The findings from this research underscore the importance of strategic alloy design in optimizing catalyst performance and advancing catalytic reaction processes.

4.4.4 Photochemical method of production

Photochemical methods for NH_3 synthesis are inspired by the natural nitrogen fixation process that occurs during photosynthesis, which is catalyzed by enzymes in plants. However, due to its inherently slow reaction rate, this biological process is unsuitable for large-scale NH_3 production. Instead, photochemical synthesis employs a semiconductor to facilitate the conversion of protons (H^+) and nitrogen (N_2) into NH_3 (Ithisuphalap et al., 2019; Olabi et al., 2023; Yu et al., 2023). In this approach, the semiconductor absorbs solar or ultraviolet radiation, exciting electrons from the valence band (VB) to the conduction band (CB). This process generates electron-hole pairs, with the remaining holes (h^+) in the VB enabling photo-oxidation of water, while the excited electrons in the CB drive the photo-reduction of N_2, leading to NH_3 formation (Olabi et al., 2023). Therefore, identifying an efficient semiconductor photocatalyst capable of effectively harvesting solar energy and utilizing it to reduce N_2 to NH_3 under ambient conditions remains a critical research focus (Xue et al., 2019).

Ideal photocatalysts for green NH_3 production should exhibit key characteristics such as high intrinsic activity, abundant active sites, strong stability, an optimal band gap for efficient photon absorption, suppression of hydrogen evolution, and favorable reaction kinetics (Olabi et al., 2023; Zhang et al., 2019). Titanium dioxide (TiO_2) has been extensively studied as a photocatalyst for nitrogen reduction and hydrogen evolution reactions (NRR and HER) due to its affordability, excellent chemical durability, and high stability. However, a major challenge is the need to create active sites that facilitate the reduction of nitrogen at lower temperatures. One effective strategy to address this issue is carbon doping in TiO_2, which has been demonstrated to enhance nitrogen triple bond cleavage and reduce the band gap of TiO_2 (Shen et al., 2022). For instance, Han et al. (2021) reported highly active Ti^{3+} sites in porous carbon-doped TiO_x (C-TiO_x) nanosheets, achieving an NH_3 production rate of 303.6×10^{-10} mol g^{-1} s^{-1} under visible light irradiation. Similarly, Schrauzer and Guth (2002) introduced Fe doping in photocatalysts, demonstrating a solar-to-chemical conversion efficiency of 0.02% and an over 0.2% reduction in electron-hole recombination. In addition to Fe, transition metals such as Ni, Mo, and Co have also exhibited strong nitrogen reduction capabilities (Olabi et al., 2023). Furthermore, alternative photocatalysts, including g-C_3N_4, CdS, Ta_3N_5, MXene, and $BiVO_4$, have been explored for NH_3 synthesis, reinforcing the pursuit of highly efficient photocatalytic materials (Olabi et al., 2023).

The photochemical approach to NH_3 production holds promise for enhancing energy efficiency, promoting economic growth, and supporting environmental sustainability. A key advantage of this method is its complete reliance on renewable energy sources, making it a carbon-free alternative for NH_3 synthesis.

However, its practical implementation is hindered by challenges related to efficiency and stability. Overcoming these limitations requires significant advancements in photocatalyst development, particularly in optimizing band gap properties. Future research efforts should prioritize the discovery of cost-effective, stable, and highly active photocatalysts to facilitate large-scale NH_3 production and maximize the utilization of solar energy (Han et al., 2021; Zhang et al., 2019).

4.4.5 Biochemical method of production

The efficiency of the H-B process in atmospheric N_2 conversion to NH_3 is significantly lower than that achieved by natural nitrogen-fixing bacteria, which operate efficiently at ambient temperature and pressure conditions. It is estimated that approximately 48% of the nitrogen present on the surface of the Earth is fixed through bacterial activity, while the remaining 52% is obtained from NH_3-based products (Ahmed et al., 2024). Biological nitrogen fixation occurs through mechanisms such as bacterial biological nitrogen fixation and biomimetic nitrogen fixation (He et al., 2022).

Biomimetic nitrogen fixation seeks to replicate the natural nitrogenase enzyme's functionality by developing an artificial catalyst capable of synthesizing NH_3 under similar ambient conditions. This process necessitates the presence of energy carriers like nitrogenase, magnesium adenosine triphosphate ($MgATP^{2-}$), and reduced ferredoxin or flavin redox proteins, which are fundamental to microbial nitrogen fixation (Ahmed et al., 2024). The core protein involved in this process, known as MoFe, contains a specialized cluster that facilitates nitrogen reduction within the nitrogenase complex. This complex is composed of two key proteins: MoFe and Fe. The reduction of N_2 to NH_3 involves an electron donor, energy provided by ATP hydrolysis, and the coordinated interaction of these proteins (Yan et al., 2019).

During the nitrogen fixation process, hydrogen gas (H_2) is produced as a by-product when nitrogen undergoes reduction to form NH_3. The reaction mechanism, which follows a specific stoichiometric relationship, is presented in Eq. 4.3. This equation illustrates the process by which 16 molecules of MgATP and eight electrons are required to convert atmospheric nitrogen into NH_3, mimicking the energy-efficient mechanism of biological nitrogen fixation.

$$N_2 + 8H^+ + 16MgATP + 8e^- \rightarrow 2NH_3 + H_2 + 16MgADP + 16Pi \quad (4.3)$$

where ADP and Pi stand for adenosine diphosphate and phosphate, respectively, from inorganic sources.

Biomimetic chemical nitrogen fixation seeks to replicate the mechanism of nitrogenase enzymes in a controlled chemical reaction. This process has been explored through the synthesis of catalysts, such as MoFe proteins, which can form biohybrid systems capable of producing NH_3 when exposed to visible light, followed by ATP hydrolysis. Researchers have demonstrated that iron (Fe)

TABLE 4.2 Different ammonia production techniques, with their respective benefits and drawbacks highlighted.

Method	Advantages	Disadvantages
Thermocatalytic method	High production efficiency Reliable and robust Widely used in industry	Energy-intensive Reliance on fossil fuels
Electrochemical method	Renewable energy utilization Operates under milder conditions	Cost-effective electrocatalysts Limited scalability
Photochemical method	Operates under milder conditions Potential to use solar energy	Availability of sunlight affects efficiency Challenges in developing efficient photocatalysts
Biochemical method	Operates at lower temperatures and pressures Potential for reduced energy consumption	Challenges in developing efficient and stable catalysts Complex biological processes
Plasma-assisted method	Operates at lower temperatures and pressure Potential advantages in terms of energy efficiency	Complex and costly setup Potential issues related to plasma stability and control
Chemical looping method	Allows for better control of the separation and synthesis steps Potential for improved efficiency	Complexity in the design and operation of chemical looping systems Limited practical implementation

can function as an effective active site for nitrogen reduction by incorporating artificial MoFe and Fe centers within a polymer network, thereby mimicking natural nitrogenase activity (Yao et al., 2020). Although these developments are promising, significant challenges remain, including the need for more advanced modeling techniques and a better understanding of the intricate mechanisms governing this process. Generating accurate isotopic calibration data, especially concerning atmospheric ammonium (NH_4^+) pollution, is a critical area for further investigation (Ling et al., 2019).

The diverse array of NH_3 production methods each presents unique advantages and challenges, reflecting a trade-off between efficiency, sustainability, and technical feasibility (Table 4.2). The thermocatalytic (H-B) method remains the industrial cornerstone due to its high production efficiency and reliability, yet its heavy reliance on fossil fuels and energy-intensive operation—requiring extreme temperatures and pressures—contributes significantly to greenhouse gas emissions. In contrast, emerging methods aim to mitigate these drawbacks.

Electrochemical synthesis leverages renewable energy and operates under milder conditions, offering a pathway to CO_2-free NH_3, though it faces hurdles in scalability and the development of cost-effective electrocatalysts. Similarly, photochemical methods harness solar energy for ambient-condition synthesis but struggle with photocatalyst inefficiency and sunlight dependency, limiting consistent large-scale application. Biochemical approaches, inspired by natural nitrogen fixation, operate at benign temperatures and pressures with reduced energy demands, yet their complexity and the instability of biohybrid systems pose challenges for industrial adaptation. Plasma-assisted synthesis disrupts traditional energy requirements by enabling N_2 activation at lower temperatures, yet its complex setup and plasma stability issues hinder practicality. Meanwhile, chemical looping promises enhanced efficiency through controlled reaction-separation steps but remains constrained by system complexity and limited real-world implementation. Collectively, these methods underscore the need for balanced innovation: optimizing traditional processes for sustainability while advancing emerging technologies to address scalability, cost, and durability.

4.5 Ammonia as a fuel in internal combustion engines

NH_3 is gaining traction as a carbon-neutral fuel for ICEs, offering a pathway to decarbonize sectors such as maritime transport, power generation, and heavy-duty machinery. With over 250 million tons produced annually, NH_3 benefits from a mature global infrastructure for storage and transportation, originally developed for agricultural applications. However, its adoption in ICEs faces challenges due to its inherent combustion properties, including low flame speed, high ignition temperature, and nitrogen-bound emissions. Despite these limitations, advances in engine design and fuel blending strategies have enabled progress in both SI and CI engines, positioning NH_3 as a viable alternative to fossil fuels in specific applications (Chavando et al., 2024; Cheng et al., 2025).

4.5.1 Utilizing ammonia in spark-ignition engines

SI engines are particularly suited to leverage the high-octane rating of NH_3 ($\approx$110), which resists knocking and allows for elevated CR. To address the slow flame propagation and narrow flammability range of NH_3, modifications such as increased spark energy, advanced ignition timing, and higher CR (10–18:1) are essential. Onboard NH_3 cracking systems, which dissociate NH_3 into hydrogen (H_2) and nitrogen (N_2), further enhance combustion stability by introducing the rapid flame speed of hydrogen. Blending NH_3 with hydrogen or natural gas (CH_4) has proven effective: for instance, a 50:50 NH_3-CH_4 mixture reduced CO_2 emissions by 28% compared to pure natural gas in a heavy-duty SI engine (Oh et al., 2021). However, the low volumetric energy density of NH_3 results in a ~20% power loss relative to gasoline, necessitating auxiliary systems like supercharging to compensate. Emissions remain a critical concern, with NO_x

levels reaching 4000–8000 ppm due to fuel-bound nitrogen, alongside unburned NH_3 at low dissociation rates. Catalytic after-treatment systems, such as selective catalytic reduction (SCR), are required to mitigate these emissions. Recent studies demonstrate stable engine operation (COV_{IMEP} <5%) under optimized conditions, highlighting the potential of NH_3 in light-duty vehicles and hybrid systems when paired with renewable hydrogen (Mounaïm-Rousselle et al., 2021; Sharma et al., 2025; Uddeen et al., 2023).

4.5.2 Utilizing ammonia in compression-ignition engines

CI engines face greater challenges due to the near-zero CN of NH_3, which impedes autoignition. To overcome this, researchers have explored ultrahigh CR (>35:1), though such extremes are impractical for conventional engines. Instead, dual-fuel systems using pilot ignition agents like diesel, biodiesel, or dimethyl ether (DME) are prioritized. For example, NH_3-DME blends achieve stable combustion at 60% NH_3 fractions, though cycle-to-cycle variability increases with higher NH_3 content. Intake air preheating and high-pressure direct injections improve reactivity, while homogeneous charge compression-ignition (HCCI) modes require combustion temperatures exceeding 1300 K for efficiency. In marine applications, NH_3-diesel blends reduce carbon emissions but exacerbate NO_x levels due to fuel-bound nitrogen. Recent experiments with aqueous NH_3 injections in dual-fuel CI engines revealed prolonged ignition delays and elevated CO/HC emissions under lean conditions, underscoring the need for exhaust after-treatment. Despite these hurdles, the compatibility of NH_3 with large-bore engines positions it as a critical candidate for decarbonizing maritime shipping, where torque output and fuel storage capacity are paramount (Breeze, 2017; Dimitriou & Javaid, 2020; Lion et al., 2020).

As in Table 4.3, the application of NH_3 in SI and CI engines reflects distinct advantages and challenges tailored to their operational frameworks. SI engines capitalize on the antiknock properties of NH_3 and compatibility with hydrogen cocombustion, making them suitable for light-duty vehicles and renewable energy systems. In contrast, CI engines, though constrained by the low reactivity of NH_3, offer promise in heavy-duty and maritime sectors through dual-fuel systems and pilot ignition. Both engine types grapple with high NO_x emissions, necessitating advanced catalytic converters or SCR systems.

Key barriers to widespread adoption include the low energy density of NH_3, which demands larger fuel storage systems, and the energy-intensive nature of NH_3 cracking. Future advancements hinge on integrating renewable hydrogen production to ensure carbon-neutral NH_3 synthesis, alongside innovations in combustion strategies such as jet ignition or HCCI modes. Catalyst development for efficient NH_3 dissociation and material science breakthroughs to mitigate corrosion risks are equally critical.

In conclusion, NH_3 represents a transformative fuel for ICEs in hard-to-abate sectors, bridging the gap between existing infrastructure and decarbonization

TABLE 4.3 Comparison of ammonia combustion engines (Bro & Pedersen, 1977; Cornelius et al., 1966; Hu et al., 2024; Heywood, 1981; Niki et al., 2019; Pan et al., 2023; Rouwenhorst et al., 2020; Ryu et al., 2014; Zhang et al., 2023; Zhou et al., 2018).

Aspect	Spark-Ignition (SI) engines	Compression-Ignition (CI) engines
Combustion method	Forced ignition (spark plugs) with H_2 coinjection for flame acceleration	Autoignition via pilot fuels (diesel, DME) or extreme compression ratios (>35:1)
Modifications	Higher compression ratios, NH_3 cracking systems, corrosion-resistant materials	Dual-fuel injection, intake preheating, ultrahigh compression ratios (rarely practical)
Advantages	Antiknock performance Lower CO_2 with NH_3-CH_4 blends Renewable H_2 compatibility	Suitable for maritime/heavy-duty use Torque output comparable to diesel
Challenges	High NO_x emissions Power loss vs. gasoline Hydrogen backfire risk	Fuel-bound nitrogen → NO_x spikes Cycle instability at high NH_3 fractions
Emissions	NO_x: 4000–8000 ppm Unburned NH_3 at low loads	NO_x increases with NH_3 content Elevated CO/HC under lean combustion
Applications	Light-duty vehicles, hybrid systems, and power generation	Marine engines, heavy machinery, and industrial power plants

goals. While technical challenges persist, sustained research into hybrid combustion systems, emissions control, and renewable integration could unlock the full potential of NH_3 as a cornerstone of sustainable energy systems.

4.6 Life cycle assessment of ammonia production methods

NH_3 production is a critical industrial process, yet traditional routes such as the H-B method are energy-intensive and heavily reliant on fossil fuels. Table 4.4 summarizes five NH_3 production pathways: thermocatalytic (blue), electrochemical (green), plasma-assisted, photochemical, and biochemical, highlighting raw materials, energy requirements, CO_2 emissions, and estimated costs. This section provides a review-style discussion, integrating life cycle assessment (LCA) insights, recent technological advances, and ongoing challenges.

TABLE 4.4 The raw materials, energy required, carbon emissions, and estimated cost of different methods for ammonia production (Bicer et al., 2016; Green ammonia synthesis, 2023; Smith et al., 2020; Ye et al., 2022; Zhang et al., 2022).

Production method	Raw materials	Energy required (MJ/kg NH_3)	CO_2 emissions (kg CO_2/kg NH_3)	Estimated cost (USD/ton)
Thermocatalytic (Blue)	Natural gas and air	28–35	0.5–1.5 (with CCS)	USD 300–600
Electrochemical (Green)	Water and air	30–45	0–0.34	USD 500–800
Plasma-Assisted	Air and methane/electrolysis H_2	40–60	0.5–2.0	USD 600–1000
Photochemical	Water, air, and sunlight	50–70	Near-zero	USD 800–1200
Biochemical	Biomass (e.g., bagasse)	25–35	0.34–0.85	USD 400–600

Thermocatalytic production, often referred to as blue NH_3 when paired with CCS, requires 28–35 MJ/kg NH_3 plus an additional 3–8 MJ/kg NH_3 for CCS. Although CO_2 emissions can drop to 0.5–1.5 kg CO_2/kg NH_3 (a 60%–90% reduction versus conventional H-B), upstream methane leakage during natural gas extraction can offset these gains, elevating global warming potential (GWP). Estimated costs range from 300 to 600 USD/ton NH_3, largely influenced by fossil fuel prices and CCS infrastructure, while long-term sustainability is hampered by continued dependence on fossil feedstocks and the energy penalty of compressing and storing captured CO_2 (Royal, 2020; Smith et al., 2020; Saygin et al., 2023; Ye et al., 2022; Wang et al., 2021; Zhang et al., 2025).

Electrochemical or NH_3 uses water electrolysis powered by renewable electricity to generate hydrogen, which is then reacted with nitrogen, cutting fossil inputs and lowering CO_2 emissions to 0–0.34 kg CO_2/kg NH_3 (cradle-to-gate). However, the process demands 30–45 MJ/kg NH_3, mostly for electrolysis (~50 kWh/kg H_2), with capital and operational costs reaching 500–800 USD/ton NH_3. High-cost electrolyzers, grid integration challenges, and the need for durable catalysts (e.g., Ru) limit the commercial scale-up of the technology, despite its potential for near-zero fossil fuel consumption (Liu et al., 2020).

Plasma-assisted NH_3 targets a lower-temperature pathway by ionizing nitrogen in a plasma discharge, yet this approach remains energy-intensive (40–60 MJ/kg NH_3). If hydrogen is sourced from renewables, CO_2 emissions can drop to around 0.5 kg CO_2/kg NH_3, but reliance on methane-based hydrogen pushes it closer to 2.0 kg. Costs of 600–1000 USD/ton NH_3 reflect the low readiness level of the technology and the limited durability data on plasma reactors. Challenges surrounding scalability, reactor stability, and uncertain long-term environmental impacts also constrain widespread adoption (Peng et al., 2023; Wang et al., 2024).

Photochemical NH_3 production aims to exploit sunlight directly for nitrogen reduction, theoretically achieving near-zero operational emissions. Practical barriers include low conversion efficiencies (5%–15%), high land requirements for solar arrays, and photocatalyst degradation (e.g., TiO_2), driving energy usage estimates to 50–70 MJ/kg NH_3. Current cost projections (800–1200 USD/ton NH_3) are further complicated by material waste and inconsistent pilot-scale data, indicating that significant R&D is necessary before photochemical routes become viable at commercial scales (Choe et al., 2021; Souissi, 2024).

Biochemical methods rely on biomass feedstocks—often agricultural residues—and microbial or thermochemical pathways to form NH_3, leading to 25–35 MJ/kg NH_3. Resulting CO_2 emissions range from 0.34 to 0.85 kg CO_2/kg NH_3, representing up to a 78% reduction compared with conventional natural gas–based processes when biomass is sustainably sourced. Although costs of 400–600 USD/ton NH_3 appear competitive, feedstock seasonality, competing uses for biomass (e.g., biofuels), and logistics of collection and preprocessing create uncertainties. Coproducts such as biochar can sequester additional carbon, but the precise climate benefits vary with local agronomic practices (Bazzanella & Ausfelder, 2017; Davis et al., 2018; Raganati et al., 2021; Morgan et al., 2019).

Comparatively, thermocatalytic (blue) NH_3 reduces direct emissions through CCS but remains susceptible to methane leakage and fossil-fuel reliance, while electrochemical (green) NH_3 has near-zero direct emissions but faces steep capital costs and demanding electricity needs. Plasma-assisted and photochemical pathways could, in principle, avoid fossil inputs and lower operational temperatures, yet high energy consumption and limited commercial data cast doubt on near-term viability. Biochemical routes show promise when integrated into a circular economy framework, although feedstock availability and seasonal variability still pose limits. Overall, each environmental and economic performance of the technology hinges on regional resource availability, policy support (e.g., carbon pricing), and technical breakthroughs that can reduce operational costs and lifecycle impacts.

4.7 Current challenges in sustainable ammonia production

Sustainable NH_3 production faces a range of interconnected technical, economic, and social barriers that limit rapid adoption. A central challenge is energy intensity, notably for electrochemical processes, which can require up to 50 kWh/kg of hydrogen—enough to power mid-sized cities. Such heavy demand may overwhelm regions lacking robust renewable energy infrastructures and intensify land-use conflicts, as large solar farms or wind projects can encroach on agricultural zones or marine ecosystems. This raises doubts over whether "green" NH_3 truly minimizes ecological impacts once broader infrastructure footprints are considered.

Economic viability also remains elusive, as capital expenditures for technologies like electrolyzers (1200–2000 USD/kW) or plasma reactors can exceed those of traditional systems by 50%–200%. Catalyst constraints worsen this cost differential; while substituting rare metals such as Ru with iron or nickel lowers material expenses, it can require operating temperatures of 400–500°C and lead to frequent catalyst degradation, thereby offsetting efficiency gains. Moreover, infrastructure inertia poses a structural hurdle. Established H-B plants, honed over decades to use fossil fuels, account for most of the current NH_3 output. In contrast, decentralized approaches like modular photochemical systems or biomass gasification grapple with fragmented logistics, challenging storage requirements (given the toxicity of NH_3), and a lack of established market pathways. Retrofitting existing facilities with carbon capture (CCS) can offer interim emissions reductions yet risks perpetuating fossil fuel use if policymakers and investors consider CCS a permanent rather than transitional option.

Sociopolitical factors compound these issues. Communities often resist renewable energy initiatives where extensive land or offshore areas are required. Meanwhile, labor and economic interests in fossil-dependent regions may impede large-scale transitions. Further complicating global markets is the scarcity of unified standards for "green" NH_3 certification, making it difficult to ensure genuine reductions in carbon intensity and opening the door for questionable "greenwashing" based on unverifiable credits.

4.8 Future directions: Innovations and systemic risks

Moving forward, catalyst innovation is poised to play a pivotal role in overcoming many of these barriers. Substituting rare-metal catalysts (e.g., Ru) with iron or nickel could reduce capital expenditures for electrochemical systems by 30%–40%, though real-world challenges like corrosion in humid electrolyzers or maintaining catalyst stability under high-energy plasma conditions remain. More ambitious approaches include bio-inspired nitrogenase enzymes, known for their ability to fix nitrogen at ambient conditions, yet such enzymes typically require expensive stabilization. Process redesigns like electrochemical looping or membrane reactors could reshape production entirely, but commercializing these architectures calls for rethinking plant configurations and supply networks.

Hybrid systems offer a second avenue of innovation. Combining plasma-assisted nitrogen dissociation with green hydrogen may reduce energy needs below 35 MJ/kg NH_3, especially if it leverages surplus electricity from intermittent renewables. Pairing NH_3 synthesis directly with renewable sources (e.g., offshore wind) could also tap into periods of excess power generation, although the volatility of NH_3 poses safety and transport challenges. Similarly, waste-to-NH_3 strategies, such as gasifying municipal refuse or agricultural residues, align with circular economy ideals but can be derailed by inconsistent feedstock composition and high-temperature processing requirements that produce additional toxic byproducts unless carefully managed.

Circular economy approaches thus require scrutiny. While biochar from biomass gasification can sequester carbon, it may gradually release CO_2 over time due to microbial decomposition. Likewise, promoting NH_3 as a hydrogen carrier raises questions about toxicity, leakage, and the energy penalty of reconverting NH_3 to hydrogen. Additionally, policy and market mechanisms must extend beyond theoretical carbon pricing; they must integrate binding mandates and transparent certification frameworks to ensure genuinely low-carbon NH_3. Startups proposing "disruptive" solutions may thrive in niche contexts but often struggle to scale-up against the entrenched economics of gigaton-scale H-B plants.

Ultimately, the success of sustainable NH_3 hinges on systemic alignment, including strategic grid integration for hydrogen and NH_3, closer collaboration between industries (e.g., NH_3 producers and waste management firms), and decisive policymaking that addresses fossil fuel subsidies. Without cohesive efforts to harmonize technological innovation, market structures, and governance, even the most promising breakthroughs risk stalling in the gap between laboratory success and widespread commercial adoption.

4.9 Conclusion

The transition to NH_3 as a sustainable fuel requires a combination of technological innovation, infrastructure adaptation, and strategic policy support. This chapter has explored key production pathways including thermocatalytic, electrochemical, plasma-assisted, photochemical, and biochemical methods, each presenting distinct advantages and limitations in terms of efficiency, scalability, and energy requirements. While significant progress has been made, challenges such as high energy consumption, catalyst optimization, and process integration with renewable energy sources remain critical areas for further research.

Beyond production, direct utilization of NH_3 in ICEs offers a promising route for decarbonizing the transportation sector. However, low flame speed, high ignition temperature, and NO_x emissions present engineering challenges that must be addressed through fuel blending strategies, ignition enhancement techniques, and advanced engine designs Despite these limitations, ongoing advancements indicate that NH_3 could serve as a viable, carbon-free alternative for both SI and CI engines, reinforcing its role in the broader clean energy transition.

Moving forward, the successful implementation of NH_3 will depend on further improvements in production efficiency, optimization of combustion technologies, and the development of NH_3-compatible fuel infrastructure. While policy frameworks and market incentives will play a role in scaling adoption, technological advancements remain the key driver in determining the long-term viability of NH_3 as a sustainable fuel. By addressing these challenges through multidisciplinary research and industry collaboration, NH_3 can emerge as a practical, scalable solution for reducing carbon emissions across multiple sectors.

Acknowledgment

The author(s) would like to acknowledge the support provided by the Deanship of Scientific Research (DSR) at King Fahd University of Petroleum & Minerals (KFUPM) for funding this work through project No. INRC 2605.

References

Ahmed, H. S., Yahya, Z., Ali khan, W., & Faraz, A (2024). Sustainable pathways to ammonia: a comprehensive review of green production approaches. *Clean Energy, 8*(2), 60–72.

Amar, I. A., Lan, R., Petit, C. T., & Tao, S. (2011). Solid-state electrochemical synthesis of ammonia: a review. *Journal of solid state electrochemistry, 15*(9), 1845–1860.

Andersen, J. A., Holm, M. C., van't Veer, K., Christensen, J. M., Østberg, M., Bogaerts, A., & Jensen, A. D. (2023). Plasma-catalytic ammonia synthesis in a dielectric barrier discharge reactor: A combined experimental study and kinetic modeling. *Chemical Engineering Journal, 457*, 141294.

Ashik, U.P.M., Viswan, A., Kudo, S., & Hayashi, J.I. (2018). Nanomaterials as catalysts. Applications of nanomaterials (pp. 45–82). Woodhead Publishing.

Bañares-Alcántara, R., Dericks, G., III, Fiaschetti, M., Grünewald, P., Lopez, J. M., Tsang, E., ... Zhao, S. (2015). *Analysis of islanded ammonia-based energy storage systems* (pp. 1–150). University of Oxford.

Bazzanella, A., & Ausfelder, F. (2017).Low carbon energy and feedstock for the European chemical industry: Technology Study. DECHEMA, Gesellschaft für Chemische Technik und Biotechnologie eV.

Bicer, Y., Dincer, I., Vezina, G., & Raso, F. (2017). Impact assessment and environmental evaluation of various ammonia production processes. *Environmental management, 59*(5), 842–855.

Bicer, Y., Dincer, I., Zamfirescu, C., Vezina, G., & Raso, F. (2016). Comparative life cycle assessment of various ammonia production methods. *Journal of Cleaner Production, 135*, 1379–1395.

Breeze, P. (2017). *Piston engine-based power plants*. Academic Press.

Bro, K., & Pedersen, P. S. (1977). *Alternative diesel engine fuels: An experimental investigation of methanol, ethanol, methane and ammonia in a DI diesel engine with pilot injection*. SAE Technical Paper.

Cai, T., Zhao, D., & Gutmark, E. (2023). Overview of fundamental kinetic mechanisms and emission mitigation in ammonia combustion. *Chemical Engineering Journal, 458*, 141391.

Cardoso, J. S., Silva, V., Rocha, R. C., Hall, M. J., Costa, M., & Eusébio, D. (2021). Ammonia as an energy vector: Current and future prospects for low-carbon fuel applications in internal combustion engines. *Journal of Cleaner Production, 296*, 126562.

Chai, W. S., Bao, Y., Jin, P., Tang, G., & Zhou, L. (2021). A review on ammonia, ammonia-hydrogen and ammonia-methane fuels. *Renewable and Sustainable Energy Reviews, 147*, 111254.

Chavando, A., Silva, V., Cardoso, J., & Eusebio, D. (2024). Advancements and challenges of ammonia as a sustainable fuel for the maritime industry. *Energies, 17*(13), 3183.

Chen, S., Perathoner, S., Ampelli, C., Mebrahtu, C., Su, D., & Centi, G. (2017). Electrocatalytic synthesis of ammonia at room temperature and atmospheric pressure from water and nitrogen on a carbon-nanotube-based electrocatalyst. *Angewandte Chemie International Edition, 56*(10), 2699–2703.

Cheng, Q., Muhammad, A., Kaario, O., Ahmad, Z., & Martti, L. (2025). Ammonia as a sustainable fuel: Review and novel strategies. *Renewable and Sustainable Energy Reviews, 207*, 114995.

Choe, S., Kim, S. M., Lee, Y., Seok, J., Jung, J., Lee, J.S., & Jang, Y.J. (2021). Rational design of photocatalysts for ammonia production from water and nitrogen gas. *Nano Convergence, 8*(1), 22.

Christensen, M. (2016). Laminar Burning Velocity and Development of a Chemical Kinetic Model for Small Oxygenated Fuels. [Doctoral Thesis (compilation), Combustion Physics].

Chonggen, P. A. N., Ying, L. I., & Huazhang, L. I. U. (2011). Effects of reaction conditions on performance of Ru catalyst and iron catalyst for ammonia synthesis. *Chinese Journal of Chemical Engineering, 19*(2), 273–277.

Clarke, R. E., Giddey, S., Ciacchi, F. T., Badwal, S. P. S., Paul, B., & Andrews, J. (2009). Direct coupling of an electrolyser to a solar PV system for generating hydrogen. *International journal of hydrogen energy, 34*(6), 2531–2542.

Davis, R. E., et al. (2018). *Process design and economics for the conversion of lignocellulosic biomass to hydrocarbon fuels and coproducts: 2018 biochemical design case update; biochemical deconstruction and conversion of biomass to fuels and products via integrated biorefinery pathways*. National Renewable Energy Lab (NREL).

Department, S. R. (2025). Forecast demand for ammonia worldwide from 2021 to 2050. *Chemicals & Resources*. https://www.statista.com/statistics/1345802/forecast-value-global-ammonia-demand/.

Dimitriou, P., & Javaid, R. (2020). A review of ammonia as a compression ignition engine fuel. *International Journal of Hydrogen Energy, 45*(11), 7098–7118.

El-Adawy, M., Nemitallah, M. A., & Abdelhafez, A. (2024). Towards sustainable hydrogen and ammonia internal combustion engines: Challenges and opportunities. *Fuel, 364*, 131090.

Engelmann, Y., van't Veer, K., Gorbanev, Y., Neyts, E. C., Schneider, W. F., & Bogaerts, A. (2021). Plasma catalysis for ammonia synthesis: a microkinetic modeling study on the contributions of Eley–Rideal reactions. *ACS Sustainable Chemistry & Engineering, 9*(39), 13151–13163.

Giddey, S., Badwal, S., & Kulkarni, A. (2013). Review of electrochemical ammonia production technologies and materials. *International Journal of Hydrogen Energy, 38*(34), 14576–14594.

Gielen, D., Boshell, F., Saygin, D., Bazilian, M. D., Wagner, N., & Gorini, R. (2019). The role of renewable energy in the global energy transformation. *Energy strategy reviews, 24*, 38–50.

Grasham, O., Dupont, V., Camargo-Valero, M. A., García-Gutiérrez, P., & Cockerill, T. (2019). Combined ammonia recovery and solid oxide fuel cell use at wastewater treatment plants for energy and greenhouse gas emission improvements. *Applied energy, 240*, 698–708.

Green ammonia synthesis. (2023). Nat. Synth, • 2, 581–582. https://doi.org/10.1038/s44160-023-00362-y.

Hagen, S., Barfod, R., Fehrmann, R., Jacobsen, C. J., Teunissen, H. T., & Chorkendorff, I. (2003). Ammonia synthesis with barium-promoted iron–cobalt alloys supported on carbon. *Journal of Catalysis, 214*(2), 327–335.

Han, Q., et al. (2021a). Progress and challenges in photocatalytic ammonia synthesis. Materials Advances, 2(2), 564–581.

Han, Q., et al. (2021a). Progress and challenges in photocatalytic ammonia synthesis. Materials Advances, 2(2), 564–581.

Hartfield, G., Blunden, J., & Arndt, D. S. (2018). State of the Climate in 2017. *Bulletin of the American Meteorological Society, 99*(8), Si–S310.

He, H., Zhu, Q. Q., Yan, Y., Zhang, H. W., Han, Z. Y., Sun, H., … Du, M. (2022). Metal–organic framework supported Au nanoparticles with organosilicone coating for high-efficiency electrocatalytic N2 reduction to NH3. *Applied Catalysis B: Environmental, 302*, 120840.

Heywood, J. B. (1981). Automotive engines and fuels: A review of future options. *Progress in Energy and Combustion Science, 7*(3), 155–184.

Hong, J., Zhang, T., Zhou, R., Dou, L., Zhang, S., Zhou, R., … Cullen, P. J. (2022). Green chemical pathway of plasma synthesis of ammonia from nitrogen and water: a comparative kinetic study with a N2/H2 system. *Green Chemistry, 24*(19), 7458–7468.

Hu, X., Li, J., Pan, J., Zhang, R., Wei, H., & Shu, G. (2024). On combustion and emission characteristics of ammonia/hydrogen engines: emphasis on energy ratio and equivalence ratio. *Fuel, 365*, 131183.

Imhoff, T. B., Gkantonas, S., & Mastorakos, E. (2021). Analysing the performance of ammonia powertrains in the marine environment. *Energies, 14*(21), 7447.

Ishaq, H., & Crawford, C. (2024). Review and evaluation of sustainable ammonia production, storage and utilization. *Energy Conversion and Management, 300*, 117869.

Ithisuphalap, K., Zhang, H., Guo, L., Yang, Q., Yang, H., & Wu, G. (2019). Photocatalysis and photoelectrocatalysis methods of nitrogen reduction for sustainable ammonia synthesis. *Small Methods, 3*(6), 1800352.

Izumi, Y., Iwata, Y., & Aika, K.-i. (1996). Catalysis on ruthenium clusters supported on CeO2 or Ni-doped CeO2: Adsorption behavior of H_2 and ammonia synthesis. *The Journal of Physical Chemistry, 100*(22), 9421–9428.

Jin, S., Wu, B., Zi, Z., Yang, P., Shi, T., & Zhang, J. (2023). Effects of fuel injection strategy and ammonia energy ratio on combustion and emissions of ammonia-diesel dual-fuel engine. *Fuel,. 341*, 127668.

Kitano, M., Inoue, Y., Yamazaki, Y., Hayashi, F., Kanbara, S., Matsuishi, S., … Hosono, H. (2012). Ammonia synthesis using a stable electride as an electron donor and reversible hydrogen store. *Nature chemistry, 4*(11), 934–940.

Kurien, C., & Mittal, M. (2022). Review on the production and utilization of green ammonia as an alternate fuel in dual-fuel compression ignition engines. *Energy Conversion and Management, 251*, 114990.

Li, K., Chen, S., Wang, H., & Wang, F. (2023). Plasma-assisted ammonia synthesis over Ni/LaOF: dual active centers consisting of oxygen vacancies and Ni. *Applied Catalysis A: General, 650*, 118983.

Ling, C., Zhang, Y., Li, Q., Bai, X., Shi, L., & Wang, J. (2019). New mechanism for N2 reduction: the essential role of surface hydrogenation. *Journal of the American Chemical Society, 141*(45), 18264–18270.

Lion, S., Vlaskos, I., & Taccani, R. (2020). A review of emissions reduction technologies for low and medium speed marine Diesel engines and their potential for waste heat recovery. *Energy Conversion and Management, 207*, 112553.

Liu, X., Elgowainy, A., & Wang, M. (2020). Life cycle energy use and greenhouse gas emissions of ammonia production from renewable resources and industrial by-products. *Green Chemistry, 22*(17), 5751–5761.

Loulergue, L., Schilt, A., Spahni, R., Masson-Delmotte, V., Blunier, T., Lemieux, B., … Chappellaz, J. (2008). Orbital and millennial-scale features of atmospheric CH4 over the past 800,000 years. *Nature, 453*(7193), 383–386.

Ma, F., Guo, L., Li, Z., Zeng, X., Zheng, Z., Li, W., … Yu, W. (2023). A review of current advances in ammonia combustion from the fundamentals to applications in internal combustion engines. *Energies, 16*(17), 6304.

Lhuillier, C., Brequigny, P., Contino, F., & Rousselle, C. (2019). *Combustion characteristics of ammonia in a modern spark-ignition engine* (No. 2019-24-0237). SAE Technical Paper.

Cornelius, W., Huellmantel, L.W., & Mitchell, H.R. (1965). *Ammonia as an engine fuel* (No. 650052). SAE Technical Paper.

Cornelius, W., Huellmantel, L.W., & Mitchell, H.R. (1965). Ammonia as an engine fuel (SAE Technical Paper No. 650052). SAE International. https://doi.org/10.4271/650052

Morgan, T. J., Youkhana, A., Turn, S. Q., Ogoshi, R., & Garcia-Peŕez, M. (2019). Review of biomass resources and conversion technologies for alternative jet fuel production in Hawai'i and tropical regions. *Energy & fuels, 33*(4), 2699–2762.

Mounaïm-Rousselle, C., Brequigny, P., Dumand, C., & Houillé, S. (2021). Operating limits for ammonia fuel spark-ignition engine. *Energies, 14*(14), 4141.

Nanba, T., Javaid, R., & Matsumoto, H. (2019). Ammonia synthesis by using hydrogen produced from renewable energy. *Catalysts Catal, 61*(2), 66–71.

Nazemi, M., Panikkanvalappil, S. R., & El-Sayed, M. A. (2018). Enhancing the rate of electrochemical nitrogen reduction reaction for ammonia synthesis under ambient conditions using hollow gold nanocages. *Nano Energy, 49*, 316–323.

Niki, Y., Nitta, Y., Sekiguchi, H., & Hirata, K. (2019). Diesel fuel multiple injection effects on emission characteristics of diesel engine mixed ammonia gas into intake air. *Journal of Engineering for Gas Turbines and Power, 141*(6), 061020.

Oh, S., Park, C., Kim, S., Kim, Y., Choi, Y., & Kim, C. (2021). Natural gas–ammonia dual-fuel combustion in spark-ignited engine with various air–fuel ratios and split ratios of ammonia under part load condition. *Fuel, 290*, 120095.

Mounaïm-Rousselle, C., Bréquigny, P., Medina, A. V., Boulet, E., Emberson, D., & Løvås, T. (2021). Ammonia as fuel for transportation to mitigate zero carbon impact. In *Engines and fuels for future transport* (pp. 257-279). Singapore: Springer Singapore.

Oreskes, N. (2018). The scientific consensus on climate change: How do we know we're not wrong? In Climate modelling: Philosophical and conceptual issues (pp. 31–64). Cham: Springer International Publishing.

Olabi, A. G., Abdelkareem, M. A., Al-Murisi, M., Shehata, N., Alami, A. H., Radwan, A., … Sayed, E. T. (2023). Recent progress in Green Ammonia: Production, applications, assessment; barriers, and its role in achieving the sustainable development goals. *Energy Conversion and Management, 277*, 116594.

Pan, J., Tang, R., Wang, Z., Gao, J., Xu, Q., Shu, G., & Wei, H. (2023). An experimental and modeling study on the oxidation of ammonia and n-heptane with JSR. *Proceedings of the Combustion Institute, 39*(1), 477–485.

Peng, P., Su, J., & Breunig, H. (2023). Benchmarking plasma and electrolysis decomposition technologies for ammonia to power generation. *Energy Conversion and Management, 288*, 117166.

Pischinger, R., Klell, M., & Sams, T. (2002). *Thermodynamik der Verbrennungskraftmaschine*.

Raganati, F., Miccio, F., & Ammendola, P. (2021). Adsorption of carbon dioxide for post-combustion capture: A review. *Energy & Fuels, 35*(16), 12845–12868.

Rehbein, M. C., Meier, C., Eilts, P., & Scholl, S. (2019). Mixtures of ammonia and organic solvents as alternative fuel for internal combustion engines. *Energy & Fuels, 33*(10), 10331–10342.

Rouwenhorst, K. H., Krzywda, P. M., Benes, N. E., Mul, G., & Lefferts, L. (2020). Ammonia, 4.Green ammonia production. *Ullmann's Encyclopedia of Industrial Chemistry, 7*, 1–20.

Royal, Society. (2020). Ammonia: zero-carbon fertiliser, fuel and energy store. https://royalsociety.org/news-resources/projects/low-carbon-energy-programme/green-ammonia/.

Ryu, K., Zacharakis-Jutz, G. E., & Kong, S.-C. (2014). Performance characteristics of compression-ignition engine using high concentration of ammonia mixed with dimethyl ether. *Applied Energy, 113*, 488–499.

Saygin, D., Blanco, H., Boshell, F., Cordonnier, J., Rouwenhorst, K., Lathwal, P., & Gielen, D. (2023). Ammonia production from clean hydrogen and the implications for global natural gas demand. *Sustainability, 15*(2), 1623.

Schrauzer, G., & Guth, T. (2002). Photolysis of water and photoreduction of nitrogen on titanium dioxide. *Journal of the American Chemical Society, 99*(22), 7189–7193.

Sharma, V., Panesar, A., de Sercey, G., & Begg, S. (2024). A review of ammonia combustion and emissions characteristics in spark-ignition engines and future road map. *Energies, 18*(1), 41.

Shen, H., Yang, M., Hao, L., Wang, J., Strunk, J., & Sun, Z. (2022). Photocatalytic nitrogen reduction to ammonia: Insights into the role of defect engineering in photocatalysts. *Nano Research, 15*(4), 2773–2809.

Singh, E., Badra, J., Mehl, M., & Sarathy, S. M. (2017). Chemical kinetic insights into the octane number and octane sensitivity of gasoline surrogate mixtures. *Energy & Fuels, 31*(2), 1945–1960.

Smith, C., Hill, A. K., & Torrente-Murciano, L. (2020). Current and future role of Haber–Bosch ammonia in a carbon-free energy landscape. *Energy & Environmental Science, 13*(2), 331–344.

Souissi, N. (2024). *Fueling the future: A techno-economic evaluation of e-ammonia production for marine applications* (No. 40). OIES Paper: ET.

Sun, J., Chen, Q., Zhao, X., Lin, H., & Qin, W. (2022). Kinetic investigation of plasma catalytic synthesis of ammonia: insights into the role of excited states and plasma-enhanced surface chemistry. *Plasma Sources Science and Technology, 31*(9), 094009.

Tiseo, I. (2025). Annual greenhouse gas emissions worldwide from 1970 to 2024. *Energy & Environment*. https://www.statista.com/statistics/1285502/annual-global-greenhouse-gas-emissions/.

Tornatore, C., Marchitto, L., Sabia, P., & De Joannon, M. (2022). Ammonia as green fuel in internal combustion engines: state-of-the-art and future perspectives. *Frontiers in Mechanical Engineering, 8*, 944201.

Uddeen, K., Tang, Q., Shi, H., Magnotti, G., & Turner, J. (2023). April. Multiple spark ignition approach to burn ammonia in a spark-ignition engine: An optical study (p. 288442). WCX SAE World Congress Experience.

Wang, D., Ji, C., Wang, S., Yang, J., & Wang, Z. (2021). Numerical study of the premixed ammonia-hydrogen combustion under engine-relevant conditions. *International Journal of Hydrogen Energy, 46*(2), 2667–2683.

Wang, L., Lee, E. W., Hussian, S. A., Yuen, A. C. Y., & Feng, W. (2021). Quantitative impact analysis of driving factors on annual residential building energy end-use combining machine learning and stochastic methods. *Applied Energy, 299*, 117303.

Wang, N., Otor, H. O., Rivera-Castro, G., & Hicks, J. C. (2024). Plasma catalysis for hydrogen production: a bright future for decarbonization. *ACS catalysis, 14*(9), 6749–6798.

Wang, Q., Guo, J., & Chen, P. (2019). Recent progress towards mild-condition ammonia synthesis. *Journal of Energy Chemistry, 36*, 25–36.

Westlye, F. R., Ivarsson, A., & Schramm, J. (2013). Experimental investigation of nitrogen based emissions from an ammonia fueled SI-engine. *Fuel, 111*, 239–247.

Xie, P., Yao, Y., Huang, Z., Liu, Z., Zhang, J., Li, T., ... Wang, C. (2019). Highly efficient decomposition of ammonia using high-entropy alloy catalysts. *Nature communications, 10*(1), 4011.

Xue, X., Chen, R., Yan, C., Zhao, P., Hu, Y., Zhang, W., ... Jin, Z. (2019). Review on photocatalytic and electrocatalytic artificial nitrogen fixation for ammonia synthesis at mild conditions: Advances, challenges and perspectives. *Nano Research, 12*(6), 1229–1249.

Yan, D., Li, H., Chen, C., Zou, Y., & Wang, S. (2019). Defect engineering strategies for nitrogen reduction reactions under ambient conditions. *Small Methods, 3*(6), 1800331.

Yao, Y., Wang, J., Shahid, U. B., Gu, M., Wang, H., Li, H., & Shao, M. (2020). Electrochemical synthesis of ammonia from nitrogen under mild conditions: current status and challenges. *Electrochemical Energy Reviews, 3*(2), 239–270.

Yapicioglu, A., & Dincer, I. (2019). A review on clean ammonia as a potential fuel for power generators. *Renewable and sustainable energy reviews, 103*, 96–108.

Ye, J., Li, Z., Kubicki, D. J., Zhang, Y., Dai, L., Otero-Martínez, C., ... Hoye, R. L. (2022). Elucidating the Role of Antisolvents on the Surface Chemistry and Optoelectronic Properties of CsPbBr x I3-x Perovskite Nanocrystals. *Journal of the American Chemical Society, 144*(27), 12102–12115.

Ye, L., Nayak-Luke, R., Banares-Alcantara, R., & Tsang, E. (2017). Reaction:"Green" ammonia production. *Chem, 3*(5), 712–714.

Yu, S., Xiang, T., Alharbi, N. S., Al-aidaroos, B. A., & Chen, C. (2023). Recent development of catalytic strategies for sustainable ammonia production. *Chinese Journal of Chemical Engineering, 62*, 65–113.

Zhang, R., Chen, L., Wei, H., Li, J., Ding, Y., Chen, R., & Pan, J. (2023). Experimental investigation on reactivity-controlled compression ignition (RCCI) combustion characteristic of n-heptane/ammonia based on an optical engine. *International Journal of Engine Research, 24*(6), 2478–2488.

Zhang, S., Zhao, Y., Shi, R., Waterhouse, G. I., & Zhang, T. (2019). Photocatalytic ammonia synthesis: Recent progress and future. *EnergyChem, 1*(2), 100013.

Zhang, Y., Liu, H., Li, J., Deng, Y., Miao, X., Xu, D., … Tian, Y. (2022). Life cycle assessment of ammonia synthesis in China. *The International Journal of Life Cycle Assessment, 27*(1), 50–61.

Zhang, Y., Zanganeh, J., & Moghtaderi, B. (2025). Advancement in neutral and negative emissions production of ammonia from carbonaceous sources: A review. *International Journal of Hydrogen Energy, 109*, 40–62.

Zheng, D., Peng, Z., He, D., Zhang, M., Li, J., Ding, Y., & Du, Y. (2024). Investigation of the NH3 consumption and H2O/NO formation during the oxidation of n-heptane and iso-octane blended with ammonia. *Fuel, 357*, 129793.

Zhou, D., Tay, K. L., Tu, Y., Li, J., Yang, W., & Zhao, D. (2018). A numerical investigation on the injection timing of boot injection rate-shapes in a kerosene-diesel engine with a clustered dynamic adaptive chemistry method. *Applied Energy, 220*, 117–126.

Zhu, H. L., Li, Z. B., Jing, S., Wu, X. L., Meng, X. B., & Jia, Y. (1999). *Ammonia synthesis technology* (p. 1229058). Pat: CN.

Zitouni, S.-E., Brequigny, P., & Mounaïm-Rousselle, C. (2022). Turbulent flame speed and morphology of pure ammonia flames and blends with methane or hydrogen. *Proceedings of the Combustion Institute*: 39. Elsevier.

Chapter 5

Methanol production from solid waste

Abdellah Zouheir[a], Anisa Emrani[b] and Asmae Berrada[a]
[a] *LERMA Lab, College of Engineering and Architecture, International University of Rabat, Sala Al Jadida, Morocco,* [b] *ENSAS/Laboratory of Materials, Processes, Environment and Quality (LMPEQ), Cadi Ayyad University (UCA), Safi, Morocco*

5.1 Introduction

In the current context of climate change, mitigating greenhouse gas emissions is of utmost importance. Two major contributors are the traditional pathways of waste disposal (notably landfilling and incineration) and the use of fossil fuels in industry. The UN Environment Programme (2012) reports that an estimated 11.2 billion tons of solid waste are collected globally each year. The emissions from fossil fuel combustion and industrial processes have amounted to 37.79 billion tons of carbon dioxide in 2023 (Methanol Institute, 2025). These two environmental challenges can be addressed simultaneously through innovative green methods, such as the waste-to-chemical processes. The idea of converting waste into valuable chemical products has gained traction with the emergence of concepts such as the "circular economy" and growing concerns regarding the environmental issues mentioned earlier. In these processes, nonrecyclable solid waste can be converted into a wide range of valuable chemicals, including ethanol, methanol, ammonia, and olefins. Effectively addressing the environmental issues linked to waste and contributing to the decarbonization of the global economy by replacing traditional chemical processes that use fossil resources.

Among waste-to-chemical products, methanol stands out as a target product. The Methanol Institute (2025) reports that global methanol consumption reached 120 million tons in 2024, with a forecasted annual increase of 3.6%. Methanol is also the building block for several derivatives, such as Acetic acid, formaldehyde, silicones, and olefins, to name a few. Currently, the use of natural gas and coal accounts for the majority of methanol production worldwide, resulting in carbon-intensive processes with emissions reaching 3.8 kg CO2eq per kilogram

Decarbonizing through Sustainable Hydrogen and its Derivatives.
DOI: https://doi.org/10.1016/B978-0-443-27632-3.00006-2

of produced methanol (Demirel, 2015). In contrast, waste and renewables have a significant carbon mitigation potential, with a negative carbon footprint in some configurations (Methanol Institute, 2022). In addition to its high energy density (22.7 MJ/kg HHV), methanol is a liquid under standard conditions, making it easier to store and transport than other energy vectors, such as hydrogen. This chapter is dedicated to an innovative waste-to-methanol process that combines anaerobic digestion and tri-reforming. Tri-reforming has garnered attention due to its distinct advantages over other reforming technologies, including improved energy efficiency, the ability to adjust the H_2/CO ratio, and a higher potential for CO_2 utilization. However, to capitalize on these advantages, the process operating conditions must be carefully selected.

In this context, the optimization of the tri-reformer has been investigated in several studies. Arab Aboosadi et al. (2011) used differential evolution (DE) to optimize hydrogen yield in a fixed-bed reactor from a mixture of prereformed gas and natural gas, whereas Khajeh et al. (2015) employed the same approach on a fluidized-bed reactor. In both studies, reactant ratios ($CH_4/CO_2/H_2O/O_2$) and inlet temperature were set as decision variables. Additionally, constraints were imposed on the H_2/CO ratio and the maximum temperature reached inside the reactor. The fixed-bed reactor increased CH_4 conversion by 3.8% when compared to the traditional steam reformer, producing syngas with a hydrogen yield of 1.84 and an H_2/CO ratio of 1.7. In contrast, the fluidized-bed tri-reformer achieved a 99.4% CH_4 conversion, a 2.1 hydrogen yield, and a 1.84 H_2/CO ratio, demonstrating superior performance compared to the fixed-bed reactor. Optimum inlet temperature and reactant ratios were 1100 K and 1/1.3/2.46/0.47, respectively, in Arab Aboosadi et al. (2011) study, whereas Khajeh et al. (2015) reported optimal conditions of 1100 K and 1/1.32/2.49/0.44. Khademi et al. (2021) investigated the use of membrane reactors with side-feed for tri-reforming, utilizing the DE algorithm to optimize the reactor length by adjusting the reactant ratios, shell-side pressure, and gas inlet temperature. The study concluded that the reactor with oxygen side-feeding outperformed the conventional membrane reactor, as well as the reactor setups with H_2O and CO_2 side-feeding, achieving a methane conversion of 99.98%, a hydrogen yield of 1.91, and a H_2/CO ratio of 1.72. CO_2 conversion was at its lower bound of 13%, and an optimal reactor length of 1.39 meters was obtained with reactant ratios of 1/1/2.73/0.62, a pressure of 21.72 bar, and an inlet temperature of 1200 K. Osat et al. (2023) optimized the tri-reforming reactor to maximize three conflicting objectives: energy efficiency, CO_2 conversion, and the H_2/CO ratio. The problem was formulated as three biobjective optimization subproblems, where the inlet temperature, the oxygen ratio, and the CO_2 ratio were chosen as the decision variables. The results show that the first scenario yields a maximal H_2/CO ratio and energy efficiency of 5.48 and 71.9%, respectively. When H_2/CO and CO_2 conversion are considered as the objective functions, the Pareto front shows that CO_2 conversion increases from 4.10% to 30.62% when the H_2/CO ratio decreases from 5.48 to 1.14. In the final scenario, the Pareto front indicates

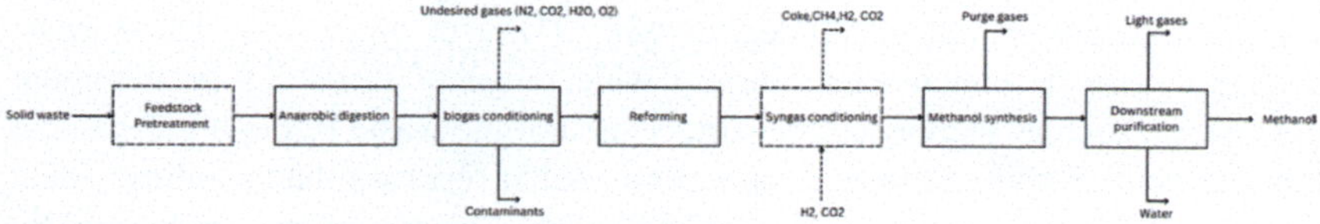

FIGURE 5.1 Block diagram of the waste-to-methanol process.

CO_2 conversions ranging from 13.6% to 30.6%, while energy efficiency varies between 65.6% and 71.9%.

Ahmed and Padhiyar (2020) addressed the multiobjective optimization (MOO) of a tri-reforming packed-bed reactor with a mixture of prereformed and natural gas as feedstock. The reactor was modeled through a mathematical model solved in MATLAB, and the integrated genetic algorithm optimization function "gamultiobj" was used. The authors formulated four optimization scenarios based on three objective functions, namely maximizing the hydrogen flow rate in the product gas, minimizing the CO_2 flow rate, and minimizing the energy loss related to the pressure drop in the reactor. Three of the scenarios had two objective functions, while the last scenario had all three objective functions. The three selected decision variables were the feed temperature, H_2O/CH_4 ratio, and $O2/CH_4$ ratio. The results showed that smaller values of reactants' flow rates lead to the minimum power loss. In contrast, a significant amount of H_2O yielded higher hydrogen and lower CO_2 emissions.

However, optimization studies using biogas as the feedstock and methanol as the end-product are limited. Furthermore, only a few studies have performed the multiobjective optimization of the process. Additionally, none of the reviewed studies implement both catalyst sintering and carbon deposition constraints into the formulation of the problem. Within this scope, this study will focus on the multiobjective optimization of the tri-reformer for methanol synthesis. The three objectives are to maximize hydrogen production, minimize the difference between the H_2/CO ratio and its optima l value for methanol synthesis, and to maximize CO_2 utilization in the reforming block. Additionally, constraints on the maximal temperature and carbon deposition in the reforming reactor will be implemented.

5.2 Process description

The solid waste-to-methanol process begins with the collection of solid waste and the extraction of its organic fraction (Fig. 5.1). It is important to note that the term solid waste is not exclusive to solid state matter; in contrast, the term is commonly used to refer to other physical states of matter, such as sludge (Chandra et al., 2015; Nagler et al., 2018). The suitability of solid waste from a specific source depends on several characteristics, such as the waste composition and its biodegradability (Hou et al., 2020). These characteristics can be enhanced through a wide range of feedstock pretreatment operations, such as increasing the surface area of the waste through mechanical pretreatment. For example, Chandra et al. (2015) reported that grinding wheat and rice straw to a particle size of 0.75 mm increased methane production by 38.7%. Different pretreatment methods can also be combined to overcome the limitations of standalone techniques. For example, Nagler et al. (2018) reported that elevated temperatures help counteract the adverse effects of sodium on sludge dewaterability in the thermochemical pretreatment of wastewater sludge. At the same time, the presence of NaOH allows the process to operate at lower temperatures (Nagler et al., 2018). Another practical approach to improving feedstock characteristics for anaerobic digestion (AD) is cofeeding in anaerobic codigestion (AcoD)—the simultaneous digestion of two or more types of organic waste. When appropriately combined, certain feedstocks can trigger synergistic effects. For example, Hou et al. (2020) report that codigesting food waste with rice straw and rice bran increased methane yield by 27.4% and improved energy recovery by 35.23%.

After the pretreatment step, biomass is fed to the AD reactor, in which a consortium of microorganisms breaks down biodegradable material through a complex mechanism consisting of an extracellular stage (hydrolysis), and intracellular stages (acidogenesis, acetogenesis, and methanogenesis), and several intermediate compounds (acetate, H_2, CO_2, …). The outflow of the digester is a biogas containing mainly methane (40%–70%), CO_2 (30%–60%), with hydrogen, oxygen, and nitrogen present at lower concentrations (1%–5%), and trace quantities of other compounds such as H_2S and NH_3. The composition of biogas depends on the feedstock, the pretreatment step, and the operating conditions of the digestion reactor. Anaerobic digesters typically operate between 20°C and 60°C, which corresponds to two temperature regimes: mesophilic AD (20°C–40°C) and thermophilic AD (50 C–65°C). The latter is often preferred for solid feedstock. However, the AD process is susceptible to temperature fluctuations in both temperature ranges, with Wang et al. (2019) reporting that a 278K fluctuation can lead to reactor failure. The organic loading rate of the feedstock also has a significant impact on waste recovery; a higher loading rate typically means a higher quantity of waste being treated in the same reactor. However, an organic loading rate that is too high may lead to process failure due to the accumulation of inhibitors, shock to microorganisms, and biomass washout resulting from low hydraulic retention times (Appels et al., 2008).

Raw biogas discharged from the digestion reactor typically contains CO_2, H_2O, and contaminants such as H_2S, NH_3, and siloxanes. These contaminants generate harmful environmental emissions, lead to material corrosion, and poison the catalyst by depositing on its surface or being adsorbed to its active sites. Gao et al. (2023) report that even a very low H_2S concentration of 1 ppm results in a significant decrease in biogas conversion in the reforming reactor. At the same time, ammonia has a lesser but still considerable inhibitory effect (Gao et al., 2023). In the same context, Elsayed et al. (2017) recommend removing at least 96% of siloxanes to ensure stable operation. Standard biogas cleaning technologies include water and organic solvent scrubbing to eliminate ammonia, hydrogen sulfide, and siloxanes. Adsorption beds employing activated carbon, zeolites, or silica gel can be used to remove H_2S and NH_3. Additionally, CO_2, N_2, and O_2 can be separated through pressure swing adsorption (PSA) or membrane separation technologies (Abdullah et al., 2018; Gkotsis et al., 2023; Werkneh, 2022).

The next unit operation in the process is biogas reforming. In this step, the methane in the biogas is catalytically converted into a mixture of hydrogen, carbon monoxide, and carbon dioxide (Syngas), in addition to water as a by-product. Various reforming processes are available, each characterized by the oxidizing agent it employs; three agents are used primarily: steam, CO_2, and O_2 (or air). Steam reforming of natural gas is the most prevalent industrial method for hydrogen production (Szablowski et al., 2025). However, the resulting synthesis gas is unsuitable for methanol synthesis due to its H_2:CO ratio of 3:1, necessitating further conditioning. Furthermore, steam reforming typically results in a net production of CO_2 through the Water-Gas-Shift (WGS) reaction and leads to a high carbon dioxide footprint of the process (Ashour et al., 2025). The use of oxygen alone as the oxidizing agent in partial oxidation is attractive because the produced syngas has a H_2:CO ratio of 2:1, and the reaction is exothermic, eliminating the need for external heat. However, the severe and rapid catalyst deactivation hinders its large-scale industrialization (Majewski & Wood, 2014). Similarly, dry reforming with CO_2 suffers from substantial coke deposition and a low H_2/CO ratio (Jabbour et al., 2019). To mitigate the disadvantages of single-reagent technologies, a mixture of oxidizing agents can be used. Bi-reforming, for example, combines steam reforming and dry reforming, which allows for adjusting the H_2:CO ratio through the reactant ratios, and mitigates coke deposition on the catalyst compared to dry reforming, in addition to the higher CO_2 utilization in the process compared to steam methane reforming. In tri-reforming, all three oxidizing agents are used. This integrated approach leverages the distinct advantages of each reforming method, effectively mitigating the limitations inherent in individual processes. Since tri-reforming is central to our case study, a detailed discussion is presented in Section 5.2.

In the methanol synthesis reactor, three catalytic reactions take place: CO hydrogenation, CO_2 hydrogenation, and the WGS reaction. Both hydrogenation reactions are exothermic and mole-reducing; thus, they are usually conducted under high pressures of 50-100 bars and a temperature range of 220°C–280°C, in the presence of a Cu-based catalyst. The suitability of syngas composition for methanol synthesis is determined primarily by the stoichiometric number ($SN = \frac{H_2 - CO_2}{CO + CO_2}$), which has a theoretical optimum value of 2. An SN higher than two means that there is an excess of hydrogen, whereas values below 2 indicate a hydrogen deficiency. On an industrial scale, methanol is synthesized from syngas: a gas mixture containing hydrogen, carbon monoxide, and a low concentration of carbon dioxide. The concentration of carbon dioxide must also be controlled, since high hydrogen content leads to low equilibrium conversions and slower kinetics (Nestler et al., 2020). Grabow and Mavrikakis (2011) suggest that the CO_2 concentration should be below 12%. Their results also demonstrated that small quantities of water inhibit methanol synthesis. Therefore water must be removed before feeding the mixture into the methanol synthesis reactor. In addition to water condensation, other conditioning unit operations can be implemented to calibrate the syngas composition and eliminate contaminants such as sulfides. For instance excess hydrogen can be balanced via the WGS reaction, which converts hydrogen and carbon dioxide into water and carbon monoxide. On the other hand, PSA can be used to remove the excess hydrogen from the process. CO_2 is also captured from syngas using PSA, or alternative technologies such as amine-based absorption (Oreggioni et al., 2015). When the syngas is hydrogen-deficient, a reverse water-gas-shift (RWGS) reactor can be introduced into the process. Alternatively, the SN can be adjusted by supplying hydrogen from an external source, such as an electrolysis process. However, this might significantly increase the cost of methanol production (Bube et al., 2024).

5.3 Tri-reforming of biogas: Analysis

Tri-reforming combines (i) Steam Methane Reforming (SMR), (ii) Dry Reforming of Methane (DRM), and (iii) Partial Oxidation Reforming (POX):

$$CH_4 + H_2O \leftrightarrow CO + 3H_2 \quad \Delta H^\circ = 206\ \text{kJ/mol} \tag{2.1}$$

$$CH_4 + CO_2 \leftrightarrow 2CO + 2H_2 \quad \Delta H^\circ = 247\ \text{kJ/mol} \tag{2.2}$$

$$CH_4 + 1/2O_2 \leftrightarrow CO + 2H_2 \quad \Delta H^\circ = -36\ \text{kJ/mol} \tag{2.3}$$

This technology offers several advantages, including the ability to adjust the syngas composition by varying the reactant ratios. This flexibility enables the direct feeding of the reformate to the methanol reactor without additional conditioning, thereby reducing a significant portion of the process cost. Another notable advantage is the lower heating demand (Konak et al., 2006); in fact, the reforming reactor in tri-reforming can be operated under an adiabatic setup. Furthermore, high concentrations of steam and oxygen in the reformer feed can suppress coke formation in the reactor (Horn et al., 1994). However, to fully capitalize on these advantages, operating parameters must be selected with careful consideration. One key parameter is the feed composition, typically expressed as reactant molar ratios (H_2O/CH_4, CO_2/CH_4, O_2/CH_4). The steam ratio (H_2O/CH_4) correlates directly to the activity of the SMR in the reactor; and increasing the steam ratio results in a feed with higher steam concentration effectively shifting the SMR equilibrium toward the products, this results in an increase in the hydrogen yield as SMR produces three moles of hydrogen for each mole of CH_4 compared to two moles in POX and one mole in DRM. Fig. 5.2 illustrates that higher steam ratios result in a higher concentration of hydrogen in the produced syngas. Additionally, an increase in CO_2 concentration

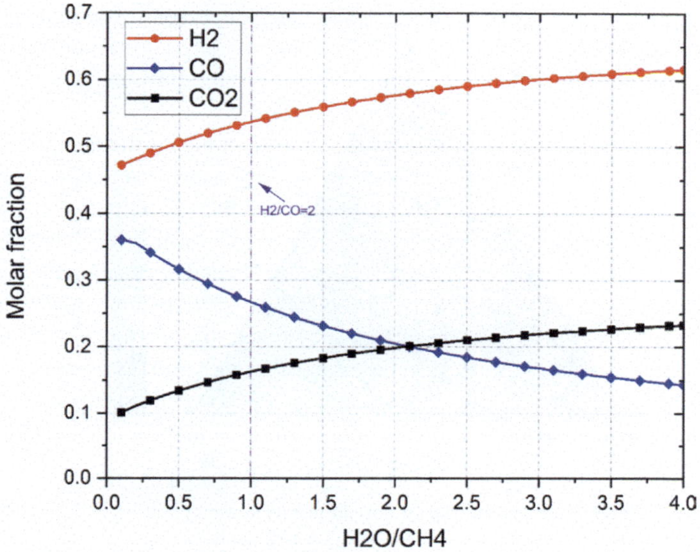

FIGURE 5.2 Effect of the steam ratio on syngas components' concentration at an inlet temperature of 850°C, 5 bar, and reactant ratios 1/0.5/0.6/0.4.

is also observed. This can be explained by the fact that increasing the steam concentration in the feed also shifts the WGS reaction's equilibrium toward producing more CO_2 and hydrogen; this phenomenon also explains H_2/CO ratios that exceed 3. Whereas the optimal H_2/CO ratio of 2 is obtained at a H_2O/CH_4 ratio of 1 in the studied case of Fig. 5.2. Moreover, increasing the CO_2/CH_4 ratio leads to higher CO_2 conversions. However, a CO_2-rich feed results in reduced hydrogen production, and the H_2/CO ratio decreases toward one as the CO_2 content increases. Higher O_2/CH_4 ratios promote the POX reaction, resulting in a H_2/CO ratio that is pushed toward 2. Additionally, when operating the reaction under adiabatic conditions, POX becomes a primary source of heat inside the reactor. Increasing the O_2/CH_4 ratio results in higher temperatures within the reactor; however, an inadequate feed composition may lead to elevated temperatures, potentially causing catalyst sintering. Another important phenomenon observed in the reactor is the deposition of coke. Fig. 5.3 shows the influence of reactant ratios on the coke molar flow and the coke deposition zones. It can be observed that high CO_2 concentrations in the feedstock can result in substantial coke deposition; however, coke formation can be suppressed with high H_2O and O_2 concentrations. Another critical parameter is temperature, since both SMR and DMR are strongly endothermic reactions, and the WGS reaction is disfavored at high temperatures; increasing the temperature leads to higher conversions, as well as increased yields of hydrogen and carbon monoxide. At low temperatures, substantial coke formation can occur due to the Boudouard and Beggs side reactions being more favorable at lower temperatures. However, coke deposition can still occur at high temperatures due to the methane cracking reaction.

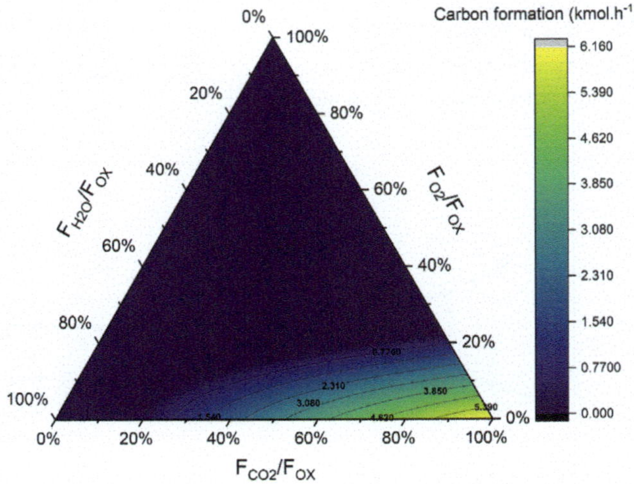

FIGURE 5.3 Effect of reactant ratio on the coke formation flow rate, with $T_{\text{in}} = 750°\text{C}$, $P = 5$ bar, FCH4 = 5 kmol·h^{-1} and $F_{\text{OX}} = F_{\text{H2O}} + F_{\text{CO2}} + F_{\text{O2}} = 10$ kmol·h^{-1}.

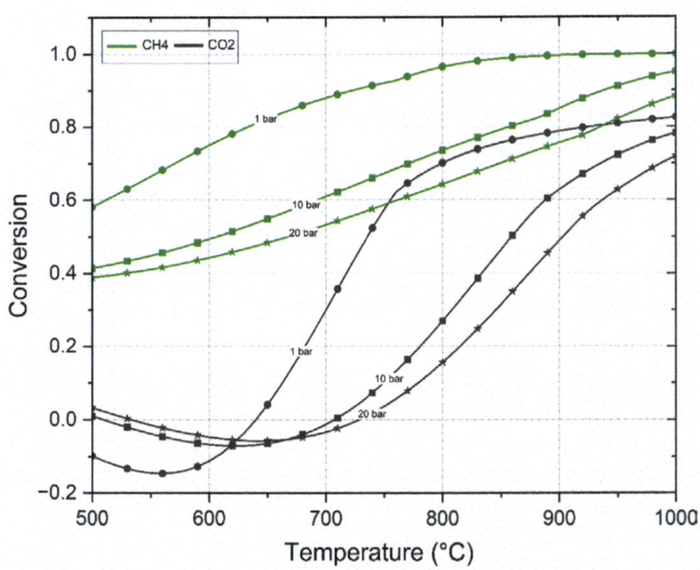

FIGURE 5.4 Effect of temperature and pressure on methane and carbon dioxide conversion at reactant ratios 1/0.3/0.3/0.3.

Fig. 5.4 illustrates that increasing temperatures above 650°C, in the studied case, leads to an increase in CH_4 and CO_2 conversions, reaching 99% and 82%, respectively. In contrast, a drop in CO_2 conversion is observed between 500°C and 650°C. This can be attributed to the side reactions producing CO_2, which are favored at lower temperatures, indicating that additional CO_2 is produced within the reactor, primarily due to the WGS reaction and other side reactions at this temperature range. Fig. 5.4 also demonstrates that increasing pressure leads to lower conversions for the same feed composition and temperature. This can be explained by the fact that all three reforming reactions (SMR, DRM, POX) are mole-increasing reactions, and increasing pressure shifts the equilibrium toward the reactants according to the Le Chatelier principle. The complex and interrelated nature of these effects on the independent variables of the process highlights the need to optimize operating parameters when aiming for a specific outcome.

5.4 Materials and methods

5.4.1 Reactor model

5.4.1.1 Gibbs free energy minimization model

The tri-reforming reactor in this work is simulated under the assumption of chemical equilibrium. The model computes the reactor's equilibrium composition by minimizing Gibbs free energy for a set of species, without explicitly specifying reaction stoichiometry. In this case study, the following species are considered: H_2, CO, CO_2, H_2O, N_2, O_2, CH_4, and C(s).

The total Gibbs free energy of the reactor is defined by the sum of the chemical potential of the species and is expressed as

$$G^t = \sum_{i=1}^{N} n_i \, \mu_i \tag{4.1}$$

$$\sum_{i=1}^{N} n_i \cdot a_{i,j} = A_j, \quad j = 1, 2, \ldots, k$$

where μ_i is the chemical potential of species i and n_i is the number of moles. The equilibrium composition is obtained by solving the constrained minimization problem of finding the mole numbers n_i that minimize the total Gibbs free energy. Optimization is typically performed using the method of Lagrange multipliers, and the problem is constrained by the elemental balance:

$$\sum_{i=1}^{N} n_i \cdot a_{i,j} = A_j, \ j = 1, 2, \ldots, k \tag{4.2}$$

with A_j being the total number of atoms in the jth specie, and a_{ij} the number of atoms of element j in the ith specie, Lagrange multipliers are then implemented in Eq. (4.3), and the constrained optimization is reduced to solving Eq. (4.4):

$$L = G^t - \sum_{j=1}^{k} \lambda_j \left(\sum_{i=1}^{N} n_i \cdot a_{i,j} - A_j \right) \tag{4.3}$$

$$\frac{\partial L}{\partial n_i} = 0 \quad (4.4)$$

The Gibbs free energy minimization reactor offers several advantages over commonly used models from the literature, including its ability to estimate carbon formation under equilibrium conditions, enabling us to identify optimal conditions with no coke deposition and to evaluate the catalyst deactivation associated with this phenomenon. However, an additional constraint must be implemented to assess thermal deactivation and catalyst sintering. Although the additional partial oxidation reaction in the tri-reformer is exothermic and enables adiabatic operation, a temperature peak occurs at the reactor entrance. At this point, the temperature can surpass the catalyst deactivation threshold, commonly set at 1700 K (Arab Aboosadi et al., 2011). However, the Gibbs free energy minimization model does not compute the temperature profile in the reactor, as the model does not contain energy balances. Consequently, the constraint related to thermal sintering of the catalyst cannot be assessed, and an additional equation is required to estimate the maximum temperature.

5.4.1.2 Kinetic model

This section is dedicated to developing a model that links the maximum temperature to reactant ratios, inlet temperature, and reactor pressure. This enables the reactor model to evaluate the maximum temperature reached inside the reactor. For this purpose, the discretized 1-D plug flow reactor (Rplug) model from Aspen Plus simulation software was used. Additionally, Langmuir-Hinshelwood-Hougen-Watson (LHHW) kinetics were incorporated into the model to calculate the temperature at each cell along the reactor length. The kinetic model used in this study is the LHHW model developed by Arab Aboosadi et al. (2011) and presented in Eqs. (4.5–4.8). Furthermore, De Groote and Froment's effectiveness factors ($\eta 1 = 0.07$, $\eta 2 = 0.06$, $\eta 3 = 0.7$, and $\eta 4 = 0.05$) are introduced to account for intraparticle transfer limitations in the formation and consumption rates (Wang et al., 2019). With Ki being the kinetic constants presented in Table 5.1, and P_j the partial pressure of species j:

$$r_1 = \frac{k_1}{p_{\mathrm{H_2}}^{2.5}}\left(p_{\mathrm{CH_4}}p_{\mathrm{H_2O}} - \frac{p_{\mathrm{H_2}}^3 p_{\mathrm{CO}}}{K_I}\right) \times \frac{1}{\phi^2} \quad (4.5)$$

$$r_2 = \frac{k_2}{p_{\mathrm{H_2}}^{3.5}}\left(p_{\mathrm{CH_4}}p_{\mathrm{H_2O}}^2 - \frac{p_{\mathrm{H_2}}^4 p_{\mathrm{CO_2}}}{K_{II}}\right) \times \frac{1}{\phi^2} \quad (4.6)$$

$$r_3 = \frac{k_3}{p_{\mathrm{CO}}p_{\mathrm{H_2O}}}\left(p_{\mathrm{CO}}p_{\mathrm{H_2O}} - \frac{p_{\mathrm{H_2}} p_{\mathrm{CO_2}}}{K_{III}}\right) \times \frac{1}{\phi^2} \quad (4.7)$$

TABLE 5.1 Kinetic constants of the tri-reforming LHHW model.

Parameter	*A*	Unit	*B* (J/mol)
k_1	1.17×10^{12}	(kmol $bar^{0.5}$)(kg s)$^{-1}$	240,100
k_2	2.83×10^{11}	(kmol $bar^{0.5}$)(kg s)$^{-1}$	243,900
k_3	5.43×10^{2}	(kmol bar^{-1})(kg s)$^{-1}$	67,130
k_{4a}	8.11×10^{2}	(kmol bar^{-2})(kg s)$^{-1}$	86,000
k_{4b}	6.82×10^{2}	(kmol bar^{-2})(kg s)$^{-1}$	86,000
K_I	1.20×10^{13}	bar^2	223,076
K_{II}	2.12×10^{11}	bar^2	186,492
K_{III}	1.77×10^{-2}	bar^2	−36,584
K_{CO}	8.23×10^{-5}	bar^{-1}	−70,650
K_{H_2}	6.12×10^{-9}	bar^{-1}	−82,900
K_{CH_4}	6.65×10^{-4}	bar^{-1}	−38,280
K_{H_2O}	1.77×10^{5}	–	88,680
$K^C_{CH_4}$	1.26×10^{-1}	bar^{-1}	−27,300
$K^C_{O_2}$	7.78×10^{-7}	bar^{-1}	−92,800

$$r_4 = \frac{k_{4a} p_{CH_4} p_{O_2}}{\left(1 + K^C_{CH_4} p_{CH_4} + K^C_{O_2} p_{O_2}\right)} + \frac{k_{4b} p_{CH_4} p_{O_2}}{\left(1 + K^C_{CH_4} p_{CH_4} + K^C_{O_2} p_{O_2}\right)} \tag{4.8}$$

With $\phi = 1 + K_{CO} p_{CO} + K_{H_2} p_{H_2} + K_{CH_4} p_{CH_4} + K_{H_2O} \frac{p_{H_2O}}{p_{H_2}}$

$$X_\alpha(T) = \boldsymbol{A}_\alpha \exp\left(-\boldsymbol{B}_\alpha/(RT)\right)$$

The correct implementation of the model was first verified through the cross-comparison of the obtained syngas' molar fractions with the results from the Gibbs equilibrium reactor model. This comparison was performed using a dataset of 100 points generated to fill the design space (Sobol sequence) uniformly. Fig. 5.5 illustrates the validation results for 10 points from the dataset. The results show good agreement between the Gibbs model and the plug flow reactor model with an average root mean squared error (RMSE) of 0.72%.

The kinetic model was then validated by comparing the temperature and components' molar fraction profiles with the results from Rezaei and Catalan's (2020) paper. Figs. 5.6 and 5.7 present the molar fraction and temperature profiles derived from the literature and those obtained from the Aspen simulation, and show that both results are in good agreement; therefore, the tri-reforming kinetic model is considered validated for this study.

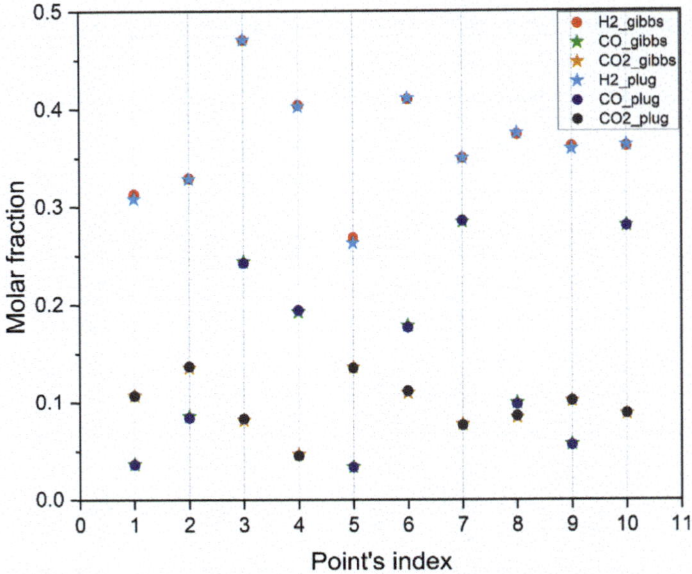

FIGURE 5.5 Validation simulation results of the Gibbs and kinetic model of the tri-reforming reactor.

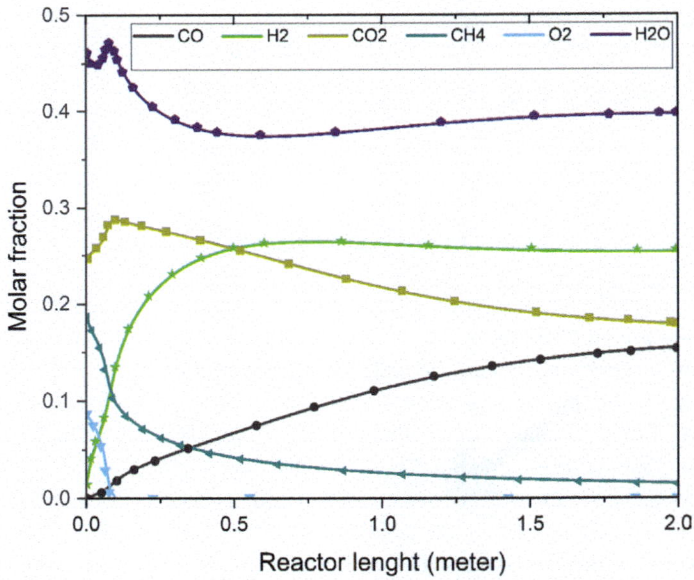

FIGURE 5.6 Mole-fraction profiles of components in the tri-reforming reactor. Solid lines = simulation results.

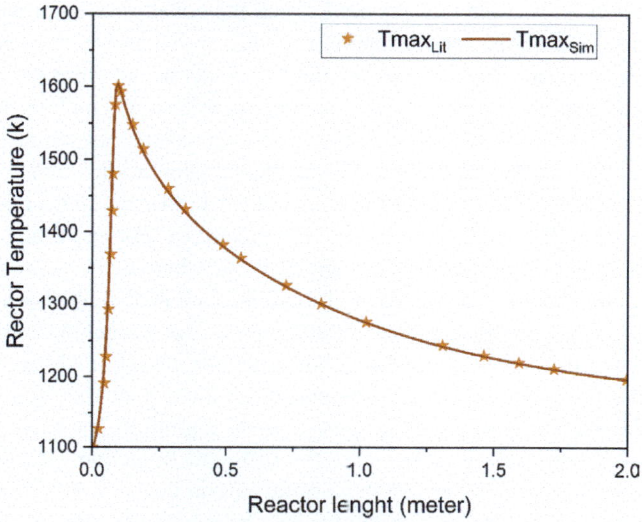

FIGURE 5.7 Temperature profile in the tri-reforming reactor. Solid lines = simulation results.

5.4.2 Genetic algorithm

A genetic algorithm (GA) is an evolutionary algorithm that can be used to solve both constrained and unconstrained optimization problems by emulating biological evolution. The algorithm iteratively updates a set of individual solutions, with the initial population typically being randomly initialized. At each generation, genetic operators are applied: crossover recombines parts of two parent solutions to create new offspring, and mutation introduces random small changes to the candidate solutions. The iterative application of these concepts drives the search toward improved solutions, and over time, the population "evolves" toward an optimal solution.

GAs have been widely applied in various fields of research and are particularly well-suited for solving multiobjective optimization problems (MOOP), where two or more conflicting objectives must be optimized simultaneously (Konak et al., 2006). In fact, the generic single-objective optimization GA can be adapted to solve MOOPs and generate a set of nondominated solutions in a single GA run. Several authors have developed different strategies to extend GA into MOOPs, such as the weight-based GA (WBGA) and the niched Pareto genetic algorithm (NPGA), with each MOO algorithm having its own advantages and disadvantages (Hajela & Lin, 1992; Horn et al., 1994).

In this study, we employ the nondominated sorting GA II (NSGA-II), an elitist GA that addresses nonlinear constrained problems while being computationally more efficient than other algorithms (Deb et al., 2000). Additionally, it offers ease of implementation through MATLAB's "gamultiobj" built-in function from the Global Optimization Toolbox. The algorithm was implemented in MATLAB alongside the process model, and the equation system was solved iteratively to evaluate the fitness function of each individual in the population. After which, genetic operators are applied to generate an improved pool of solutions, and the process repeats until a stopping criterion is met. In this study, two stopping criteria were implemented: first, a convergence criterion based on stagnation of the Pareto front over successive generations, and a criterion related to the acceptable computational cost of the optimization process, which is determined by the maximum number of generations. For our case study, the maximum number of generations was set to 1000, whereas the population size was set to 200. The other parameters of the GA, such as the crossover fraction and the elite count, remained at their default values. Upon reaching convergence or another stopping criterion, the algorithm generates the set of 70 Pareto points. These points present the best nondominated solutions that the algorithm has found in the search space, corresponding to the optimum for each objective as well as trade-off solutions between the different objectives. To determine the best solution among these points, a multicriteria decision analysis method must be employed. We use the Technique for Order of Preference by Similarity to Ideal Solution (TOPSIS). In this method, the Pareto points are ranked by their proximity to the ideal solution (a hypothetical solution that simultaneously achieves the best possible value for all three objectives), with the best point being the one closest to the ideal solution and the farthest from the worst point.

5.5 Optimization results and discussion

In this section, the GA and the model described in Section 5.4.1 were used to optimize the reforming reactor for methanol synthesis. The case study was based on biogas with a composition of 65% CH_4 and 35% CO_2 at a molar flow rate of 1200 $Nm^3 \cdot h^{-1}$, consistent with biogas from the organic fraction of municipal waste via AD, as reported in the biorefinery study by Valentino et al. (2018). The independent parameters of the optimization study were the inlet temperature, reaction pressure, steam ratio, oxygen ratio, and the carbon dioxide ratio, whereas the dependent parameters were hydrogen production, the H_2/CO ratio, and the CO_2 utilization ratio, corresponding to the fraction of biogenic CO_2 in the biogas consumed during the reforming step. It should be noted that the CO_2 utilization ratio may take negative values if CO_2 is produced in the reactor. Eq. (5.1) presents the formula for the CO_2 utilization ratio:

$$\eta_{CO_2} = \frac{F_{CO_2}^{in} - \left(F_{CO_2}^{sep,out} + F_{CO_2}^{syngas}\right)}{F_{CO_2}^{in}} \tag{5.1}$$

where $F_{CO_2}^{in}$ is the molar flow of CO_2 in the feed entering the reactor, and $F_{CO_2}^{sep,out}$ is the CO_2 separated from the feed, additionally $F_{CO_2}^{syngas}$ is the CO_2 content of the syngas at the reformer outlet. The H_2/CO ratio is defined by Eq. (5.2). Additionally, hydrogen yield, CH_4 conversion, and CO_2 conversion are described in Eqs. (5.3–5.5) and will be used to analyze the results:

$$H2/CO = \frac{F_{H_2}^{syngas}}{F_{CO}^{syngas}} \tag{5.2}$$

$$Y_{H_2} = \frac{F_{H_2}^{syngas} - F_{H_2}^{in}}{F_{CH_4}^{in}} \tag{5.3}$$

$$X_{CH_4} = \frac{F_{CH_4}^{in} - F_{CH_4}^{syngas}}{F_{CH_4}^{in}} \tag{5.4}$$

$$X_{CO_2} = \frac{F_{CO_2}^{in} - F_{CO_2}^{syngas}}{F_{CO_2}^{in}} \tag{5.5}$$

The three objectives optimized in this section (Eq. 5.6) are: maximizing hydrogen production, maximizing CO_2 utilization, and minimizing the difference between H_2/CO and its optimal value for methanol synthesis. Additionally, two nonlinear constraints were implemented into the problem (Eq. 5.7), the first is set to avoid catalyst thermal deactivation by ensuring the temperature inside the reactor does not exceed 1700 K, and the second is to prevent carbon deposition on the catalyst by imposing that the molar fraction of solid carbon in the syngas is zero. The bounds of the four dependent parameters were taken from the literature and are presented in Eqs. (5.8–5.12). In contrast, the lower bound of the CO_2/CH_4 corresponds to the residual CO_2 fraction after a CO_2 separation step with 90% efficiency, while the upper bound reflects the scenario without separation. The main assumptions used in this section are the adiabatic conditions in the reforming reactor and thermodynamic equilibrium at its output:

$$\text{Maximize}\, f(x) = \left[f_1(x) = \text{H}_2 \text{ production},\ f_2(x) = \eta_{CO_2},\ f_3(x) = -\left|\frac{\text{H}_2}{\text{CO}} - 2\right|\right] \tag{5.6}$$

$$\text{Constraints} = \left[F_{CO_2}^{syngas} = 0;\ T_{max} < 1700\ K\right] \tag{5.7}$$

$$800\ k < T_{in} < 1300\ k \tag{5.8}$$

$$1\ bar < P < 20\ bar \tag{5.9}$$

$$0.5 < \frac{\text{H}_2\text{O}}{\text{CH}_4} < 4 \tag{5.10}$$

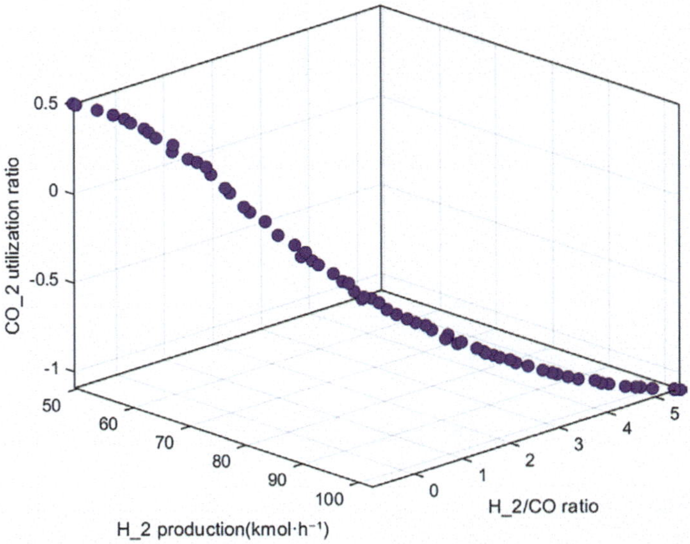

FIGURE 5.8 Pareto front generated by the genetic algorithm.

$$0.05 < \frac{CO_2}{CH_4} < 0.54 \tag{5.11}$$

$$0.2 < \frac{O_2}{CH_4} < 0.8 \tag{5.12}$$

Once the problem was formulated and implemented in MATLAB, the GA was executed until one of the stopping criteria was satisfied (convergence or maximum number of iterations). The values of these criteria were selected based on the trade-off between convergence toward the true Pareto front and the computational cost of the problem. After the final iteration, the algorithm generates a set of 70 nondominated Pareto points, as shown in Fig. 5.8. A point is said to be nondominated when there is no other feasible point that is strictly better in all objectives at the same time. The obtained Pareto points reveal substantial trade-offs among the three variables, such that improving one objective requires a compromise in the other two. The Pareto front (Figs. 5.9–5.11) indicates that hydrogen production lies between 49.58 and 102.59 kmol·h^{-1}, whereas the H_2/CO ratio ranges from 1.11 to 7.50, and the CO_2 utilization ratio varies between −1.09 and 0.511. Furthermore, only a few Pareto points yield a reformate composition suitable for methanol production. At the same time, a significant portion of the biogas CO_2 can be utilized efficiently in the reforming step of the process. However, Fig. 5.9 shows that several points in the Pareto front result in CO_2 production in the reformer. Table 5.2 presents the key points of the Pareto front.

TABLE 5.2 Key Pareto points of the problem.

Point	Tin (K)	Pin (bar)	H_2O/CH_4	CO_2/CH_4	O_2/CH_4	Hydrogen production ($kmol \cdot h^{-1}$)	CO_2 utilization ratio	$\left\lvert \frac{H_2}{CO} - 2 \right\rvert$
Point E1	1232.271	4.296	0.51815	0.537	0.649	49.583	0.512	−0.8831
Point E2	1181.096	1.562	3.859004	0.055	0.294	102.598	−1.095	5.501
Point E3	1218.957	2.575	0.855508	0.407	0.421	72.312	0.091	0
Point T	1219.558	2.524	0.567	0.529	0.475	63.867	0.394	−0.485

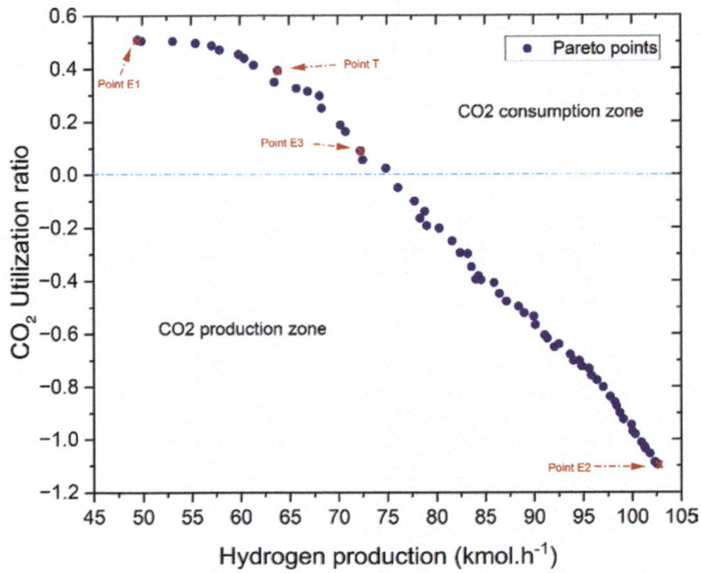

FIGURE 5.9 2D-projection of the Pareto front showing the trade-off between CO_2 utilization and hydrogen production.

Point E1 presents the case with the highest CO_2 utilization rate. This point should be used when CO_2 utilization is the sole critical design criterion, as it results in 51.2% of biogenic CO_2 from biogas being consumed in the reforming block. Furthermore, this scenario does not require CO_2 separation from the biogas upstream of the reformer, since all the biogenic CO_2 is fed directly to the reactor. Point E1 also results in a low carbon dioxide content in syngas, with a molar fraction of 0.083, potentially removing the requirement for a CO_2 separation step downstream of the reformer. However, the low hydrogen yield of 1.42 and the H_2/CO ratio of 1.11 make this syngas unsuitable for methanol synthesis. Moreover, low hydrogen production directly translates to low methanol production, negatively impacting the economic feasibility of the process. Moreover, the H_2/CO ratio at this point indicates a hydrogen deficiency in the process, necessitating a conditioning method to increase the hydrogen concentration in the syngas (e.g., WGS reactor, hydrogen supply, CO separation). Point E1 corresponds to a temperature of 1232.27 K and a pressure of 4.29 bar. The feed composition is 1/0.51/0.53/0.64 ($CH_4/H_2O/CO_2/O_2$). Under these conditions, the CH_4 and CO_2 conversions are 0.99 and 0.51, respectively, and both constraints related to catalyst deactivation have been satisfied. Additionally, the Pareto points highlight that high CO_2 utilization is achieved when the steam ratio is low and the CO_2 ratio is high, which can be explained by the fact that under these conditions, CO_2 production by the WGS reaction is suppressed. In contrast, CO_2 conversion through the DRM reaction is favored.

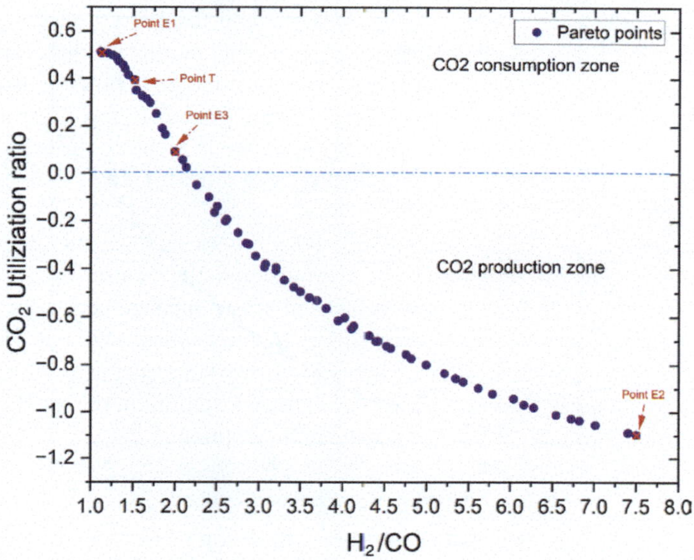

FIGURE 5.10 2D-projection of the Pareto front showing the trade-off between H_2/CO ratio and the CO_2 utilization ratio.

Point E2 corresponds to the highest hydrogen production flow rate in the studied space at 102.58 $kmol \cdot h^{-1}$ with a hydrogen yield of 2.94. However, this is accompanied by the lowest CO_2 utilization ratio at −1.09 and a H_2/CO ratio of 7.5. The operating conditions of point E2 are commonly used for hydrogen production when other objectives are not critical in the process design. Under this setup, the produced syngas is not suitable for methanol synthesis, and the reforming block emits more than twice the CO_2 it receives at a flow rate of 39.25 $kmol \cdot h^{-1}$ and a CO_2 molar fraction of 0.162 in the produced syngas. The high H_2/CO ratio indicates that the syngas presents a significant hydrogen surplus. Its hydrogenation into methanol typically requires the implementation of a conditioning unit (RWGS reactor, hydrogen separation, ...) to lower the hydrogen concentration in the syngas and adjust the H_2/CO ratio. Point E2 corresponds to an inlet temperature of 1181.09 K, a pressure of 1.56 bar, and the feed composition is 1/3.86/0.05/0.29 ($CH_4/H_2O/CO_2/O_2$). The CH_4 and CO_2 conversions at this point are 0.98 and −10.66, respectively, with 22.44 $kmol \cdot h^{-1}$ of CO_2 exiting the reactor, indicating substantial CO_2 production in the reformer. The Pareto front (Fig. 5.8) demonstrates that an increased hydrogen yield is obtained with high steam ratios and a CO_2 ratio close to its minimum value. This is because increasing the concentration of steam and decreasing the CO_2 concentration in the feed promotes hydrogen production through the WGS and SMR reactions.

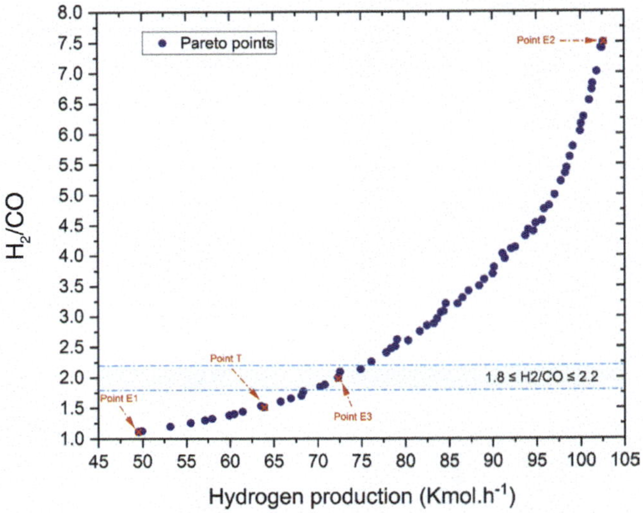

FIGURE 5.11 2D-projection of the Pareto front showing the trade-off between hydrogen production and H_2/CO ratio.

Point E3 presents the scenario that is most favorable for direct CO hydrogenation to methanol, as the H_2/CO ratio is at its optimal value. Using its corresponding optimal conditions provides flexibility to bypass the conditioning required for H_2/CO ratio adjustment in the other scenarios. However, the Pareto front shows that necessary compromises in hydrogen production and CO_2 utilization should be made to obtain the optimal H_2/CO ratio; the hydrogen yield is 26.87% lower than its maximum value in point E2, at a hydrogen production of 74.91 $kmol{\cdot}h^{-1}$, and the CO_2 utilization ratio is 0.091, presenting an 82.2% decrease compared to point E1. The inlet temperature of point U1 corresponds to 1218.87 K, the reactor pressure is 2.575 bar, and the reactant ratios are 1/0.855/0.407/0.4. The CH_4 and CO_2 conversions are 0.98 and 0.12, respectively. Point E3 and the other Pareto points demonstrate that the process favors higher temperatures and lower pressures ($P < 5$ bar) for all the optimization scenarios studied. This can be attributed to the fact that both SMR and DMR are strongly endothermic reactions that are favored by high temperatures. Furthermore, all three reforming reactions are mole-increasing and disfavored by high pressures.

Point T is the best Pareto point selected by the TOPSIS method, and results in a hydrogen yield of 1.83, a CO_2 utilization ratio of 0.394, and an H_2/CO ratio of 1.51. The operating conditions of this point are used when all three objectives are equally important from a design perspective. The inlet temperature in point T is 1219 K, the reactor pressure is 2.52 bar, and the reactant ratios are 1/0.567/0.475/0.529. The CH_4 and CO_2 conversions are 0.99 and 0.40, respectively, with a CO_2 molar flow of 11.02 $kmol{\cdot}h^{-1}$ exiting the reactor, and the maximum temperature inside the reactor is 1243.47 K. Point U2 presents a compromise of 37.74% in the hydrogen yield, 23.02% in CO_2 utilization, and 24.5% in the H_2/CO ratio compared to the utopia point. The results at this point demonstrate that necessary compromises must be made in each objective to obtain reasonable trade-off solutions.

The Pareto front also illustrates that a 1% compromise in the optimal value of hydrogen yield leads to a 4.81% increase in CO_2 utilization (from −1.095 to −1.042) and improves the H_2/CO ratio by 10.92% (from 5.5 to 4.9). On the other hand, compromising 1 % in the optimal value of the H_2/CO ratio ($\left|\frac{H_2}{CO} - 2\right| = 0.02$) only results in a 0.152% increase in hydrogen production (from 72.31 to 72.42 $kmol{\cdot}h^{-1}$) and no notable improvement in the CO_2 utilization ratio. A similar trend is observed for the CO_2 utilization objective, where a compromise of 1% results in an increase of 0.819% in hydrogen production (from 49.57 to 50.36 $kmol{\cdot}h^{-1}$) and an improvement of 1.312% in the H_2/CO ratio (from 0.883 to 0.871). The results reveal that small compromises in the value of hydrogen production can lead to considerable improvement in the other two objectives. In contrast, important compromises in the H_2/CO ratio and the CO_2 utilization ratio must be made to improve the other objectives.

The optimization results demonstrate that the steam ratio varies between 0.52 and 3.86, with high steam ratios corresponding to a high hydrogen yield and a negative CO_2 utilization ratio. On the other hand, the oxygen ratio ranges from 0.29 to 0.64, whereas the CO_2 ratio varies between 0.05 and 0.53. High O_2 and CO_2 ratios correspond to high CO_2 utilization ratios. This can be attributed to the shift in equilibrium of dry reforming with the increase in CO_2 concentration in the feed. A high O_2 ratio leads to higher operating temperatures and compensates for the low H_2O ratio in the total flow rate of the oxidizing agent. Furthermore, in these scenarios, coke deposition is inhibited primarily by the presence of oxygen in high concentrations. The inlet temperature of the reactor ranges from 1181.096 to 1232.271 K, indicating that the reactor favors high temperatures. Moreover, the temperature range suggests that the maximal temperature in the reactor can be controlled by adjusting the reactant ratios. Finally, the reactor pressure varies between 1.56 and 4.29 bar, which indicates that the process favors low pressures due to the three reforming reactions being mole-increasing reactions.

5.6 Conclusion

This chapter presents an innovative green process for producing methanol from waste, with a focus on methanol synthesis from OMSFW-based biogas with a 65/35 composition. Tri-reforming was first investigated to assess the effect of different parameters on the dependent variables of the process and the catalyst deactivation phenomena. The GA NSGA-II was then used to optimize the process for three conflicting objectives simultaneously, while accounting for the causes of catalyst deactivation. The decision variables considered were the inlet temperature, reactor pressure, and the ratios of steam, oxygen, and carbon dioxide in the feed. Moreover, the optimization objectives were maximizing hydrogen production, maximizing CO_2 utilization, and minimizing the difference between H_2/CO and its optimal value for methanol synthesis. Additionally, two constraints related to catalyst deactivation were implemented in the optimization problem formulation. The optimization algorithm generated a set of 70 nondominated points, and four representative points were used to analyze the results. The main conclusions drawn from this study are as follows.

- High steam ratios result in the highest hydrogen yield, while high CO_2 ratios result in the highest CO_2 utilization in the reformer.
- Coke formation typically occurs when high CO_2 ratios are coupled with low H_2O and O_2 ratios, while temperature and pressure also play a role.
- In the case study, the point corresponding to the highest hydrogen yield exhibits a hydrogen yield of 2.91, an H_2/CO ratio of 7.5, and results in extensive CO_2 production in the reactor.
- Up to 51.2% of the CO_2 in biogas can be consumed in the reforming reactor; however, this is accompanied by a significant compromise in hydrogen production.
- The target H_2/CO ratio of 2 is obtained under the following conditions: inlet temperature 1218.9 K, pressure 2.57 bar, and feed ratios of 1/0.85/0.40/0.42. and corresponds to a hydrogen yield of 2.07 and 9.1% CO_2 utilization in the reformer.
- The best Pareto point obtained with TOPSIS corresponds to a hydrogen yield of 1.83, 39.4% CO_2 utilization, and a H_2/CO ratio of 1.515.
- Small compromises in the optimal value of hydrogen production can lead to significant improvements in the two other objectives. In contrast, important compromises in CO_2 utilization and the H_2/CO ratio must be made to improve the other objectives.
- The reforming reactor tends to favor elevated temperatures and reduced pressures, with the reactant ratios varying within their bounds across the Pareto points.

Ultimately, the findings demonstrate that the waste-to-methanol process offers a promising alternative to fossil-based methods, providing a practical pathway for incorporating significant quantities of waste into chemical processes. Furthermore, the study illustrates the potential of tri-reforming, highlighting its key advantages such as the suppression of carbon formation and the flexible adjustment of the H_2/CO ratio. Building on these results, Future work may involve optimizing additional dependent variables (e.g., energy efficiency), evaluating economic performance under diverse conditions, and exploring alternative process configurations.

References

Abdullah, A. H., Mat, R., Somderam, S., Abd Aziz, A. S., & Mohamed, A. (2018). Hydrogen sulfide adsorption by zinc oxide-impregnated zeolite (synthesized from Malaysian kaolin) for biogas desulfurization. *Journal of Industrial and Engineering Chemistry, 65*, 334–342. doi:10.1016/j.jiec.2018.05.003.

Ahmed, M. Z., & Padhiyar, N. (2020). Multi objective optimization of a tri-reforming process with the maximization of H_2 production and minimization of CO_2 emission & power loss. *International Journal of Hydrogen Energy, 45*(43), 22480–22491. doi:10.1016/j.ijhydene.2020.06.210.

Appels, L., Baeyens, J., Degrève, J., & Dewil, R. (2008). Principles and potential of the anaerobic digestion of waste-activated sludge. *Progress in Energy and Combustion Science, 34*(6), 755–781. doi:10.1016/j.pecs.2008.06.002.

Arab Aboosadi, Z., Jahanmiri, A. H., & Rahimpour, M. R. (2011). Optimization of tri-reformer reactor to produce synthesis gas for methanol production using differential evolution (DE) method. *Applied Energy, 88*(8), 2691–2701. doi:10.1016/j.apenergy.2011.02.017.

Ashour, A., Challiwala, M. S., Musa, T., Wilhite, B., & Elbashir, N. (2025). Modeling tri-reforming of methane for carbon dioxide utilization and hydrogen production. *Energy, 335*, 137977. doi:10.1016/j.energy.2025.137977.

Bube, S., Sens, L., Drawer, C., & Kaltschmitt, M. (2024). Power and biogas to methanol – A techno-economic analysis of carbon-maximized green methanol production via two reforming approaches. *Energy Conversion and Management, 304*, 118220. doi:10.1016/j.enconman.2024.118220.

Chandra, R., Takeuchi, H., Hasegawa, T., & Vijay, V. K. (2015). Experimental evaluation of substrate's particle size of wheat and rice straw biomass on methane production yield. *Agricultural Engineering International: CIGR Journal, 17*(2), 2. https://cigrjournal.org/index.php/Ejounral/article/view/2844.

Deb, K., Agrawal, S., Pratap, A., & Meyarivan, T. (2000). A fast elitist non-dominated sorting genetic algorithm for multi-objective optimization: NSGA-II. In M. Schoenauer, K. Deb, G. Rudolph, X. Yao, E. Lutton, J. J. Merelo, & H.-P. Schwefel (Eds.), *Parallel problem solving from nature PPSN VI* (pp. 849–858). Springer. doi:10.1007/3-540-45356-3_83.

Demirel, Y. (2015). Technoeconomics and sustainability of renewable methanol and ammonia productions using wind power-based hydrogen. *Journal of Advanced Chemical Engineering, 5*(3). doi:10.4172/2090-4568.1000128.

Elsayed, N. H., Elwell, A., Joseph, B., & Kuhn, J. N. (2017). Effect of silicon poisoning on catalytic dry reforming of simulated biogas. *Applied Catalysis A: General, 538*, 157–164. doi:10.1016/j.apcata.2017.03.024.

Gao, Y., Jiang, J., Meng, Y., Ju, T., & Han, S. (2023). Influence of H2S and NH3 on biogas dry reforming using Ni catalyst: A study on single and synergetic effect. *Frontiers of Environmental Science & Engineering, 17*(3), 32. doi:10.1007/s11783-023-1632-1.

Gkotsis, P., Kougias, P., Mitrakas, M., & Zouboulis, A. (2023). Biogas upgrading technologies – Recent advances in membrane-based processes. *International Journal of Hydrogen Energy, 48*(10), 3965–3993. doi:10.1016/j.ijhydene.2022.10.228.

Grabow, L. C., & Mavrikakis, M. (2011). Mechanism of methanol synthesis on Cu through CO_2 and CO hydrogenation. *ACS Catalysis, 1*(4), 365–384. doi:10.1021/cs200055d.

Hajela, P., & Lin, C.-Y. (1992). Genetic search strategies in multicriterion optimal design. *Structural Optimization, 4*(2), 99–107. doi:10.1007/BF01759923.

Horn, J., Nafpliotis, N., & Goldberg, D. E. (1994). A niched Pareto genetic algorithm for multiobjective optimization. In *Proceedings of the First IEEE Conference on Evolutionary Computation.*

IEEE World Congress on Computational Intelligence: 1 (pp. 82–87). doi:10.1109/ICEC.1994.350037.

Hou, T., Zhao, J., Lei, Z., Shimizu, K., & Zhang, Z. (2020). Synergistic effects of rice straw and rice bran on enhanced methane production and process stability of anaerobic digestion of food waste. *Bioresource Technology, 314*, 123775. doi:10.1016/j.biortech.2020.123775.

Jabbour, K., Saad, A., Inaty, L., Davidson, A., Massiani, P., & El Hassan, N. (2019). Ordered mesoporous Fe-Al2O3 based-catalysts synthesized via a direct 'one-pot' method for the dry reforming of a model biogas mixture. *International Journal of Hydrogen Energy, 44*(29), 14889–14907. doi:10.1016/j.ijhydene.2019.04.110.

Khademi, M. H., Alipour-Dehkordi, A., & Tabesh, M. (2021). Optimal design of methane tri-reforming reactor to produce proper syngas for Fischer-Tropsch and methanol synthesis processes: A comparative analysis between different side-feeding strategies. *International Journal of Hydrogen Energy, 46*(27), 14441–14454. doi:10.1016/j.ijhydene.2021.01.215.

Khajeh, S., Arab Aboosadi, Z., & Honarvar, B. (2015). Optimizing the fluidized-bed reactor for synthesis gas production by tri-reforming. *Chemical Engineering Research and Design, 94*, 407–416. doi:10.1016/j.cherd.2014.08.018.

Konak, A., Coit, D. W., & Smith, A. E. (2006). Multi-objective optimization using genetic algorithms: A tutorial. *Reliability Engineering & System Safety, 91*(9), 992–1007. doi:10.1016/j.ress.2005.11.018.

Majewski, A. J., & Wood, J. (2014). Tri-reforming of methane over Ni@SiO 2 catalyst. *International Journal of Hydrogen Energy, 39*(24), 12578–12585. doi:10.1016/j.ijhydene.2014.06.071.

Methanol Institute. (2022). *Carbon footprint of methanol*. Methanol Institute https://www.methanol.org/wp-content/uploads/2022/01/CARBON-FOOTPRINT-OF-METHANOL-PAPER_1-31-22.pdf.

Methanol Institute. (2025). *Methanol price and supply/demand*. Methanol Institute https://www.methanol.org/methanol-price-supply-demand/.

Nagler, M., Aichinger, P., Kuprian, M., Pümpel, T., Insam, H., & Ebner, C. (2018). A case study for a cost-benefit-based, stepwise optimization of thermo-chemical WAS pre-treatment for anaerobic digestion. *Journal of Material Cycles and Waste Management, 20*(1), 266–273. doi:10.1007/s10163-016-0577-x.

Nestler, F., Schütze, A. R., Ouda, M., Hadrich, M. J., Schaadt, A., Bajohr, S., & Kolb, T. (2020). Kinetic modelling of methanol synthesis over commercial catalysts: A critical assessment. *Chemical Engineering Journal, 394*, 124881. doi:10.1016/j.cej.2020.124881.

Oreggioni, G. D., Brandani, S., Luberti, M., Baykan, Y., Friedrich, D., & Ahn, H. (2015). CO_2 capture from syngas by an adsorption process at a biomass gasification CHP plant: Its comparison with amine-based CO_2 capture. *International Journal of Greenhouse Gas Control, 35*, 71–81. doi:10.1016/j.ijggc.2015.01.008.

Osat, M., Shojaati, F., & Hafizi, A. (2023). A multi-objective optimization of three conflicting criteria in a methane tri-reforming reactor. *International Journal of Hydrogen Energy, 48*(16), 6275–6287. doi:10.1016/j.ijhydene.2022.06.221.

Rezaei, E., & Catalan, L. J. J. (2020). Evaluation of CO2 utilization for methanol production via tri-reforming of methane. *Journal of CO_2 Utilization, 42*, 101272. doi:10.1016/j.jcou.2020.101272.

Szablowski, L., Wojcik, M., & Dybinski, O. (2025). Review of steam methane reforming as a method of hydrogen production. *Energy, 316*, 134540. doi:10.1016/j.energy.2025.134540.

UN Environment Programme. (2012). *Sustainable, resource efficient cities: making it happen*. UNEP.

Valentino, F., Gottardo, M., Micolucci, F., Pavan, P., Bolzonella, D., Rossetti, S., & Majone, M. (2018). Organic fraction of municipal solid waste recovery by conversion into added-value polyhydroxyalkanoates and biogas. *ACS, 6*(12), 16375–16385 Sustainable Chemistry & Engineering. doi:10.1021/acssuschemeng.8b03454.

Wang, N., Lu, X., Tsang, Y., Mao, Y., Tsang, C., & Yueng, V. A. (2019). A comprehensive review of anaerobic digestion of organic solid wastes in relation to microbial community and enhancement process. *Journal of the Science of Food and Agriculture, 99*(2), 507–516. doi:10.1002/jsfa.9315.

Werkneh, A. A. (2022). Biogas impurities: environmental and health implications, removal technologies and future perspectives. *Heliyon, 8*(10), e10929. doi:10.1016/j.heliyon.2022.e10929.

Part III

Green and sustainable hydrogen economy

Chapter 6

Long-, medium-, and short-term hydrogen storage

Youssef Achour[a], Salma Boujmiraz[b], Asmae El Mokrini[c] and Asmae Berrada[a]

[a]*LERMA Lab, College of Engineering and Architecture, International University of Rabat, Sala Al Jadida, Morocco,* [b]*LRIT, Faculty of Science, Mohammed V University in Rabat, Rabat, Morocco,* [c]*BEAR Lab, Rabat Business School, International University of Rabat, Sale, Morocco*

6.1 Introduction

A continuous research effort is carried out worldwide in order to resolve climatic problems and preserve the environment. (Climate Change and Renewable Energy, 2019). In the midst, hydrogen possesses several interesting properties that make it an outstanding energy vector. All of which explains why hydrogen-related technologies have undergone a global shift in recent years. Nonetheless, hydrogen still faces several challenges in terms of production, storage, and application. The aim of sustainability has driven the scientific community to concentrate on green hydrogen for its great potential as an energy carrier. Compared to conventional fossil fuels, hydrogen exhibits nearly three times the energy content per unit mass (Ball & Weeda, 2015). However, hydrogen's low density makes it challenging to find efficient means for its storage. In this context, hydrogen storage technologies have seen great improvement over the last few years. The current literature offers a variety of options depending on the application in terms of hydrogen storage. The applications of hydrogen storage spread widely across the domestic and international markets, for small, medium, and large scales (Achour et al., 2023; Achour et al., 2025b; Mrabet & Berrada, 2021).

The main areas of research regarding hydrogen storage technologies consist of the cryogenic liquification method, the pressurized gas means, and the solid-state storage in metal hydrides. The transportation field favors pressurized gas hydrogen storage over the other methods (Johnson et al., 2015; Li et al., 2023; Ye et al., 2023). Liu et al. (2010) and Zheng et al. (2013) in 2010 and 2013, respectively, conducted thorough experimental studies on pressurized hydrogen storage. Their work demonstrated that the rapid refueling in this latter storage

Decarbonizing through Sustainable Hydrogen and its Derivatives.
DOI: https://doi.org/10.1016/B978-0-443-27632-3.00007-4

technology is accompanied by some issues concerning the thermodynamics of the process. The main issues resume with the heat generated by the kinetic energy that follows the fast refueling. The tank's inner temperature was also noticed to increase significantly due to the Joule–Thomson effect and the rise in temperature caused by the hydrogen compression. In most reviews, there are three important aspects of high-pressure gaseous hydrogen safety, i.e., hydrogen-induced cracking of metals at room temperature, temperature rise in hydrogen during fast refilling, and potential risks such as diffusion, leakage, and deflagration (Dwivedi & Vishwakarma, 2018; Zheng et al., 2012).

Liquid hydrogen (LH_2) storage, on the other hand, particularly formic acid and alcohols, is highly attractive as a hydrogen carrier, as they can be made from carbon dioxide (CO_2) or other materials. They can be used in stationary power storage units such as hydrogen filling stations, along with their direct use as transportation fuels. The challenge with this technology lies not only in the complexity of achieving it but also in the amount of energy and time that are consumed. To liquefy hydrogen, a temperature of 20.15 K must be reached, a value near absolute zero (0 K). This storage technology, despite having high liquid density (high gravimetric and volumetric efficiency), reduces the tank volume. It remains incredibly expensive with extreme losses (40% in the best cases) (Valenti, 2016). Other drawbacks are highlighted, including the technical complexity of hydrogen uptake and release, and hydrogen braze (Cardozo Soares Amaral, Oh, Do, & Choi, 2024). Zhang et al. (2024) studied LH2 release and cloud formation through a series of experiments. The study included a large-scale application with different release hole sizes and ground materials. The findings clearly indicate the rapid spread of hydrogen clouds. Similar studies raised the risk of explosion and helped improve safety measurements. LH_2 evaporation due to some technical issues is a major problem that accompanies this storage method (Statharas et al., 2000).

Light elements such as B, C, Li, Na, N, Mg, and many more constitute what are known as hydrides. They are remarkable candidates for high volumetric and gravimetric density hydrogen storage. Nonetheless, these hydrides are susceptible to some challenges concerning reversibility, expensive prices, degradation, and the rarity of materials. The solid-state hydrogen storage technology consists of a physical adsorption-desorption cycle based on the Van der Waals force (Principi et al., 2009; Zlotea & Latroche, 2013). Magnesium-based alloys, in particular, have shown promising hydrogenation properties. Researchers' extensive examination of these alloys demonstrated their potential as they are lightweight materials, present favorable kinetics, and have a high storage capacity of up to 7.6 wt%. Manganese is a cheap and abundant material, on top of its advantageous properties. These alloys may also form solid solutions and compounds, in addition to having strong recyclability and heat resistance (Verga et al., 2009). Furthermore, they present fast absorption and desorption rates of hydrogen. Zeolites, carbon-based materials, and metal-organic framework (MOF) materials are only a few of the wide varieties of nanoporous materials that have been researched (Barthelemy et al., 2017; Ren et al., 2017). Nevertheless, at room temperature and low pressure, microporous materials have poor hydrogen storage ability. Activated carbon can only store 0.5–2.0 wt% of hydrogen at room temperature, even at pressures of up to 20 MPa. Activated carbon with a 3000 m^2/g specific surface area can store approximately 2.0 wt% of hydrogen at 77 K at 0.1 MPa and 0.44 wt% at 298 K at 5 MPa (Jordá-Beneyto et al., 2007). Only 1.0 wt% of hydrogen can be stored in MOF-5 at 298 K and 2 MPa, whereas only 1.25 wt% can be stored in MOF-177 at standard pressure (Li et al., 2009).

This chapter compares different hydrogen storage solutions for short-, medium-, and long-term applications to fill a significant research gap. This is achieved by thoroughly assessing each storage option. This study makes a novel contribution by thoroughly comparing technical feasibility, as well as the economic and environmental ramifications. This will help to provide important insights into the effective use of hydrogen as a significant energy vector. This chapter lists some of the most developed hydrogen storage technologies with their respective recent developments. The aim is directed toward comprehensiveness rather than technical modeling.

The first section of this chapter comprises a comprehensive literature review regarding hydrogen storage technologies. The second section is dedicated to presenting general assumptions regarding the different hydrogen storage technologies. The follow-up sections are a summary of the recent developments and the associated mathematical modeling for numerical validation. Furthermore, a thorough discussion of these technologies and the discussion of short, medium-, and long-term hydrogen storage solutions. Finally, a conclusion with final remarks and perspectives.

6.2 General assumptions

This chapter compares different methods of hydrogen storage, including gaseous, solid, liquid, etc. The aim of this chapter is to identify the most suitable hydrogen storage technology for short-, medium-, and long-term applications. Fig. 6.1 displays the different hydrogen storage technologies. Numerous safety measures for hydrogen storage and transportation have been established and detailed in Table 6.1.

From an energy perspective, hydrogen has several important properties. It has a low molecular weight, which gives it an excellent ability to disperse quickly. This high diffusion is useful in certain chemical processes and promotes good mixing with other gases. Table 6.2 summarizes the main chemical and thermal characteristics of hydrogen (Levikhin & Boryaev, 2024).

Storing hydrogen requires that the inlet quantity be equal to the outlet quantity from any device used for storage. Eqs. (6.1) and (6.2) represent the fundamental rules of physics and engineering, as they govern the behavior of

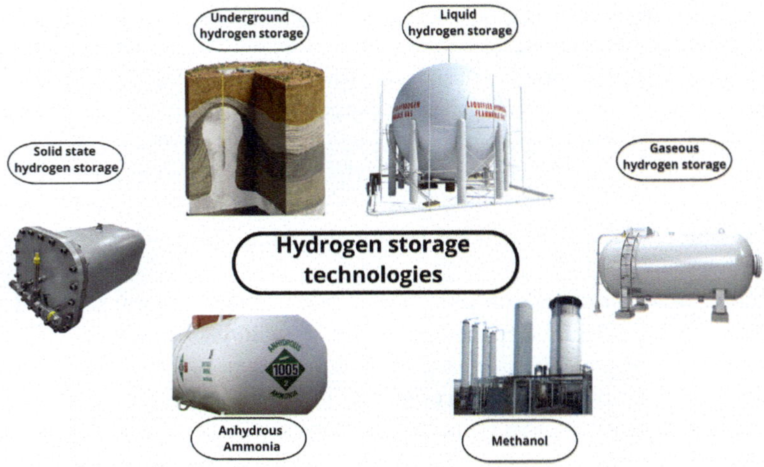

FIGURE 6.1 Illustration of the various hydrogen storage technologies.

TABLE 6.1 Regulatory documents for hydrogen-related technologies.

ISO 13984	Liquid hydrogen—land vehicle fueling system interface
ISO 14687	Hydrogen fuel—product specification.
ISO/CD 13985	Liquid hydrogen—land vehicle fuel tanks.
ISO/WD 13986	Tank containers for multimodal transportation of liquid hydrogen.
ISO/WD 15594	Airport hydrogen fueling facility.
ISO/WD 15866	Gaseous hydrogen blends and hydrogen fuel—service stations.
ISO/WD 15869	Gaseous hydrogen and hydrogen blends—land vehicle fuel tanks.
ISO/WD 15916	Basic requirements for the safety of hydrogen systems.
ISO/AWI 17268	Gaseous hydrogen—land vehicle fueling connectors.
ISO/TR 15916	2004(E) Basic considerations for the safety of hydrogen systems.

mass and energy in all processes. They ensure that the total mass and energy of the closed systems (hydrogen storage facilities in this case) remain constant. These principles are essential for understanding and designing energy-efficient systems and processes. The conservation of mass is given by Eq. (6.1), while the conservation of energy is formulated in Eq. (6.2):

$$\frac{dm}{dt} = \dot{m}_i - \dot{m}_e \tag{6.1}$$

TABLE 6.2 Chemical and thermal characteristics of hydrogen.

Characteristic	Value
PCI (lower heating value)	10,800 kJ/Nm^3–119,930 kJ/kg 3.00 kWh/Nm^3–33.33 kWh/kg
HHV (higher heating value)	12,770 kJ/Nm^3–141,860 kJ/kg 3.55 kWh/Nm^3–39.41 kWh/kg
Gas density at 20.3K	1.34 kg/m^3
Gas density at 273K	0.08988 kg/Nm^3
Liquid density at 20.3K	70.79 kg/m^3
Theoretical energy of liquefaction	14,112 J/g (3.92 kWh/kg)
Autoignition temperature in air	858K
Flame temperature in air at 300K	2318K
Flammability limits in air (vol %)	4–75
Detonation limits in air (vol %)	13–65
Detonation velocity in air	2.0 km/s
Stoichiometric mixture in air (vol)	29.53%

PCI, *pressure-composition isotherm.*

$$\frac{dE}{dt} = \dot{Q} - \dot{W} + \dot{m}_i \times h_i - \dot{m}_e \times h_e \tag{6.2}$$

where $\dot{m}$ is the mass flow rate entering the system, and E is the rate of change of energy. Q is the heat-transfer rate into the system. W is the work done by the system, $\dot{m}_i \times h_i$ is the enthalpy flow rate entering the system while $\dot{m}_e \times h_e$ is the enthalpy flow rate leaving the system. The levelized cost of hydrogen storage (LCOHS) can be best described as the ratio of the net present cost of the storage system to the cumulative stored mass of hydrogen:

$$LCOHS = \frac{\text{Cost}_{\text{net}}}{\sum_1^N rM_{H_2}} \tag{6.3}$$

where r is the discount rate, $M_{H_2}$is the cumulative mass stored in the system, is Cost_{net} is expressed as

$$\text{Cost}_{\text{net}} = \text{Cap}Ex_s + \sum_1^N OpEx_s \tag{6.4}$$

6.3 Gaseous hydrogen storage

Compressed gas hydrogen storage is a mature, common storage technology. It is the most considered technology for fuel-cell electric vehicles (Achour et al., 2024). Fig. 6.2 displays the 3-D geometrical design of the gaseous

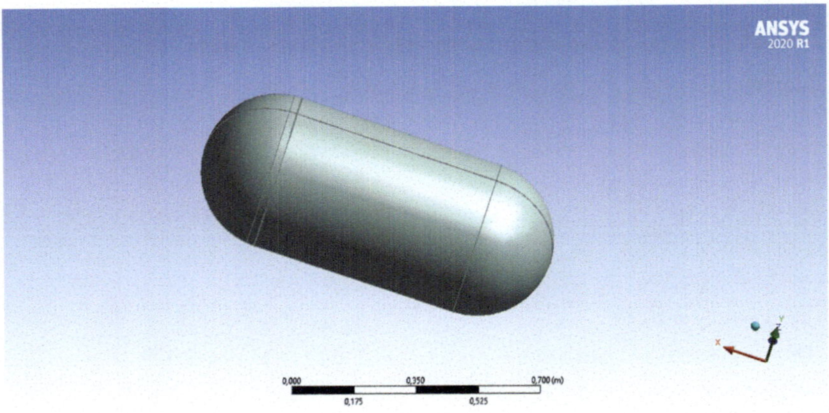

FIGURE 6.2 Three-dimensional geometrical design of the gaseous hydrogen storage tank.

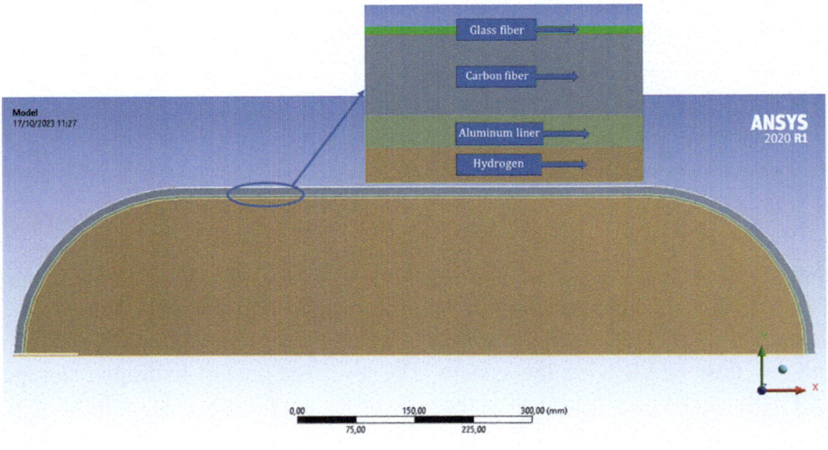

FIGURE 6.3 Cross-section of the hydrogen gaseous tank with its different layers.

hydrogen storage tank, while Fig. 6.3 represents the different layers used in the manufacturing of the hydrogen tank. Each layer adds certain parameters to the device. Table 6.3 presents the main physical and thermodynamic properties of the cylinder material (Liu et al., 2010).

The CFD simulation is based on a 2-D axisymmetric model. Slow filling results in laminar flow, characterized by smooth, orderly layers of gas moving parallel to each other. This contrasts with the turbulent flow observed in high-speed jet flows (Liu et al., 2010; Zheng et al., 2013). All thermodynamic properties and state-governing equations were calculated by the National Institute of Standards and Technology. The model contains two distinct computational domains: the fluid domain inside the tank, filled with gaseous hydrogen, and the

TABLE 6.3 Main properties of cylinder materials.

	Density (kg/m^3)	Specified heat [J/(kg·K)]	Thermal conductivity [W/(m·K)]
Glass fiber	1570	840	0.612
Carbon fiber	1750	800	5.95
Aluminum liner	2700	902	238

domain consisting of the tank itself and the inner tube inlet. The heat transfer between the gaseous hydrogen and the environment consists of three parts: convection between the hydrogen and the inner wall, conduction between the two walls, and finally convection between the outer wall and the outside air. The governing equations for the filling process are the unsteady Favre-averaged Navier–Stokes equations (Suryan et al., 2013), as represented by Eqs. (6.5) and (6.6):

$$\frac{\partial \rho}{\partial t} + \nabla \cdot (\rho \vec{v}) = 0 \tag{6.5}$$

$$\frac{\partial}{\partial t}(\rho \vec{v}) + \nabla \cdot (\rho \vec{v} \vec{v}) = -\nabla p + \nabla \cdot (\bar{\tau}) + \rho g \tag{6.6}$$

$$\frac{\partial}{\partial t}(\rho E) + \nabla \cdot (\vec{v}(\rho E + p)) = \nabla \cdot \left(k_{\text{geff}} \nabla T + \left(\bar{\bar{\tau}}_{eff} \cdot \vec{v}\right)\right) \tag{6.7}$$

where t represents the time, ρ is the gas's density, $\vec{v}$ is Favre's average velocity, p is the static pressure, $\bar{\bar{\tau}}$ is the stress tensor, and ρg is the gravitational body force, respectively. The stress tensor is given by Eq. (6.8):

$$\bar{\bar{\tau}} = \mu \left[\nabla \vec{v} - \left(\frac{2}{3}\right) \nabla \cdot \vec{v} I \right] \tag{6.8}$$

With μ being the dynamic viscosity and I is the unit tensor. E is the gas energy in J/k, and k_{geff} is the effective thermal conductivity (Eqs. 6.9 and 6.10):

$$E = h - \left(\frac{p}{\rho}\right) + \left(\frac{1}{2}\right) v^2 \tag{6.9}$$

$$k_{\text{geff}} = k_g + k_t \tag{6.10}$$

$$k_t = \frac{(c_{pg} \mu_t)}{Pr_t} \tag{6.11}$$

Pr_t is the turbulent Prandtl number, and $\bar{\bar{\tau}}_{\text{eff}}$ represents the viscous heating. μ in stress tensor is replaced with μ_{eff} as shown in Eq. (6.12):

$$\mu_{\text{eff}} = \mu + \mu_t \tag{6.12}$$

The flow chosen for our study is a realizable k–epsilon (k–ε) turbulent flow. This model provides better results than standard k–ε models. It has been tested

TABLE 6.4 Initial and boundary conditions.

Boundary conditions	Values
Ambient temperature	293K
Convective heat-transfer coefficient	10 W/m^{2}K
Inlet pressure	20 MPa
Inlet mass flow	0.025 & 0.090 kg/s
Initial temperature	293K

and validated for free flows such as jets, mixing layers, rotating homogeneous shear flows, and channel and boundary layer flows (Suryan et al., 2013), etc. The initial and boundary conditions are detailed in Table 6.4.

The temperature distribution at different moments during the filling of the tank is observed in Fig. 6.4. At $\Delta t = 29.5$ s, the temperature reaches 366 K. The clear feature is the presence of high temperature gradients throughout the tank, with a concentration of hot zones along the tank. Cooling zones are identified adjacent to the tank walls (they are considered thermal dissipators), allowing heat transfer from the hydrogen to the walls. These cooling zones play a crucial role in managing heat distribution and preventing the tank from overheating. Additionally, Fig. 6.4 showcases a good distribution of heat in different areas of the tank, suggesting uniformity in temperature distribution. At $\Delta t = 3$ s, which is a much faster filling time compared to the previous one. Temperature contours reveal the presence of hot spots, where temperatures are higher compared to other points. These hot spots indicate a high concentration of heat, resulting from disturbances in the hydrogen flow within the tank. Fig. 6.5 shows the velocity contours of the flow at different moments during the filling process. At a flow period of 2 s, the appearance of velocity contours generating turbulence is observed, with a maximum velocity of 2.70 m/s. As the filling time progresses, turbulence gradually increases, and at $\Delta t = 29.5$ s, the velocity reaches 9 m/s.

Fig. 6.5 demonstrates several changes in the structure and shape of the velocity contours. These variations can be observed over time, showing the evolution of the flow. However, as time increases, vorticity structures form in some regions of the tank. The effect of viscosity on the flow velocity contours can also be noticed, with more pronounced velocity gradients near the walls due to higher viscous effects. It is noted that at a flow period of 2 s, the velocity reaches 3 m/s. This is due to the preimposed initial conditions. The initial conditions are favorable for generating turbulence contours and a maximum velocity of 3 m/s. Over time, several factors influence the velocity reached at each moment, such as interactions between hydrogen particles, pressure loss, and energy dissipation.

During the filling process, thermal energy is supplied to the tank cylinder, resulting in an increase in temperature, which is transferred to the tank walls

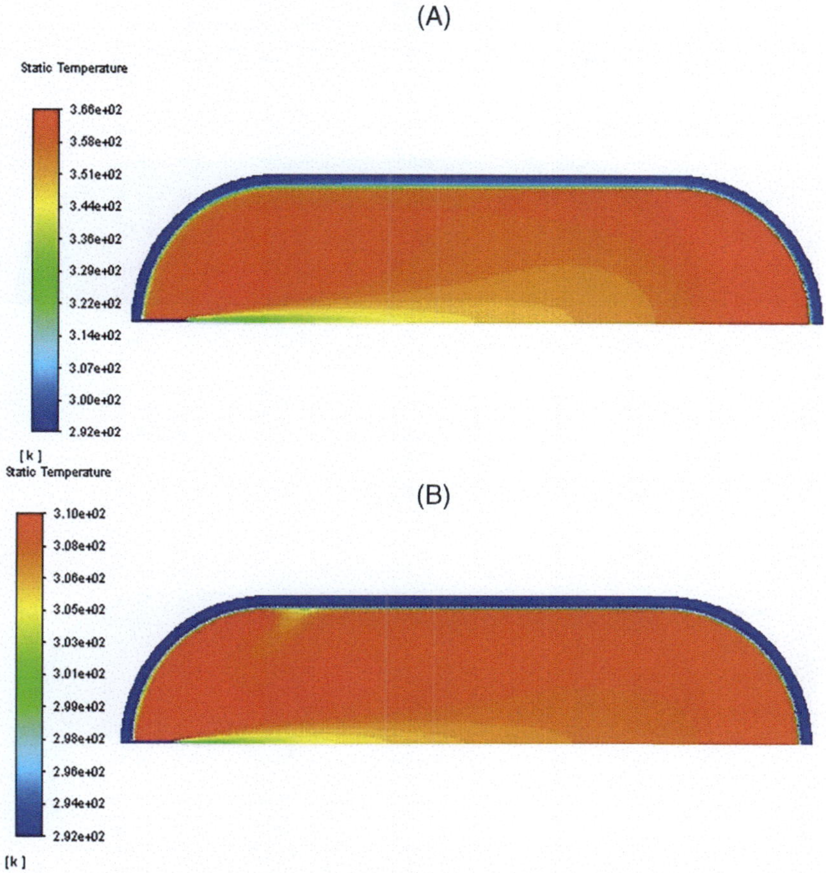

FIGURE 6.4 Contour of the isotherm at different filling periods: (A) at $\Delta t = 29.5$ s and (B) at $\Delta t = 3$ s.

through forced convection. Initially, the heat transfer is high, increasing the Nusselt number due to a significant thermal flux caused by the temperature difference between the hydrogen and the tank walls. As the heat transfer reaches its peak, energy dissipation increases, reducing the energy available for forced convection and decreasing the Nusselt number. Eventually, temperatures balance, causing the heat transfer by convection to stabilize and the Nusselt number to level off. Fig. 6.6 represents the Nusselt number evolution with time. The Nusselt number is essentially what quantifies the efficiency of heat transfer in the system.

6.4 Metal-hydride storage

Solid-state reactions are characterized by the fraction of a material that has reacted. The process typically begins with the host metal absorbing some of

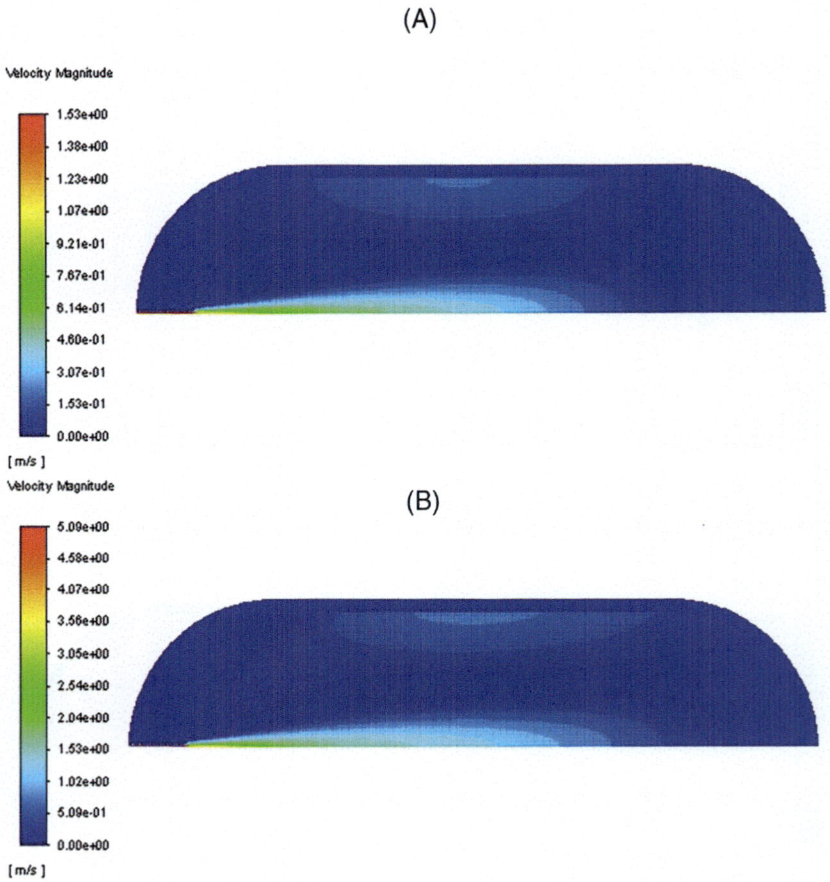

FIGURE 6.5 Contour of velocity magnitudes at different filling periods: (A) at $\Delta t = 29.5$ s and (B) at $\Delta t = 2$ s.

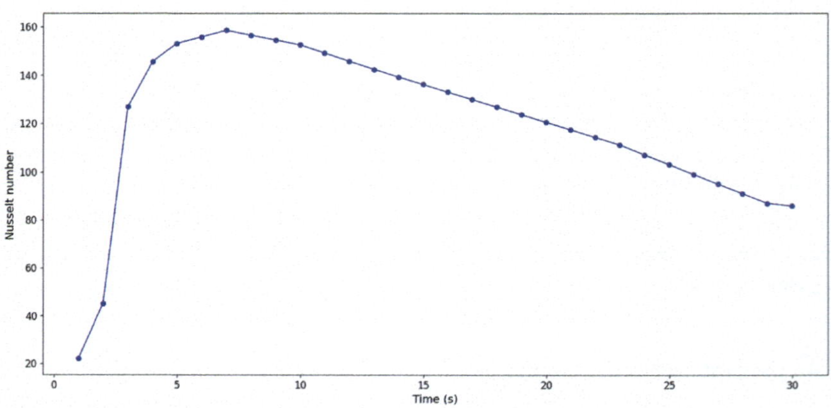

FIGURE 6.6 Nusselt number evolution with time.

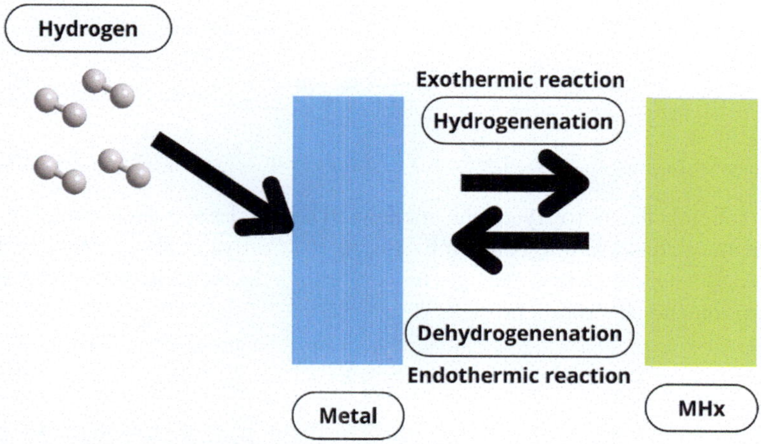

FIGURE 6.7 Schematic of the metal-hydride H_2 storage concept.

the hydrogen by dissolving it into the metal (α-phase). The concentration of hydrogen gradually rises inside the metal, which explains the local interactions created between hydrogen atoms. These interactions lead to nucleation and hydride formation (β-phase). For a continuous transition between the two phases (α-to-β), the metal must be at a steady temperature T_c, also known as the critical temperature. For any temperature below T_c, the two phases' regions coexist. This coexistence is marked by a region where the pressure does not change much, creating a plateau on the pressure-composition isotherm (PCI) curve. The length of this plateau indicates the total storage capacity of the material at that constant temperature. Fig. 6.7 is a schematic of the metal-hydride H_2 storage concept. Metal hydrogenation is the absorption of hydrogen to form M-H_x. This chemical reaction is often exothermic. On the other hand, dehydrogenation is the release of H_2 to form the metal M. This reaction, however, is endothermic (Andersson & Grönkvist, 2019; Klopčič et al., 2023).

The energy equation for the solid-state hydrogen configuration is given by Eq. (6.13) (Broom, 2011):

$$(\rho \cdot C_p)\frac{\partial T}{\partial t} + (\rho_g \cdot C_{p_g})\vec{v}_g \cdot \nabla T = \nabla \cdot (k_e \cdot \nabla T) + m \cdot (\Delta H - T \cdot (C_{p_g} - C_{p_s})) \tag{6.13}$$

where ρ, ρ_g are the densities of the solid and gaseous phases, respectively. C_p, C_{p_g}, C_{p_s} are the specific heat capacities of the solid phase, gaseous phase, and the solid hydride, respectively. $\vec{v}_g$ is the velocity vector of the gaseous phase, k_e is the effective thermal conductivity, T is the temperature, m is the reaction rate of hydrogen absorption or desorption, ΔH is the enthalpy change of the reaction. Hydride and hydrogen mass balance are expressed by Eqs. (6.14) and (6.15), respectively (Yartys et al., 2019):

$$(1-\varepsilon)\frac{\partial \rho_s}{\partial t} = -m \tag{6.14}$$

$$\varepsilon\frac{\partial \rho_g}{\partial t} + \nabla\cdot\left(\rho_g \vec{v}_g\right) = m \tag{6.15}$$

with ε being the porosity of the medium. Darcy's Law, which explains the momentum conservation, is given by Eq. (6.16) (Yartys et al., 2019). This law helps understand how hydrogen gas moves through the metal-hydride bed during absorption and desorption processes:

$$\vec{v}_g = -\frac{K}{\mu_g}\nabla P_g \tag{6.16}$$

where K is the permeability of the porous medium, μ_g is the dynamic viscosity of the gaseous phase, P_g is the pressure of the gaseous phase. Van't Hoff Law is given by Eq. (6.17). This law relates the equilibrium pressure of hydrogen and the temperature during the absorption and desorption processes:

$$\ln P_{\text{eq}} = \frac{\Delta H}{R_g\cdot T} - \frac{\Delta S}{R_g} \tag{6.17}$$

where P_{eq} is the equilibrium pressure, ΔS is the entropy change of the reaction, and R_g is the universal gas constant. Kinetics of absorption and desorption are given by Eqs. (6.18) and (6.19):

$$m = C_a\cdot \exp\left(-\frac{E_a}{R_g\cdot T}\right)\cdot \ln\left(\frac{P_{\text{eq}}}{P_g}\cdot(\rho_s-\rho_s)\right) \tag{6.18}$$

$$m = C_d\cdot \exp\left(-\frac{E_d}{R_g\cdot T}\right)\cdot\left(\frac{P_g}{P_{\text{eq}}}-1\right)\cdot(\rho_s-\rho_0) \tag{6.19}$$

where C_a, C_d are the preexponential factors for absorption and desorption, E_a, E_d are the activation energies for absorption and desorption, and ρ_0 is the initial density of the solid phase.

Mg is one of the most promising metals for creating a hydride material. It is characterized by high hydrogen storage capacity with reversible hydrogen capacity up to 7.6 wt%. It is also an abundant material with a low cost. In addition, magnesium hydrides have good reversibility. Fig. 6.8 compares the volumetric and gravimetric energy densities of different hydrogen storage systems, including CGH_2, LH_2, metal hydrides, as well as Li-ion and vanadium-redox-flow batteries (VRFB). The US Department of Energy (DoE) 2025 and ultimate targets, and the European Clean Hydrogen Joint Undertaking (CH JU) 2030 targets are presented with a dash-dotted line (De Colvenaer & Castel, 2012; US Department of Energy, 2026). The PCIs for magnesium doped with 2% nickel are displayed in Fig. 6.9. The ideal PCI curve reveals that the plateau absorption pressure remains constant as hydrogen concentration increases, up to a temperature of 593 K. Beyond this, at higher temperatures of 623 and 653 K,

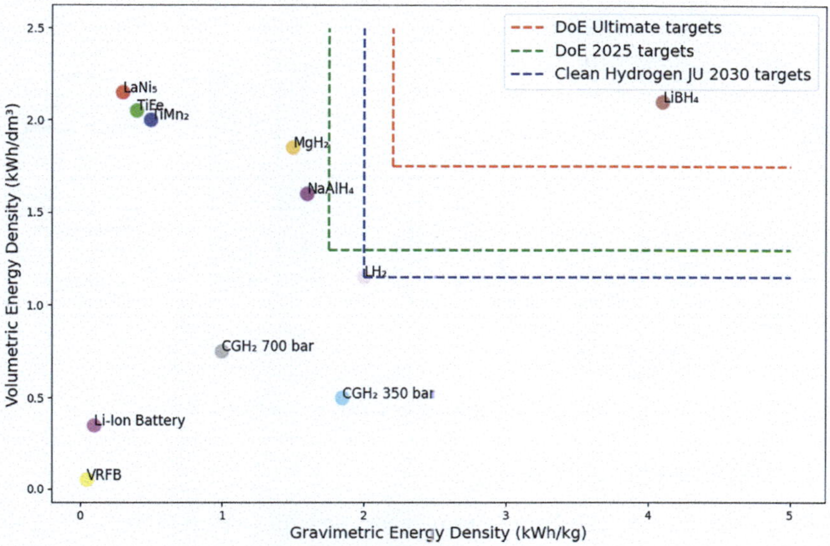

FIGURE 6.8 Volumetric and gravimetric energy densities of different hydrogen storage systems.

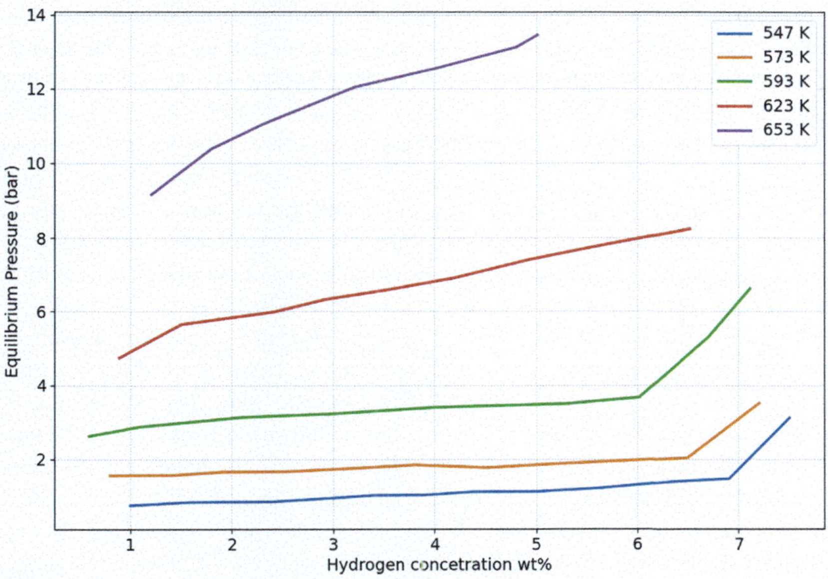

FIGURE 6.9 Pressure-composition isotherms of Mg (2% Ni).

the hydrogen absorption pressure rises proportionally with increasing hydrogen concentration. At 547 K, the material achieves a maximum hydrogen concentration W_{H2} of 7%, which declines to 5% at 653 K (Kapischke & Hapke, 1998; Berrada et al., 2025; Rkhis et al., 2023).

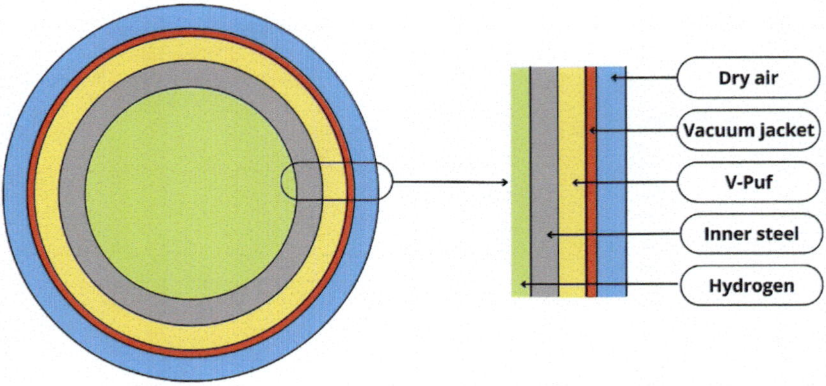

FIGURE 6.10 Cross-section of the hydrogen gaseous tank with its different layers.

6.5 Liquid hydrogen storage

LH2 storage carriers can have approximate capacities up to 50,000 m^3. The capacity varies depending on the applications, logistics, and other metrics. The tanks are equipped with an insulation-layer thickness. In the temperature range of LH_2, the air molecules condense and solidify. Safety precautions are taken to prevent air molecules (mainly oxygen and nitrogen) from solidifying, thus damaging the insulation system. Furthermore, the materials used for LH_2 storage tanks must withstand high pressures and large vacuum loads. To respond to this issue, the utilization of Vacuum-polyurethane foam (v-PUF) as the insulation layer has proven effective (Abe, 1998; Park et al., 2022). The various layers in the LH_2 storage tank are illustrated in Fig. 6.10. For optimal operation, the fluid inside the LH_2 storage tank should maintain a uniform temperature distribution. The convective heat-transfer coefficients for LH_2 and gaseous hydrogen (GH_2) within the tank are determined based on natural-convection analysis, with fluid properties sourced from Refprop provided by the NIST (Yu et al., 2025).

Stainless steel provides significant strength, resistance to hydrogen embrittlement, and toughness at very low temperatures. There is no direct contact between the vacuum jacket (VJ) and the cryogenic LH_2. The VJ is usually made of stainless steel with the appropriate composition (304L, 316L). High performance, cost effectiveness, and thermal resistance are the main parameters to consider for selecting the appropriate material. Since the v-PUF is located between the inner steel and the VJ, thermal conductivity is involved. It was determined by considering solid/gas thermal exchanges (conduction and radiation) within PUF under vacuum conditions. Convective heat-transfer coefficient for dry air is assumed to be 10 W/m^2K. as the standard value at atmospheric conditions.

Two distinct regions of fluid that interact within the LH_2 tank are observed. This phenomenon is due to the varying temperature gradients inside the tank. LH_2 typically forms at a temperature of approximately 20–25 K. In contrast,

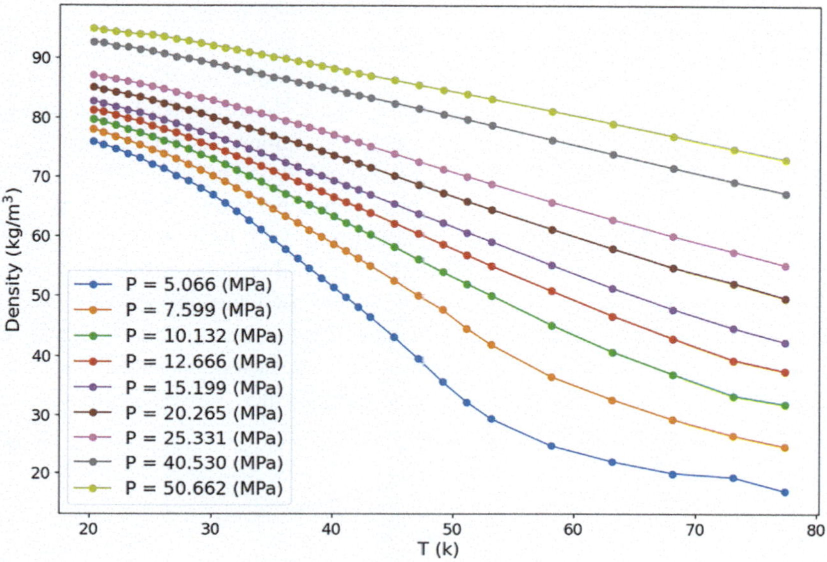

FIGURE 6.11 Hydrogen density evolution along with different temperature and pressure levels.

the gaseous hydrogen inside the tank is noticed at a higher temperature than the liquid phase, but its temperature in cryogenic conditions can range from 20 to 30 K. A significant temperature difference exists between the tank's internal region (~20–25 K) and the external environment (~297 K), which is managed by the tank's insulation. This large thermal gradient generally leads to a uniform temperature distribution across the tank's internal space, especially near the walls where heat ingress from the external environment occurs. However, despite the gradient, a notable correlation exists between the density, temperature, and pressure of the stored hydrogen. For instance the highest density is achieved under high pressures. At around 50.66 MPa and a temperature of approximately 23 K, the measured density value is close to 80 kg/m^3 for LH2. On the other hand, at a lower pressure of 5.06 MPa and a higher temperature of about 78 K, the density drops significantly to approximately 19 kg/m^3. The density variation with respect to temperature and pressure is visualized in Fig. 6.11, where a clear inverse relationship between temperature and density is shown under varying pressures, reflecting the behavior of hydrogen in both gaseous and liquid states.

The thermodynamic analysis is crucial in designing and installing LH2 systems. Maintaining cryogenic temperatures requires optimal insulation. However, cost effectiveness, materials, and other metrics should be considered. Fig. 6.12 demonstrates the heat transfer outside the tank versus the insulation-layer thickness. It is clear from the plot that the thicker the insulation layer, the more optimal the heat isolation, as the heat transfer drops.

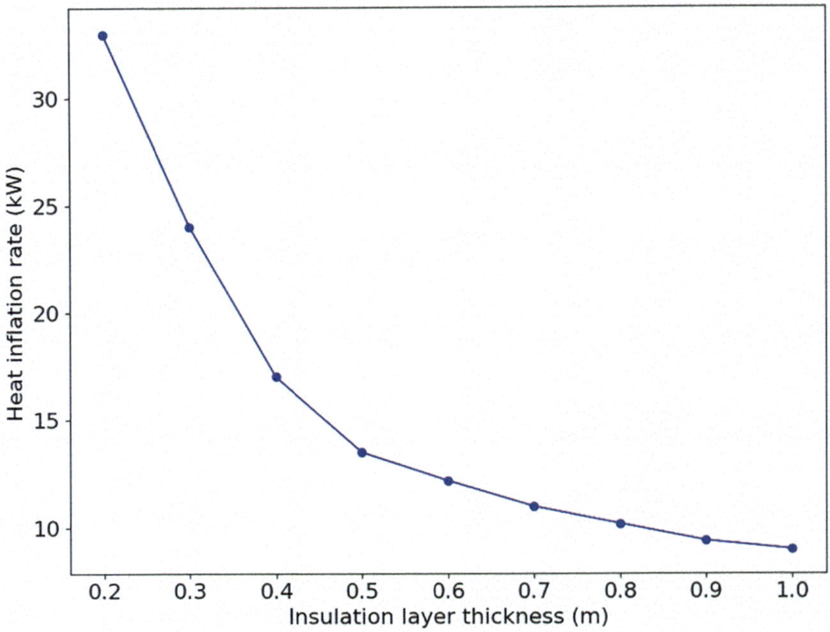

FIGURE 6.12 Heat inflation rate for different insulation-layer thicknesses.

6.6 Hydrogen storage in ammonia

Ammonia (NH3) has an excellent hydrogen storage density of 17.7 wt% gravimetrically and 123 kg/m^3 volumetrically for liquid ammonia at IO bar. Ammonia synthesis, storage, and logistics are very mature. These interesting properties of ammonia make it an attractive means for hydrogen storage (Andersson & Grönkvist, 2019). Ammonia can be produced via the Haber–Bosch process. The synthetic ammonia process, or the Haber ammonia process, is the direct combination of hydrogen and nitrogen to synthesize ammonia (Eq. 6.20) (Kyriakou et al., 2020). This process was later adapted for large-scale usage using a catalyst and high-pressure methods developed by Carl Bosch (Soloveichik, 2019). Ammonia cracking, or ammonia decomposition, which is the reversible reaction that occurs at a typical Haber–Bosch reactor, has been considered to obtain hydrogen. NH_3 offers a higher density in comparison to H_2, which makes the quantity of H_2 stored in this ammonia larger than hydrogen itself (Pourali et al., 2022):

$$3H_2 + N_2 \rightarrow 2NH_3 \tag{6.20}$$

The ammonia synthesis loop requires H_2 and N_2 in an exothermic reaction. The reactants are unchanged, which makes the ammonia production process unchanged as well. Furthermore, this important note indicates that, in principle, standard ammonia synthesis plants can easily transition from natural gas-based

TABLE 6.5 The ideal work necessary for liquefaction of various gases.

Gas	Molar fraction in the air (%vol)	Normal boiling point (°C)	Ideal work of liquification (kJ/kg)
Air	100	−194,35	738.9
Hydrogen	0.000005	−252,88	12,019
Nitrogen	78	−195,79	768.1
Oxygen	20.9	−182,97	635.6
Ammonia	–	−33,35	359.1

to renewable power-based. Hydrogen in this configuration will be produced via a water electrolyzer, thus resulting in cutting harmful emissions (Table 6.5; Pfromm, 2017).

The ammonia production process remains unchanged; however, the dehydrogenation of ammonia is very technically challenging. Dehydrogenation is the decomposition of ammonia to separate hydrogen and nitrogen molecules. Thermolysis is the most common way to obtain hydrogen from ammonia. It is worth noting that ammonia starts the decomposition process spontaneously at temperatures exceeding 200°C, but this split of the NH_3 is only partial. Additional energy (heat) at temperatures above 650°C is required for complete decomposition (Achour et al., 2025a; Giddey et al., 2017).

6.7 Underground hydrogen storage

Hydrogen can be stored underground by various means, including using the depleted gas reservoirs, salt caverns (SCs), or even saline aquifers. Each method requires a specific set of conditions, and it is accompanied by its respective challenges (Zivar et al., 2021). Depleted gas reservoirs are former natural gas fields that have been exhausted of their gas. They are considered for hydrogen storage due to their existing infrastructure and proven ability to contain gases. The specific requirements include adequate capacity, the ability to contain hydrogen, and a reliable caprock to prevent leakage (Lemieux et al., 2020). SCs are created by dissolving underground salt deposits and then removing the brine. They are highly regarded for hydrogen storage because they offer good purity and very minimal contamination risk. SCs also display adequate sealing properties, which are crucial for containing hydrogen (Pierce & Rich, 1962). Saline aquifers are porous rock formations filled with saline water (Ali et al., 2022). They are considered for hydrogen storage due to their large capacity and widespread availability. However, the containment of hydrogen in saline aquifers requires careful evaluation of the geological and reservoir properties to ensure secure storage. Underground hydrogen storage offers the advantage of storing large amounts of hydrogen at a reasonably low cost. The cost, however, should

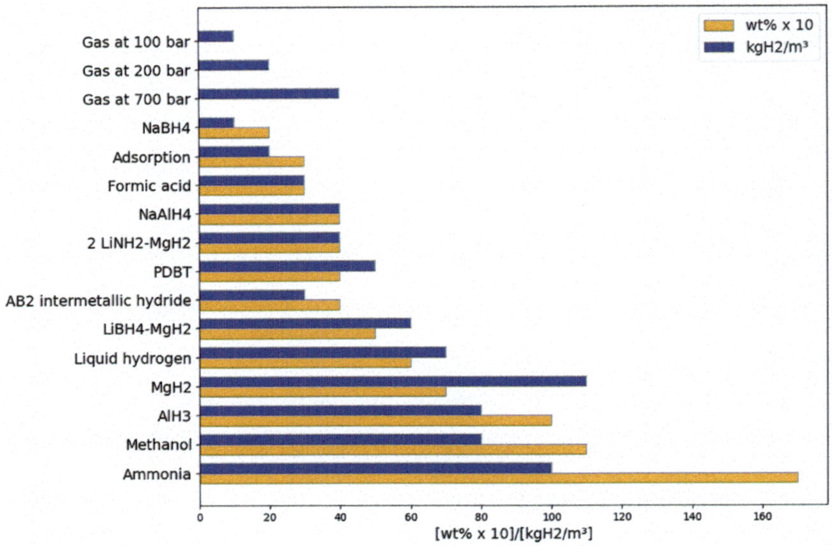

FIGURE 6.13 Volumetric and gravimetric hydrogen storage densities of various hydrogen storage technologies.

amount to pressure change, cyclic operations costs, and regular maintenance. Furthermore, these parameters depend on the geological features of the storage site (Lewandowska-Śmierzchalska et al., 2018).

6.8 General discussion

Hydrogen storage has undergone significant research and development over the last few years. This chapter reviewed the most interesting technical and economic aspects of H_2 storage. Fig. 6.13 presents the volumetric and gravimetric hydrogen storage densities of various hydrogen storage technologies (Wijayanta et al., 2019). The storage densities provided in Fig. 6.13 do not include the specific characteristics of the utilized tanks, including size, weight, volume, and capital investment. It is deduced that gaseous hydrogen storage resides at last in terms of volumetric hydrogen storage density. Nonetheless, the highest storage densities are recorded by methanol and ammonia, accompanied by MgH_2, AIH_3, and LH_2 (Moran et al., 2024). PDBT, two $LiNH_2$–MgH_2, and $NaAIH_4$ have close values of volumetric and gravimetric densities. The lowest volumetric densities for H_2 are compressed gas. For NaBH4, 20% solution in 10% NaOH is assumed. Table 6.6 summarizes the main findings and provides a holistic discussion on the optimal short-, medium-, and long-term hydrogen storage technology. Compressed gas and LH_2 storage are the favorable options for short-term storage (up to 2 weeks). LH_2 and metal hydrides are most suitable for medium-term storage. Ammonia and underground hydrogen storage display good qualities for long-term hydrogen storage.

TABLE 6.6 Advantages, drawbacks, and costs of the various hydrogen storage methods.

Methods	Advantages	Drawbacks	Discussion	LCOHS	References
Compressed gas	Storing hydrogen as a gas at high pressures (100, 200, and up to 700 bar) is a proven, mature method for H_2 storage.	Although higher pressures allow evidently more hydrogen into the tank, this approach requires stronger, more expensive tanks and very serious safety measures.	Compressed gas storage is customizable depending on the required application (Fuel Cell Electric Vehicle (FCEV), Jet fuel tank, etc.). Nonetheless, damage can occur to the tank if hydrogen filling encounters the slightest issues. In addition, the energy consumed by the compressor is significant. This technology is favorable for short-term storage.	~0.65 $/kg$H_2$	Rivard et al. (2019)
Liquid hydrogen	This method allows for high-density storage (up to 80 kgH_2). A mature technology with a compact regulatory framework and established safety measures.	Storing hydrogen as a liquid requires extremely low temperatures (−253°C). This method allows for high-density storage (up to 80 kgH_2). Nonetheless, LH_2 requires high energy costs for maintaining the low temperatures.	Liquid hydrogen storage is suitable for applications requiring high energy density and long-term storage. However, the energy-intensive cooling/compressing process and potential boil-off losses are significant drawbacks.	~1.68 $/kg$H_2$	Yu et al. (2025)

(*continued on next page*)

TABLE 6.6 Advantages, drawbacks, and costs of the various hydrogen storage methods—cont'd

Methods	Advantages	Drawbacks	Discussion	LCOHS	References
Metal hydrides	Chemical alloys such as $NaBH_4$, $NaAlH_4$, and MgH_2 can store large amounts of hydrogen by wt%. These methods are undergoing rigorous research and testing, and new findings are discovered every day.	However, these methods for H_2 storage often require high temperatures or catalysts to release the hydrogen, which can be a drawback, especially in terms of price. In addition, weight is an added problematic factor for some applications, such as FCEVs or aviation.	Metal hydrides such as LH_2 offer high volumetric hydrogen density and are suitable for stationary applications. However, the need for high temperatures or catalysts to release hydrogen and the added weight make them unsuitable for mobile applications. They are most likely a medium- to long-term solution.	~2.5 $/kgH_2	Irham et al. (2024)

(*continued on next page*)

Underground	Underground hydrogen storage technology allows for the storage of large amounts of hydrogen. It is known to have one of the lowest levelized costs of hydrogen storage. This technology benefits from naturally occurring vessels, such as salt caverns and depleted gas reservoirs, which hold hydrogen with minimal leakage and pose no health or safety risks.	Geochemical reactions can occur, which affect storage operations. Additionally, the repeated injection and withdrawal of hydrogen can cause stress on the subsurface, potentially leading to geomechanical issues such as induced seismicity. Moreover, limited experience with large-scale storage further complicates the development and implementation of this technology.	This technology is favorable for medium and long-term hydrogen storage. Storage can go from 15 to 120 days with minimal leakage. Although this technology is rather immature, it is dedicated to large-scale storage, given the specific nature of the vessels.	~1.2 $/kgH_2$	Le Duigou et al. (2017), Mishra et al. (2024)

(*continued on next page*)

TABLE 6.6 Advantages, drawbacks, and costs of the various hydrogen storage methods—cont'd

Methods	Advantages	Drawbacks	Discussion	LCOHS	References
Ammonia	Ammonia offers the highest weight percentage for hydrogen storage (17 wt%). This energy carrier provides hydrogen in a stable form, easier to handle. The ammonia market is very active, and recent research shows several advancements in transitioning from natural gas-fueled ammonia production plants to renewable-powered plants.	98% of ammonia production plants rely on fossil fuels, which makes them emit harmful substances into the atmosphere. Prices for renewable-based ammonia production plants are substantially more expensive than fossil fuel-based ones. Furthermore, extracting hydrogen from ammonia requires thermolysis, a chemical process that consumes a significant amount of heat (energy).	Ammonia is a promising hydrogen carrier due to its high hydrogen density and well-developed synthesis and distribution technology. However, the energy required for hydrogen release and toxicity issues need further assessment. Anhydrous ammonia is easier to carry and store, which makes it an excellent hydrogen carrier, and it is a favorable medium-to-long-term hydrogen storage solution.	~1.5 $/kgH_2$[a]	Armijo and Philibert (2020), Elishav et al. (2020)

[a] *The estimated LCOHS from ammonia is considered gray ammonia with a production price of 250 $/t.LCOHS, levelized cost of hydrogen storage; LH2, liquid hydrogen.*

6.9 Conclusion

This chapter enlightens the important topic of hydrogen storage by carefully comparing the various technologies and their respective recent developments over the last few years. The various technologies were reviewed, and some of them were simulated for a better understanding of the mechanisms behind the storage and release processes. Each technology has specific characteristics that make it suitable for its respective applications. Furthermore, they are actively undergoing further research and development.

Underground storage and some other types are not yet mature, ready-to-use technologies for storage, despite presenting some interesting properties (high densities) and an attractive LCOHS. Moreover, ammonia and methanol, despite having substantially high energy demands, could possibly surpass the other means of storage, such as compressed gas or liquefication.

After reviewing the physical, thermodynamic, economic, and ecological aspects of various hydrogen storage technologies, it turned out that compressed gas and LH_2 storage are the favorable options for short-term storage (up to 2 weeks). LH_2 and metal hydrides are most suitable for medium-term storage (from 2 weeks to 3 months). Ammonia and underground hydrogen storage display good qualities for long-term hydrogen storage (more than a season). Nonetheless, the key argument beyond all of these considerations is the end goal for hydrogen usage. The type of application dictates the storage type.

Nomenclature

C_a/C_d	Preexponential factors for absorption and desorption
C_p	Specific heat capacity [J/(kg·K)]
E	Gas energy (J/k)
E	Rate of change of energy (J/s)
ΔH	Enthalpy change (J)
ΔS	Entropy change (J/K)
h	Enthalpy (J)
I	Unit tensor
k_{geff}	Effective thermal conductivity [W/(m ● K)]
$\dot{m}$	Mass flow rate (kg/s)
p	Static pressure (Pa)
Pr_t	Turbulent Prandtl number
Q	Heat-transfer rate (W)
r	Discount rate (%)
R_g	Universal gas constant [J/(mol·K)]
T	Temperature (K)
t	Time (s)
$\overline{\overline{\tau}}$	Stress tensor (Pa)
$\overline{\overline{\tau}}_{eff}$	Viscous heating
$\vec{v}$	Favre's average velocity (m/s)
W	Work done by the system (J)

ε	Medium porosity
μ	Dynamic viscosity (Pa·s)
ρ	Density (kg/m^3)
ρg	Gravitational body force (N)

References

Abe, A. (1998). Studies of the large-scale sea transportation of liquid hydrogen. *International Journal of Hydrogen Energy, 23*(2), 115–121. doi:10.1016/s0360-3199(97)00032-3.

Achour, Y., Berrada, A., Arechkik, A., & El Mrabet, R. (2023). Techno-economic assessment of hydrogen production from three different solar photovoltaic technologies. *International Journal of Hydrogen Energy, 48*(83), 32261–32276. http://www.journals.elsevier.com/international-journal-of-hydrogen-energy/. doi:10.1016/j.ijhydene.2023.05.017.

Achour, Y., El Mokrini, A., & Berrada, A. (2025a). Techno-economic assessment of large-scale power-to-ammonia plant: Integrating hybrid energy storage system (gravity and battery) and advanced day-ahead forecasting algorithms. *International Journal of Hydrogen Energy, 136*, 202–216. doi:10.1016/j.ijhydene.2025.04.102.

Achour, Y., El Mokrini, A., El Mrabet, R., & Berrada, A. (2025b). Optimizing power-to-ammonia plant: Sizing, operation, and production forecasting using deep learning approach. *Renewable Energy, 240*, 122234. doi:10.1016/j.renene.2024.122234.

Achour, Y., Emrani, A., El Moukrini, A., & Berrada, A. (2024). *Electric vehicles as an active energy storage system battery-integrated residential energy systems* (pp. 210–228). CRC Press. https://www.taylorfrancis.com/books/9781040148129. doi:10.1201/9781003441236-11.

Ali, M., Yekeen, N., Pal, N., Keshavarz, A., Iglauer, S., & Hoteit, H. (2022). Influence of organic molecules on wetting characteristics of mica/H2/brine systems: Implications for hydrogen structural trapping capacities. *Journal of Colloid and Interface Science, 608*, 1739–1749. http://www.elsevier.com/inca/publications/store/6/2/2/8/6/1/index.htt. doi:10.1016/j.jcis.2021.10.080.

Andersson, J., & Grönkvist, S. (2019). Large-scale storage of hydrogen. *International Journal of Hydrogen Energy, 44*(23), 11901–11919. http://www.journals.elsevier.com/international-journal-of-hydrogen-energy/. doi:10.1016/j.ijhydene.2019.03.063.

Armijo, J., & Philibert, C. (2020). Flexible production of green hydrogen and ammonia from variable solar and wind energy: Case study of Chile and Argentina. *International Journal of Hydrogen Energy, 45*(3), 1541–1558. http://www.journals.elsevier.com/international-journal-of-hydrogen-energy/. doi:10.1016/j.ijhydene.2019.11.028.

Ball, M., & Weeda, M. (2015). The hydrogen economy - vision or reality? *International Journal of Hydrogen Energy, 40*(25), 7903–7919. http://www.journals.elsevier.com/international-journal-of-hydrogen-energy/. doi:10.1016/j.ijhydene.2015.04.032.

Barthelemy, H., Weber, M., & Barbier, F. (2017). Hydrogen storage: Recent improvements and industrial perspectives. *International Journal of Hydrogen Energy, 42*(11), 7254–7262. http://www.journals.elsevier.com/international-journal-of-hydrogen-energy/. doi:10.1016/j.ijhydene.2016.03.178.

Berrada, A., Sanjari, M. J., & El Mrabet, R. (2025). Techno-economic assessment of hydrogen production: Comparative analysis of electrolyser technologies in a hybrid PV/wind system. *International Journal of Hydrogen Energy, 141*, 193–211. doi:10.1016/j.ijhydene.2025.05.384.

Broom, D. P. (2011). Hydrogen storage materials. *International Journal of Hydrogen Energy, 204*, 153160.

Cardozo Soares Amaral, P., Oh, C.B., Do, K.H., & Choi, B.I. (2024). Risk assessment of hydrogen leakage and explosion in a liquid hydrogen facility using computational analysis. International Journal of Hydrogen Energy (91, pp. 950-964). Elsevier. https://doi.org/10.1016/j.ijhydene.2024.10.129.

Climate Change and Renewable Energy. National policies and the role of communities, cities and regions. Report to the G20 Climate Sustainability Working Group (CSWG). (2019), 1–60, www.irena.org.

De colvenaer, B., & Castel, C. (2012). The fuel cells and hydrogen joint undertaking (FCH JU) in Europe. *International Journal of Low-Carbon Technologies, 7*(1), 5–9. doi:10.1093/ijlct/ctr035.

Dwivedi, S. K., & Vishwakarma, M. (2018). Hydrogen embrittlement in different materials: A review. *International Journal of Hydrogen Energy, 43*(46), 21603–21616. http://www.journals.elsevier.com/international-journal-of-hydrogen-energy/. doi:10.1016/j.ijhydene.2018.09.201.

Elishav, O., Mosevitzky Lis, B., Valera-Medina, A., & Grader, G. S. (2020). *Storage and distribution of ammonia techno-economic challenges of green ammonia as an energy vector* (pp. 85–103). Elsevier. https://www.elsevier.com/books/techno-economic-challenges-of-green-ammonia-as-an-energy-vector/valera-medina/978-0-12-820560-0. doi:10.1016/B978-0-12-820560-0.00005-9.

Giddey, S., Badwal, S. P. S., Munnings, C., & Dolan, M. (2017). Ammonia as a renewable energy transportation media. *ACS Sustainable Chemistry and Engineering, 5*(11), 10231–10239. http://pubs.acs.org/journal/ascecg. doi:10.1021/acssuschemeng.7b02219.

Irham, A., Hannan, M. A., Rahman, S. A., Ker, P. J., Wong, R. T., Roslan, M. F., Begum, R. A., & Jang, G. (2024). Cost-effectiveness and reliability evaluation of hydrogen storage-based hybrid energy systems for unreliable grid. *International Journal of Hydrogen Energy, 94*, 1314–1328. https://www.sciencedirect.com/science/journal/03603199. doi:10.1016/j.ijhydene.2024.11.119.

Johnson, T., Bozinoski, R., Ye, J., Sartor, G., Zheng, J., & Yang, J. (2015). Thermal model development and validation for rapid filling of high pressure hydrogen tanks. *International Journal of Hydrogen Energy, 40*(31), 9803–9814. http://www.journals.elsevier.com/international-journal-of-hydrogen-energy/. doi:10.1016/j.ijhydene.2015.05.157.

Jordá-Beneyto, M., Suárez-García, F., Lozano-Castelló, D., Cazorla-Amorós, D., & Linares-Solano, A. (2007). Hydrogen storage on chemically activated carbons and carbon nanomaterials at high pressures. *Carbon, 45*(2), 293–303. doi:10.1016/j.carbon.2006.09.022.

Kapischke, J., & Hapke, J. (1998). Measurement of the pressure-composition isotherms of high-temperature and low-temperature metal hydrides. *Experimental Thermal and Fluid Science, 18*(1), 70–81. doi:10.1016/S0894-1777(98)10007-9.

Klopčič, N., Grimmer, I., Winkler, F., Sartory, M., & Trattner, A. (2023). A review on metal hydride materials for hydrogen storage. *Journal of Energy Storage, 72*, 108456. doi:10.1016/j.est.2023.108456.

Kyriakou, V., Garagounis, I., Vourros, A., Vasileiou, E., & Stoukides, M. (2020). An electrochemical Haber-Bosch process. *Joule, 4*(1), 142–158. https://www.journals.elsevier.com/joule. doi:10.1016/j.joule.2019.10.006.

Le Duigou, A., Bader, A. G., Lanoix, J. C., & Nadau, L. (2017). Relevance and costs of large scale underground hydrogen storage in France. *International Journal of Hydrogen Energy, 42*(36), 22987–23003. http://www.journals.elsevier.com/international-journal-of-hydrogen-energy/. doi:10.1016/j.ijhydene.2017.06.239.

Lemieux, A., Shkarupin, A., & Sharp, K. (2020). Geologic feasibility of underground hydrogen storage in Canada. *International Journal of Hydrogen Energy, 45*(56), 32243–32259. http://www.journals.elsevier.com/international-journal-of-hydrogen-energy/. doi:10.1016/j.ijhydene.2020.08.244.

Levikhin, A. A., & Boryaev, A. A. (2024). Physical properties and thermodynamic characteristics of hydrogen. *Heliyon, 10*(17), e36414. doi:10.1016/j.heliyon.2024.e36414.

Lewandowska-Śmierzchalska, J., Tarkowski, R., & Uliasz-Misiak, B. (2018). Screening and ranking framework for underground hydrogen storage site selection in Poland. *International Journal of Hydrogen Energy, 43*(9), 4401–4414. http://www.journals.elsevier.com/international-journal-of-hydrogen-energy/. doi:10.1016/j.ijhydene.2018.01.089.

Li, J., Cheng, S., Zhao, Q., Long, P., & Dong, J. (2009). Synthesis and hydrogen-storage behavior of metal-organic framework MOF-5. *International Journal of Hydrogen Energy, 34*(3), 1377–1382. doi:10.1016/j.ijhydene.2008.11.048.

Li, M., Yang, Q., Zhang, C., Huang, S., Zhang, M., Zhang, G., Zhao, L., & Jiang, S. (2023). Experimental and numerical study of the temperature evolution in hydrogen cylinder under fast-refueling process. *International Journal of Heat & Mass Transfer, 211*. doi:10.1016/j.ijheatmasstransfer.2023.124220.

Liu, Y. L., Zhao, Y. Z., Zhao, L., Li, X., Chen, H. G., Zhang, L. F., Zhao, H., Sheng, R. H., Xie, T., Hu, D. H., & Zheng, J. Y. (2010). Experimental studies on temperature rise within a hydrogen cylinder during refueling. *International Journal of Hydrogen Energy, 35*(7), 2627–2632. doi:10.1016/j.ijhydene.2009.04.042.

Mishra, S. K., Freeman, G. M., Ganguli, S., & Huerta, N. J. (2024). Which factors dominate the levelized costs of subsurface hydrogen storage in Pennsylvania, United States. *International Journal of Hydrogen Energy, 91*, 814–821. https://www.sciencedirect.com/science/journal/03603199. doi:10.1016/j.ijhydene.2024.10.189.

Moran, C., Deane, P., Yousefian, S., & Monaghan, R. F. D. (2024). The hydrogen storage challenge: Does storage method and size affect the cost and operational flexibility of hydrogen supply chains? *International Journal of Hydrogen Energy, 52*, 1090–1100. https://www.sciencedirect.com/science/journal/03603199. doi:10.1016/j.ijhydene.2023.06.269.

Mrabet, R. E., & Berrada, A. (2021). *Hydrogen production and derivatives from renewable energy systems for a best valorization of sustainable resources* (pp. 343–363). Elsevier BV. doi:10.1016/b978-0-12-821403-9.00010-x.

Park, H., Kim, J., Bergan, P. G., & Chang, D. (2022). Structural design of flexible vacuum insulation system for large-scale LH2 storage. *International Journal of Hydrogen Energy, 47*(92), 39179–39192. http://www.journals.elsevier.com/international-journal-of-hydrogen-energy/. doi:10.1016/j.ijhydene.2022.09.063.

Pfromm, P. H. (2017). Towards sustainable agriculture: Fossil-free ammonia. *Journal of Renewable and Sustainable Energy, 9*(3). http://scitation.aip.org/content/aip/journal/jrse. doi:10.1063/1.4985090.

Pierce, W. G., Rich, E. I., (1962). Summary of rock salt deposits in the United States as possible storage sites for radioactive waste materials. U.S. Geological Survey. doi: 10.3133/b1148.

Pourali, M., Esfahani, J. A., Jahangir, H., Farzaneh, A., & Kim, K. C. (2022). Ammonia decomposition in a porous catalytic reactor to enable hydrogen storage: Numerical simulation, machine learning, and response surface methodology. *Journal of Energy Storage, 55*. http://www.journals.elsevier.com/journal-of-energy-storage/. doi:10.1016/j.est.2022.105804.

Principi, G., Agresti, F., Maddalena, A., & Russo, S. L. (2009). The problem of solid state hydrogen storage. *Energy, 34*(12), 2087–2091. doi:10.1016/j.energy.2008.08.027.

Ren, J., Musyoka, N. M., Langmi, H. W., Mathe, M., & Liao, S. (2017). Current research trends and perspectives on materials-based hydrogen storage solutions: A critical review. *International Journal of Hydrogen Energy, 42*(1), 289–311. http://www.journals.elsevier.com/international-journal-of-hydrogen-energy/. doi:10.1016/j.ijhydene.2016.11.195.

Rivard, E., Trudeau, M., & Zaghib, K. (2019). Hydrogen storage for mobility: A review. *Materials, 12*(12). https://res.mdpi.com/materials/materials-12-01884/article_deploy/materials-12-01973.pdf?filename=&attachment=1. doi:10.3390/ma12121973.

Rkhis, M., Laasri, S., Touhtouh, S., Belhora, F., Hlil, E. K., Zaidat, K., Obbade, S., & Hajjaji, A. (2023). Recent advances in magnesium hydride for solid-state hydrogen storage by mechanical treatment: A DFT study. *International Journal of Hydrogen Energy, 48*(91), 35650–35660. http://www.journals.elsevier.com/international-journal-of-hydrogen-energy/. doi:10.1016/j.ijhydene.2023.05.267.

Soloveichik, G. (2019). Electrochemical synthesis of ammonia as a potential alternative to the Haber–Bosch process. *Nature Catalysis, 2*(5), 377–380. https://www.nature.com/natcatal/. doi:10.1038/s41929-019-0280-0.

Statharas, J. C., Venetsanos, A. G., Bartzis, J. G., Würtz, J., & Schmidtchen, U. (2000). Analysis of data from spilling experiments performed with liquid hydrogen. *Journal of Hazardous Materials, 77*(1-3), 57–75. doi:10.1016/S0304-3894(00)00252-1.

Suryan, A., Kim, H. D., & Setoguchi, T. (2013). Comparative study of turbulence models performance for refueling of compressed hydrogen tanks. *International Journal of Hydrogen Energy, 38*(22), 9562–9569. doi:10.1016/j.ijhydene.2012.07.055.

US Department of Energy, DOE technical targets for onboard hydrogen storage for light-duty vehicles. US Department of Energy. https://www.energy.gov/eere/fuelcells/doe-technical-targets-onboard-hydrogen-storage-light-duty-vehicles.

Valenti, G. (2016). *Hydrogen liquefaction and liquid hydrogen storage* (pp. 27–51). Elsevier BV. doi:10.1016/b978-1-78242-362-1.00002-x.

Verga, M., Armanasco, F., Guardamagna, C., Valli, C., Bianchin, A., Agresti, F., Russo, S. L., Maddalena, A., & Principi, G. (2009). Scaling up effects of Mg hydride in a temperature and pressure-controlled hydrogen storage device. *International Journal of Hydrogen Energy, 34*(10), 4602–4610. doi:10.1016/j.ijhydene.2008.08.043.

Wijayanta, A. T., Oda, T., Purnomo, C. W., Kashiwagi, T., & Aziz, M. (2019). Liquid hydrogen, methylcyclohexane, and ammonia as potential hydrogen storage: Comparison review. *International Journal of Hydrogen Energy, 44*(29), 15026–15044. http://www.journals.elsevier.com/international-journal-of-hydrogen-energy/. doi:10.1016/j.ijhydene.2019.04.112.

Yartys, V. A., Lototskyy, M. V., Akiba, E., Albert, R., Antonov, V. E., Ares, J. R., Baricco, M., Bourgeois, N., Buckley, C. E., Bellosta von Colbe, J. M., Crivello, J. C., Cuevas, F., Denys, R. V., Dornheim, M., Felderhoff, M., Grant, D. M., Hauback, B. C., Humphries, T. D., Jacob, I., … Zhu, M. (2019). Magnesium based materials for hydrogen based energy storage: Past, present and future. *International Journal of Hydrogen Energy, 44*(15), 7809–7859. http://www.journals.elsevier.com/international-journal-of-hydrogen-energy/. doi:10.1016/j.ijhydene.2018.12.212.

Ye, J., Cui, J., Hua, Z., Xie, J., Peng, W., & Wang, W. (2023). Study on the high-pressure hydrogen gas flow characteristics of the needle valve with different spool shapes. *International Journal of Hydrogen Energy, 48*(30), 11370–11381. http://www.journals.elsevier.com/international-journal-of-hydrogen-energy/. doi:10.1016/j.ijhydene.2022.04.073.

Yu, C., Choi, M., & Chang, D. (2025). Economic analysis of cargo containment systems and boil-off hydrogen handling options for large-scale liquid hydrogen carriers considering insulation-layer thickness of liquid-hydrogen storage tank. *International Journal of Hydrogen Energy, 97*,

1055–1067. https://www.sciencedirect.com/science/journal/03603199. doi:10.1016/j.ijhydene.2024.11.334.

Zhang, Z., Lei, G., Sun, R., Pu, L., Wang, T., Dong, W., Chen, Q., Wei, Q., Liu, M., He, Y., Zheng, Z., & Zhang, S. (2024). An experimental study on the large-volume liquid hydrogen release in an open space. *Applied Sciences, 14*(9), 3645. doi:10.3390/app14093645.

Zheng, J., Guo, J., Yang, J., Zhao, Y., Zhao, L., Pan, X., Ma, J., & Zhang, L. (2013). Experimental and numerical study on temperature rise within a 70 MPa type III cylinder during fast refueling. *International Journal of Hydrogen Energy, 25*, 10956–10962. http://www.journals.elsevier.com/international-journal-of-hydrogen-energy/38. doi:10.1016/j.ijhydene.2013.02.053.

Zheng, J., Liu, X., Xu, P., Liu, P., Zhao, Y., & Yang, J. (2012). Development of high pressure gaseous hydrogen storage technologies. *International Journal of Hydrogen Energy, 37*(1), 1048–1057. doi:10.1016/j.ijhydene.2011.02.125.

Zivar, D., Kumar, S., & Foroozesh, J. (2021). Underground hydrogen storage: A comprehensive review. *International Journal of Hydrogen Energy, 46*(45), 23436–23462. http://www.journals.elsevier.com/international-journal-of-hydrogen-energy/. doi:10.1016/j.ijhydene.2020.08.138.

Zlotea, C., & Latroche, M. (2013). Role of nanoconfinement on hydrogen sorption properties of metal nanoparticles hybrids. *Colloids and Surfaces A: Physicochemical and Engineering Aspects, 439*, 117–130. www.elsevier.com/locate/colsurfa. doi:10.1016/j.colsurfa.2012.11.043.

Chapter 7

Life cycle assessment of green hydrogen in the industrial sector

Khadija Jihar[a], Asmae Berrada[a], Ilham Kirm[b], Abdellah Benzaouak[c] and Khalid Bouziane[a]

[a]*LERMA Lab, College of Engineering and Architecture, International University of Rabat, Sala Al Jadida, Morocco,* [b]*Laboratory of Chemistry-Biology Applied to the Environment, URL-CNRST-13, Applied Materials & Catalysis, Department of Chemistry, Faculty of Science, Moulay Ismail University (UMI), Meknes, Morocco,* [c]*ENSAM, Laboratory of Spectroscopy, Molecular Modeling, Nanomaterials, Water and Environment, Environmental Materials Team, Mohammed V University in Rabat, Rabat, Morocco*

7.1 Introduction

In response to the escalating urgency of climate change and the commitments outlined in the Paris Agreement, green hydrogen (H_2) produced from renewable energy sources with no direct greenhouse gas (GHG) emissions has emerged as a pivotal pillar of the global energy transition. It holds strong potential for decarbonizing "hard-to-abate" sectors such as heavy industry (steel, cement, chemicals), long-distance transportation, and seasonal energy storage (IEA, 2022). Over 30 countries have already released national hydrogen strategies, and investment in hydrogen-related projects is expected to exceed USD 300 billion by 2030 (IRENA, 2020). However, the simple switch to renewable inputs does not automatically ensure comprehensive environmental sustainability. A robust methodological framework is needed to assess environmental trade-offs across the entire value chain. This is where life cycle assessment (LCA) becomes indispensable. LCA is a comprehensive approach used to quantify the environmental impacts across all life cycle stages from raw material extraction to end-use application of hydrogen production technologies. It allows for a comparative, multicriteria, and system-based evaluation of production pathways, identifying hidden burdens and trade-offs (Maniscalco et al., 2024). Several important LCA reviews have emerged in recent years. For instance (Mahfoud et al. (2023)) conducted a meta-analysis of emissions across various hydrogen production routes, Dincer and Acar (2015) highlighted the energy and exergy performance of conventional and emerging technologies, and

Decarbonizing through Sustainable Hydrogen and its Derivatives.
DOI: https://doi.org/10.1016/B978-0-443-27632-3.00008-6

Pitchaimuthu et al. (2022) specifically evaluated solar-powered electrolysis and photoelectrochemical pathways in different electricity grid mixes. Among the key hydrogen production methods assessed through LCA are: water electrolysis powered by renewable energy (photovoltaic [PV], wind, hydro, geothermal), biomass gasification, thermochemical water splitting (e.g., Cu–Cl and S–I cycles), and photocatalysis. Each of these routes presents unique environmental and technical characteristics. For example, solar PV-electrolysis systems can approach near-zero GHG emissions but face challenges related to material use, land footprint, and system efficiency (Cetinkaya et al., 2012). Biomass gasification, when integrated with carbon capture and storage (CCS), may even achieve net-negative emissions under certain configurations (Ganesh et al., 2025). Meanwhile, photocatalytic and thermochemical water-splitting cycles remain at precommercial stages but show long-term potential for low-impact hydrogen, especially when powered by high-temperature solar or nuclear energy (Ozbilen et al., 2013; Sharafi Laleh et al., 2023). Despite the methodological richness of published LCA studies, several inconsistencies and limitations persist. Variations in functional units (e.g., per kgH_2, per MJ), system boundaries (cradle-to-gate vs cradle-to-grave), impact categories, and LCA tools (e.g., SimaPro, GaBi, OpenLCA) contribute to diverging conclusions (Chen et al., 2023). Furthermore, few studies integrate regional conditions such as solar irradiance, water stress, or grid carbon intensity factors that critically influence real-world sustainability. This omission is especially significant in emerging hydrogen regions like North Africa, where contextual variables can significantly affect LCA outcomes. The present review aims to deliver a comparative, structured synthesis of LCA studies for green hydrogen production technologies. Based on several LCA studies on green H_2 production. We systematically evaluate:

- Major production methods (electrolysis, biomass gasification, thermochemical cycles, photocatalysis),
- Different renewable energy sources powering hydrogen systems (solar PV, PV-T, CPV-T, wind, hydro, geothermal),
- Cross-cutting LCA criteria such as GHG emissions, water use, land footprint, and resource depletion.

We also present a regional case study focused on Morocco, which has emerged as a strategic hub for green hydrogen development. With high solar and wind availability, along with political backing and international partnerships, Morocco holds strong potential for competitive green H_2 production. However, the lack of localized LCA studies remains a barrier to decision-making and investment planning (ONEE), (IRESEN). In this context, we aim to assess how internationally validated LCA results can be adapted to Moroccan energy systems and resource availability. Finally, this chapter highlights the pressing need to strengthen LCA methodologies for green hydrogen, particularly through international harmonization, integration with economic and circularity indicators, and improved regional adaptation. The case of Morocco, despite its strategic potential in renewable energy, remains largely underrepresented in the literature. Developing context-specific LCA frameworks could serve as a transformative tool to guide the sustainable deployment of hydrogen technologies in the region.

7.2 Literature search methodology

This chapter analyzes several LCA studies published between 2015 and 2025, focusing on the LCA of various green hydrogen production methods. A targeted bibliographic search was carried out primarily using Scopus and Web of Science, two of the most authoritative scientific databases, covering the period up to August 2025. The search strategy relied on keyword combinations such as "life cycle assessment," "green hydrogen," "electrolysis," "photocatalysis," "thermochemical water splitting," "biomass hydrogen," and "renewable hydrogen production." Boolean operators and truncations were used to optimize the breadth and relevance of the results. The review methodology followed a structured and critical approach, combining comparative tables and thematic analysis to identify key trends, methodological divergences, and existing research gaps across the literature.

7.3 Life cycle assessment: Key steps analysis

LCA is a standardized and science-based methodology used to evaluate the environmental impacts of a product, process, or service across all stages of its life cycle. This includes everything from raw material extraction and manufacturing to distribution, use, and final disposal or recycling, an approach commonly known as "cradle-to-grave" or, in circular systems, "cradle-to-cradle" (Scholz and Tietje, 2002). The framework for conducting LCA is defined by the ISO 14040 and 14044 standards, which outline four main phases: (1) Goal and Scope Definition, (2) Life Cycle Inventory (LCI), (3) Life Cycle Impact Assessment (LCIA), and (4) Interpretation (Seghetta & Goglio, 2018). The first phase establishes the objective of the study, defines the functional unit as the reference against which all inputs and outputs are normalized (e.g., per kg of hydrogen), and sets the system boundaries to ensure methodological consistency (Curran, 2017). In the LCI phase, quantitative data are gathered on energy consumption, material use, emissions, and waste flows throughout the entire life cycle of the system being assessed (Siregar et al., 2021). During the LCIA phase, inventory data are translated into potential environmental impacts such as climate change, eutrophication, acidification, and resource depletion. This step includes mandatory elements like classification and characterization and may optionally include normalization, grouping, or weighting, depending on the goal of the study (Klöpffer & Grahl, 2014; Sonnemann et al., 2018). Finally, in the Interpretation phase, the results are analyzed to identify hotspots, assess uncertainties, and provide recommendations for improving sustainability and supporting informed decision-making (Baker & Lepech, 2009).Today, LCA is a widely used decision-support tool across various sectors, including energy, agriculture, transportation, and waste management. It plays a key role in guiding product development, minimizing environmental footprints, and shaping sustainability strategies (Finnveden et al., 2009; Rieradevall et al., 1997).

7.4 Overview of production technologies: Hydrogen production from renewable energy

7.4.1 Electrolysis method

Electrolysis is a clean and versatile method for producing green hydrogen when powered by renewable electricity sources such as wind, solar, or hydropower. The process involves splitting water into hydrogen and oxygen by applying an electric current across two electrodes (anode and cathode) in an electrolyzer cell, with the global reaction represented as: $2H_2O \rightarrow 2H_2 + O_2$ (Panigrahy et al., 2022). Depending on the electrolyte and operating conditions, different electrolysis technologies are employed, including alkaline water electrolysis (AWE), proton exchange membrane (PEM), anion exchange membrane (AEM), and solid oxide electrolysis cells (SOEC), each with unique advantages and limitations (Sebbahi et al., 2024). AWE and PEM, classified as low-temperature electrolyzers, are the most commercially mature and widely implemented, whereas SOEC operates at high temperatures and offers higher efficiency by using both heat and electricity (Bühler & Möst, 2025). AEM electrolysis, still under development, combines the cost-effectiveness of alkaline systems with the compactness of PEM designs (Mulk et al., 2024). During water electrolysis, the half-reactions at the electrodes differ depending on the electrolyzer type, but the general scheme remains, as shown in Fig. 7.1.

- **Anode (oxidation):** $2H_2O \rightarrow O_2 + 4H^+ + 4e^-$ (acidic PEM) or $4OH^- \rightarrow O_2 + 2H_2O + 4e^-$ (alkaline)
- **Cathode (reduction):** $4H^+ + 4e^- \rightarrow 2H_2$ (acidic PEM) or $4H_2O + 4e^- \rightarrow 2H_2 + 4OH^-$ (alkaline)
- **Global reaction:** $2H_2O \rightarrow 2H_2 + O_2$

These reactions ensure that the only inputs are water and electricity, and the outputs are hydrogen and oxygen, making the process carbon-free (Choi et al., 2021). Despite challenges such as high energy requirements and elevated costs compared to fossil-based hydrogen production (Jang et al., 2022), electrolysis is increasingly seen as a cornerstone of the hydrogen economy thanks to falling renewable electricity costs and rapid advances in electrolyzer technology (Wang et al., 2025).

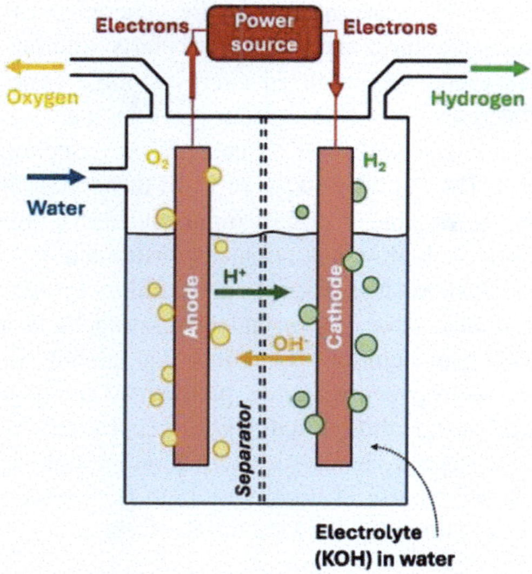

FIGURE 7.1 Schematic representation of an electrolysis cell (Sola et al., 2025).

7.4.2 Alkaline electrolysis

AWE is the most established and historically proven method for green hydrogen production. It operates using a liquid alkaline electrolyte, typically potassium hydroxide (KOH) or sodium hydroxide (NaOH), which facilitates the transport of hydroxide ions (OH^-) between the electrodes. While the overall reaction remains the electrochemical splitting of water ($2H_2O \rightarrow 2H_2 + O_2$), the half-cell reactions are specific to the alkaline environment at:

The anode, $4OH^- \rightarrow O_2 + 2H_2O + 4e^-$; and at the cathode, $4H_2O + 4e^- \rightarrow 2H_2 + 4OH^-$.

AWE systems are known for their robustness, simplicity, and relatively low capital costs, making them well-suited for large-scale, steady-state industrial applications. However, their slower response to power fluctuations makes them less ideal for coupling with highly variable renewable energy sources. This limitation has positioned AWE primarily in sectors with continuous energy demand, such as chemical and metallurgical industries (Sebbahi et al., 2024). Recent studies confirm that AWE remains a competitive option for large-scale hydrogen production, especially where cost-efficiency and system longevity are prioritized over dynamic operation flexibility (Gallandat et al., 2017).

7.4.3 Proton exchange membrane electrolysis

PEM electrolysis is a low-temperature hydrogen production technology that relies on a solid polymer membrane to conduct protons (H^+) from the anode to the cathode. This design enables the generation of high-purity hydrogen directly under pressure, making it particularly suitable for applications in hydrogen mobility and decentralized energy storage systems. The typical electrochemical reactions in a PEM electrolyzer include water oxidation at the anode ($2H_2O \rightarrow O_2 + 4H^+ + 4e^-$) and proton reduction at the cathode ($4H^+ + 4e^- \rightarrow 2H_2$). The key advantages of PEM technology lie in its compact design, fast response to variable electricity input, and high conversion efficiency. However, its widespread deployment remains constrained by the use of expensive noble-metal catalysts such as platinum and iridium. From an environmental perspective, LCA studies have shown that integrating PEM electrolysis with onshore wind power can significantly reduce the global warming potential compared to other hydrogen production pathways (Zhang et al., 2023). In response to cost and sustainability challenges, recent research has focused on the development of alternative catalysts, such as ruthenium- and iridium-based materials, which show promise in improving both performance and durability while lowering reliance on critical raw materials (Gebauer et al., 2024).

7.4.4 Anion exchange membrane electrolysis

AEM electrolysis is an emerging technology that seeks to combine the advantages of both alkaline and PEM systems. It operates using a solid polymer membrane that conducts hydroxide ions (OH^-), allowing for the use of nonprecious metal catalysts. This significantly reduces material costs while maintaining high electrochemical performance. While the half-cell reactions are similar to those in traditional alkaline electrolysis, the system architecture more closely resembles that of PEM electrolyzers, compact and responsive to dynamic operation. Despite its potential, AEM electrolysis remains in the research and development stage. Key challenges include improving membrane durability and ensuring long-term catalyst stability under real operating conditions. Nonetheless, the technology is viewed as a promising solution for low-cost, decentralized hydrogen production, particularly in regions with intermittent renewable energy sources (Mulk et al., 2024). Recent experimental studies have demonstrated that, when coupled with renewable electricity, AEM systems can achieve performance levels comparable to commercial alkaline and PEM technologies, highlighting their potential for future scale-up and commercialization (El-Shafie, 2023).

7.4.5 Solid oxide electrolysis

SOECs operate at high temperatures, typically between 600°C and 850°C, and use steam rather than liquid water as the input. In these systems, oxygen ions (O^{2-}) are transported through a solid ceramic electrolyte, enabling the integration of both thermal and electrical energy inputs. This dual-energy approach significantly enhances system efficiency, reducing electricity consumption by 20% to 25% compared to low-temperature electrolysis methods such as alkaline or PEM electrolysis. While the overall water-splitting reaction remains the same, the thermodynamic conditions differ. At the anode, oxygen ions release electrons ($O^{2-} \rightarrow \frac{1}{2} O_2 + 2e^-$), while at the cathode, steam is reduced to produce hydrogen ($H_2O + 2e^- \rightarrow H_2 + O^{2-}$). SOECs are particularly well-suited for integration with high-grade heat sources, such as industrial waste heat, concentrated solar power, or nuclear reactors. Despite their promising efficiency, the commercial readiness of SOEC technology remains limited due to challenges related to material degradation, thermal cycling, and system durability. However, recent research in South Korea has demonstrated that advances in cell materials and design can reduce energy consumption and maintain operational stability over several thousand hours (Kim et al., 2024). Comparative assessments also suggest that, under high production throughput, SOEC systems can surpass PEM electrolyzers in energy efficiency (Ferrero et al., 2013).

7.4.6 Renewable electricity sources for electrolysis

7.4.6.1 Solar photovoltaic (PV-electrolysis)

PV-powered electrolysis is currently the most widely adopted method for green hydrogen production. In this approach, solar panels convert sunlight into electricity, which is then used to drive an electrolyzer, typically either alkaline or PEM, to split water into hydrogen and oxygen. To optimize energy conversion, these systems often incorporate direct coupling strategies or power electronics, such as DC/DC converters with maximum power point tracking (MPPT). Despite its widespread use, the system's performance can be significantly impacted by solar intermittency and variations in irradiance. Recent studies highlight that careful integration of the PV system with the electrolyzer, particularly in dynamic, real-world solar conditions, can lead to substantial efficiency gains. For instance, solar-to-hydrogen (STH) conversion efficiencies ranging from 18% to 30% have been reported in well-optimized configurations (Zhang et al., 2025). Looking ahead, techno-economic projections suggest that with continued reductions in capital costs for both PV modules and electrolyzers, the levelized cost of hydrogen could fall to as low as \$0.90–1.80 per kg by 2050, making it a highly competitive clean energy vector (Vartiainen et al., 2022).

7.4.6.2 Photovoltaic-thermal hybrid

Photovoltaic-thermal (PV-T) systems combine solar electricity generation (like standard PV) with thermal energy recovery from panel waste heat. This heat can preheat the water entering the electrolyzer, reducing voltage requirements and increasing overall system efficiency. Although more complex, this integration improves hydrogen yield. A comparative study conducted in Doha, Tunis, and Stuttgart found that PV-T increased hydrogen production by 25%–78% compared to standard PV systems (Alghool et al., 2025). However, PV-T systems

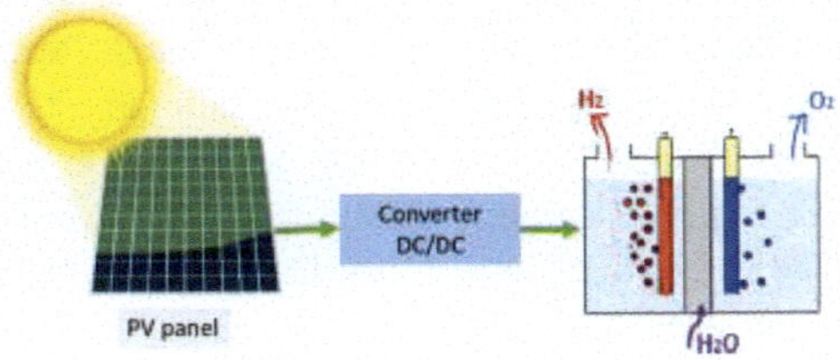

FIGURE 7.2 Green H_2 production-based PV solar (Ajeeb et al., 2024). *PV*, Photovoltaic.

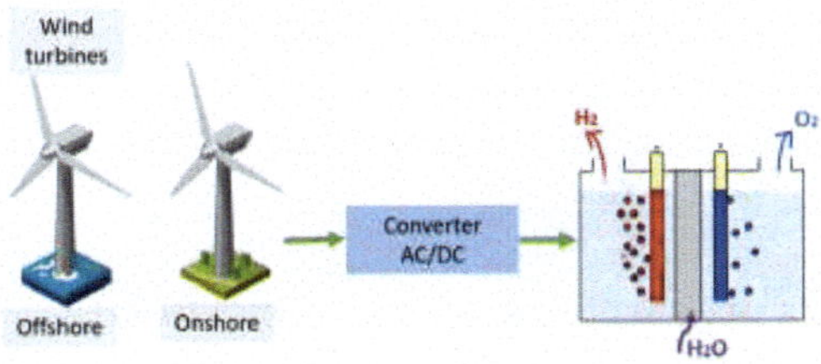

FIGURE 7.3 Green H_2 production-based wind energy (Ajeeb et al., 2024).

are more expensive, with costs ranging from \$6.66–16.80/kgH_2, compared to \$4.92–9.66/kg for PV.

7.4.6.3 Concentrated photovoltaic thermal

Concentrated PV-T (CPV-T) systems utilize optical components such as lenses or mirrors to focus sunlight onto high-efficiency PV cells, while simultaneously capturing waste heat (Fig. 7.2). This dual-energy output enables the integration of both electricity and thermal energy into hydrogen production processes, particularly high-temperature electrolysis using SOECs. The combined use of heat and power not only enhances overall system efficiency but also reduces the electrical load on the electrolyzer. Notably, a recent CPV-alkaline electrolysis system achieved a record STH efficiency of 29.1%, marking the highest performance reported to date for PV-driven electrolysis setups (Zhang et al., 2025). Techno-economic assessments (TEAs) further suggest that CPV-T systems can offer a lower hydrogen production cost per kilogram compared to conventional flat-panel PV-electrolysis configurations, making them a promising option for regions with high direct solar irradiance (Boudries, 2016).

7.4.6.4 Wind power (wind electrolysis)

Wind power is a widely available and increasingly cost-competitive renewable energy source that shows strong potential for green hydrogen production through water electrolysis (Fig. 7.3). Thanks to relatively high capacity factors in windy regions, particularly coastal and northern areas, wind energy can support steady hydrogen output when appropriately integrated with electrolyzers. Dynamic

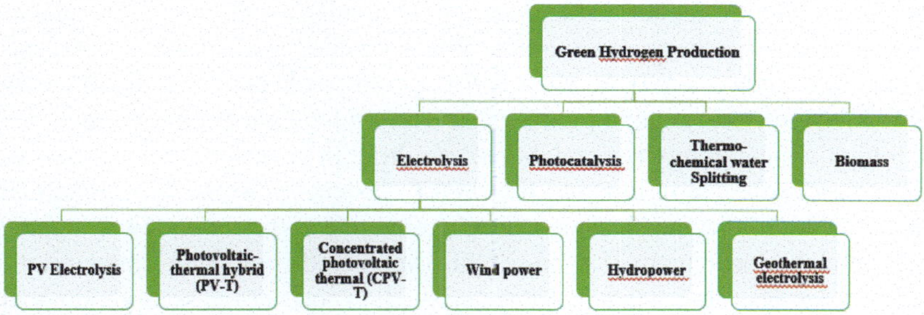

FIGURE 7.4 Different categories of green hydrogen production.

simulations in cold-climate settings such as Toronto have demonstrated that wind-powered systems significantly outperform PV-based systems under low solar irradiance, producing up to 9.03 kg of hydrogen per month compared to just 0.58 kg/month for PV, with levelized costs as low as \$0.55/kgH$_2$ (Ramadan and Gabbar, 2024). Despite its advantages, the intermittent nature of wind power introduces challenges for electrolyzer operation and system sizing. Studies have shown that integrating wind with smart energy management systems, including battery storage and load balancing, can mitigate these effects, improve electrolyzer lifespan, and optimize hydrogen yields (Elgammal & Ramlal, 2023). Furthermore, the ratio between wind power capacity and electrolyzer size plays a critical role in determining economic feasibility, with optimal configurations achieving production costs as low as €2.53/kgH$_2$ in favorable scenarios (Hofrichter et al., 2023). Overall, wind-powered electrolysis represents a technically and economically promising route for low-carbon hydrogen production, provided that system intermittency is managed through robust design, control strategies, and storage integration.

7.4.6.5 Hydropower (hydroelectricity electrolysis)

Hydropower is a continuous and dispatchable renewable energy source, ideal for powering electrolyzers with high capacity factors and minimal output curtailment. Although fewer recent studies have focused specifically on hydro-electrolysis systems, analyses indicate that these systems are among the most technologically mature and cost-effective, particularly when combined with alkaline electrolyzers (AELs) (Abdelsalam et al., 2024). The main obstacles are the geographical limitations and environmental impacts of new hydropower installations. However, existing hydropower plants can be retrofitted to efficiently power electrolyzers, thus producing low-carbon hydrogen at competitive prices.

7.4.6.6 Geothermal electrolysis

Geothermal energy presents a promising renewable source for powering electrolyzers in green hydrogen production, offering a stable and low-carbon supply of heat and electricity (Fig. 7.4). Typically operating at temperatures between

150°C and 300°C, geothermal systems are particularly well-suited for regions with abundant geothermal resources. When coupled with water electrolysis technologies such as AWE, PEM, or SOEC, geothermal energy can significantly enhance sustainability. Among these, SOEC technology is especially compatible with high-temperature geothermal heat, resulting in superior efficiency levels. For instance, one study reported that SOEC systems powered by geothermal energy achieved electrical efficiencies exceeding 80% (based on lower heating value) and produced up to 38% more hydrogen than their alkaline counterparts. Research has also shown that hydrogen production costs decrease as geothermal fluid temperature and flow rate increase, highlighting the economic advantages of high-quality geothermal resources (Yilmaz, 2017). A Franco-Icelandic project demonstrated the viability of high-temperature electrolysis (HTE) integrated with geothermal steam recovery, which significantly reduced overall electricity consumption (Sigurvinsson et al., 2006). Further efficiency gains have been realized through the HOT ELLY process, where the use of 200°C geothermal steam led to a 30% reduction in electricity demand compared to conventional electrolysis methods (Jónsson et al., 1992). From an environmental standpoint, LCA estimates indicate that geothermal-powered hydrogen production emits as little as 3.67 kg of CO_2-equivalent per kilogram of hydrogen, suggesting strong potential for long-term sustainability (Chidire et al., 2023). Finally, thermodynamic evaluations of PEM systems coupled with geothermal plants have shown the ability to produce 0.034 kg/s of hydrogen, with an electrolysis efficiency of 64% and an overall system efficiency of 6.7% (Yilmaz & Kanoglu, 2014). These findings underscore the technical, environmental, and economic viability of geothermal electrolysis integration as a robust pathway for low-carbon hydrogen generation.

Green hydrogen production via electrolysis can be categorized through two complementary lenses: the type of electrolyzer technology employed and the renewable energy source used to power the system. The main electrolyzer types, AWE, PEM, SOEC, and AEM, differ significantly in terms of efficiency, operating temperature, cost, and system adaptability. For instance, PEM electrolyzers are known for their high current densities and fast dynamic response, making them particularly well-suited for intermittent energy sources such as wind or PV solar power. In contrast, SOEC systems operate at elevated temperatures and are better aligned with high-grade thermal sources like concentrated solar power or waste heat from nuclear plants (Abdelsalam et al., 2024). Electrolyzers can be paired with various renewable electricity sources to form hybrid systems tailored to specific operational contexts. These include combinations such as PV + PEM, wind + AWE, or CPV-T (concentrated PVs with thermal recovery) + SOEC. One notable example is a PEM electrolyzer coupled with CPV, which achieved an impressive STH efficiency of 29.1%, highlighting the potential of advanced PV integration with high-performance electrolysis technologies (Zhang et al., 2025). A Canadian study comparing wind and solar power for (AWE) systems found that wind-based configurations were more cost-effective in colder climates, achieving hydrogen production costs as low as $0.55/kg (Elgammal & Ramlal, 2023). These findings underscore the interdependence between electrolyzer technology and the choice of renewable energy source. Selecting one inherently influences the optimal selection of the other, emphasizing the need for integrated design approaches that consider both dimensions simultaneously to develop green hydrogen systems that are not only efficient and scalable, but also economically viable.

7.4.7 Thermochemical water-splitting TWSC

Thermochemical water splitting is an innovative method for producing hydrogen, relying on a series of chemical reactions driven by intense heat typically provided by high-temperature energy sources such as concentrated solar power or advanced nuclear reactors. Unlike electrolysis, which requires electrical power, this approach uses exclusively thermal energy, making it particularly suitable for integration with nonelectric renewable sources. Among the most studied thermochemical cycles are the copper-chlorine (Cu–Cl) cycle and the sulfur–iodine (S–I) cycle. The Cu–Cl cycle, composed of four to five steps including hydrolysis, thermolysis, and electrochemical reactions, operates at moderate temperatures in the range of 500°C–600°C. This characteristic makes it more easily integrated into existing systems. Reported energy efficiencies for this cycle range from 32% to 58.6%, depending on operating conditions (Ghorbani et al., 2023). Conversely, the S–I cycle is based on three main steps: the decomposition of sulfuric acid, that of hydroiodic acid, and an intermediate Bunsen reaction. It requires significantly higher operating temperatures, typically between 800°C and 950°C. Despite this constraint, it can achieve comparable efficiencies, ranging from 15% to 60%, depending on the configuration (Ghorbani et al., 2023). Both cycles are the subject of extensive studies for integration with high-temperature power generation systems, including Generation IV nuclear reactors and solar thermal power plants. For example, hybrid configurations combining the Cu–Cl cycle with nuclear reactors have achieved overall energy efficiencies of up to 70% when hydrogen and electricity are produced simultaneously (Ozbilen et al., 2014). Economically, the estimated costs for hydrogen production via these cycles vary depending on the assumptions: between $1.63 and $9.47/kg for the Cu–Cl cycle, and between $5.41 and $10.40/kg for the S–I cycle (Li et al., 2022). From an environmental perspective, life cycle analyses indicate that these processes emit very little CO_2, especially when coupled with carbon-free energy sources such as nuclear or concentrated solar power, reinforcing their potential role in the transition to a green hydrogen economy (Sadeghi & Ghandehariun, 2023). In summary, thermochemical water splitting stands out as a promising sector for the production of low-carbon hydrogen. It offers interesting prospects for industrialization, particularly in integrated systems exploiting renewable or high-temperature nuclear heat sources.

7.4.7.1 *Photocatalysis*

Photocatalysis is a promising method for green hydrogen production, as it directly utilizes solar energy to split water into hydrogen and oxygen in the presence of semiconductor-based photocatalysts such as TiO_2 or graphitic carbon nitride. This pathway offers a sustainable alternative, since the only required inputs are sunlight and water, making the process intrinsically carbon-free when powered by renewable sources. However, the efficiency of photocatalytic hydrogen production strongly depends on the photocatalyst's activity, stability, and ability to harvest visible light. LCA studies have shown that the environmental footprint of photocatalytic hydrogen production is largely influenced by the extraction and fabrication of the photocatalysts, which could account for over 80% of total GHG emissions, whereas the operational phase has relatively low impacts (Maurya et al., 2023). Pilot-scale experiments confirm that photocatalytic hydrogen production systems can achieve global warming potentials as low as 4.53 kg CO_2-eq/kgH_2 under optimized conditions, outperforming conventional fossil-based hydrogen pathways (Yang et al., 2022). Nevertheless, challenges remain, particularly the short lifespan of photocatalysts and the need for rare or energy-intensive materials, which may compromise sustainability. Comparative LCA analyses emphasize that the choice of photocatalyst (e.g., TiO_2 vs. g-C_3N_4/BiOI composites) and end-of-life recycling strategies are crucial to reduce emissions and energy payback times (Ali et al., 2022). To qualify as green hydrogen, photocatalytic production must therefore rely on abundant, nontoxic photocatalysts, efficient STH conversion rates, and integration with circular economy principles for material reuse. Overall, LCA findings suggest that with material innovations and large-scale integration, photocatalysis could evolve into a competitive and environmentally superior pathway for hydrogen production.

7.4.8 Biomass feedstock

Biomass feedstock such as agricultural residues, forestry by-products, and energy crops can be converted into hydrogen through thermochemical pathways, mainly gasification. This approach offers the potential to produce renewable hydrogen with a lower carbon footprint, provided that the biomass is sourced sustainably and coupled with low-carbon energy inputs. LCA studies have shown that biomass gasification routes generally achieve lower GHG emissions compared to fossil-based hydrogen production, but the environmental impact strongly depends on the type of feedstock, cultivation methods, and conversion efficiency (Kalinci et al., 2012). However, challenges such as land use, fertilizer requirements, and transport logistics remain critical factors that can offset the environmental benefits. Comparative LCA studies highlight that hydrogen from biomass gasification has advantages in terms of global warming potential but may underperform economically and socially compared to conventional steam methane reforming, unless significant efficiency improvements are achieved

(Valente et al., 2019). More recent research confirms that biomass-based hydrogen can reach negative carbon footprints when coupled with CO_2 capture, making it a strong candidate for truly green hydrogen, although trade-offs in other impact categories, such as thermal pollution or land competition, must be carefully managed (Li et al., 2020). Furthermore, sustainability assessments emphasize that the environmental profile of this pathway is largely determined by feedstock cultivation and direct emissions during gasification, suggesting that future improvements should focus on minimizing biomass demand, optimizing energy efficiency, and integrating renewable energy inputs into the process (Iribarren et al., 2014). Overall, while biomass feedstock represents a promising route for green hydrogen, its viability depends on strict sustainability criteria and robust LCA-based evaluations to ensure that environmental, economic, and social benefits are truly achieved.

7.5 Results and discussion

The LCA can show the hydrogen production's cradle-to-grave process exhibiting a significant environmental impact. The direct impact is CO_2, which is the most harmful. It is produced using various techniques using different sources, as shown in Tables 7.1 and 7.2, representing a synthesis of the results of the LCA of hydrogen production methods. Here, efficiency refers to the percentage of energy input converted into H_2. CO_2 emissions are the amount of CO_2 emitted per unit of hydrogen produced (Ishaq & Dincer, 2021).

7.5.1 Goal and scope definition

In most LCA studies on green hydrogen, the functional unit adopted is 1 kg of H_2 produced, which enables consistent comparisons between technologies (Iribarren et al., 2014; Kim & Kim, 2024; Valente et al., 2019). However, some studies expand this definition to better reflect system-specific characteristics: Al-Ghussain et al. (2024) also incorporate electricity consumed (kWh), while Weidner et al. (2023) include the levelized cost of hydrogen (LCOH) as a complementary indicator. Hybrid PV-T and CPV-T solar systems sometimes use a surface-based unit (kgH_2/m^2) (Alghool et al., 2025), whereas photocatalysis combines hydrogen mass and catalyst lifetime to reflect the unstable performance of materials (Maurya et al., 2023; Sadeghi & Ghandehariun, 2023). System boundaries vary across studies. Most works on solar or wind-based electrolysis adopt a cradle-to-gate approach, including the manufacturing of modules, electrolyzers, and storage systems (Schropp et al., 2022; Tripathi & Subramanian, 2022). More comprehensive cradle-to-grave approaches also encompass distribution and end-of-life phases, as observed in Hunt & Nascimento, 2021, and Chen et al. (2019). Specific cases, such as electrolysis ships powered by hydropower, further extend these boundaries by including logistical impacts (Hunt & Nascimento, 2021). Geothermal systems (Ghorbani et al., 2023;

TABLE 7.1 Guidelines for key elements consideration in an LCA methodology for hydrogen production techniques.

Hydrogen production method	Goal and scope definition	Inventory analysis	Impact assessment	Interpretation and reporting
Solar PV-electrolysis	Define the objectives and scope of the investigation. 1 kg of hydrogen generated using photovoltaic electrolysis. Cradle-to-grave or cradle-to-gate life-cycle assessment systems. The importance of selecting the type of electrolyzer (PEM, AEL) (Hunt & Nascimento, 2021; Osman et al., 2024).	An analysis of inputs (photovoltaic-thermal [PV] panels, electrolyzer, water) and outputs (H_2, emissions, co-products). Primary data or LCA databases (such as Ecoinvent) (Díaz-Motta et al., 2023; Osman et al., 2024; Ravilla et al., 2024).	Evaluation of GHG emissions (1.5–3.5 kg CO_2-eq/kgH_2), water use (20-30 L/kg), and other effects using ReCiPe or TRACI (Hofrichter et al., 2023).	Highlighting regional advantages (e.g., Morocco, California), uncertainties associated with PV output and longevity, techno-economic suggestions (Osman et al., 2024; Ravilla et al., 2024).
PV-T-electrolysis	Integrating a hybrid PV-T system to evaluate efficiency gains and LCA effect (Alghool et al., 2025; Ravilla et al., 2024)	More complicated inventory: thermal and electrical integration. Information on heat exchangers, heat transfer fluids, and integration with electrolyzers (Alghool et al., 2025; Ravilla et al., 2024)	Heat recovery results in a 10%–20% reduction in effect when compared to PV alone (Alghool et al., 2025; Ravilla et al., 2024)	A comparison of PV with PV-T shows that the benefits are restricted by technological complexity. Proposals for thermodynamic optimization (Alghool et al., 2025; Ravilla et al., 2024)

(*continued on next page*)

CPV-T-electrolysis	The goal is to use solar thermal and PV concentration to increase overall system efficiency (Boudries, 2016; Farhani et al., 2024)	Data about mirrors, lenses, fluids, active cooling, and electrolyte coupling. (Boudries, 2016; Farhani et al., 2024)	GHG emissions are definitely greater than PV-T if heat losses are not accounted for. Increased exergy efficiency. (Boudries, 2016; Farhani et al., 2024)	Detailed exergy evaluations. Recommendations for ideal operating conditions (Alghool et al., 2025).
Wind-electrolysis	The goal is to measure the environmental effect of electrolysis driven by wind energy (onshore/offshore). Cradle-to-gate or cradle-to-grave method (Weidner et al., 2023)	Data concerning wind turbines, electric conversion systems, and electrolyzers. Data from the site, transportation, and maintenance have all been integrated (Weidner et al., 2023).	Offshore wind power produces GHG emissions of less than 1.5 kg CO_2-eq/kgH_2, with even lower values (<1 kg). Impacts on resources and eutrophication were investigated (Díaz-Motta et al., 2023)	Presents strong and consistent outcomes. Recommendations for appropriate coupling and network stability (Díaz-Motta et al., 2023)
Hydropower-electrolysis	Use hydroelectricity as an input. Excellent production stability and low expected emissions (Hunt & Nascimento, 2021; Jovan et al., 2021)	Inventory for hydroelectric infrastructure, environmental consequences (dams, networks), and electrolyzers (Kelman et al., 2020).	For optimal environmental performance, emissions should be less than 0.5 kg CO_2-eq/kgH_2. Without GHGs, there are less direct repercussions (Chen et al., 2023).	High reliance on hydroelectric configuration (dam, run-of-river). Analyses should be tailored to each country and season (Li et al., 2022).

(*continued on next page*)

TABLE 7.1 Guidelines for key elements consideration in an LCA methodology for hydrogen production techniques—cont'd

Hydrogen production method	Goal and scope definition	Inventory analysis	Impact assessment	Interpretation and reporting
Geothermal-electrolysis	The study of the interaction between geothermal energy and high-temperature electrolysis (mostly SOEC) (Ghorbani et al., 2023).	Complex inventory: drilling, geothermal fluids, heat exchangers, SOEC electrolysis, infrastructure (Ghorbani et al., 2023; Molho et al., 2024).	Impacts vary with geology, with GHG emissions ranging from 1 to 2 kg CO_2-eq/kgH_2. Indirect emissions from deep drilling (Ghorbani et al. 2023; Molho et al., 2024).	Uncertainties about fugitive emissions and aged wells. Recommendations: Probabilistic modeling (Ghorbani et al., 2023; Molho et al., 2024).
Thermochemical splitting	Analysis of Cu–Cl, Fe–Cl, and S–I cycles for water dissociation without external electrolysis (Ozbilen et al., 2014; Sadeghi & Ghandehariun, 2023).	Technical inventory includes active materials, reactors, heat sources (solar and nuclear), and recycling (Ghorbani et al., 2023).	The Cu–Cl cycle generates 0.3–0.9 kg CO_2-eq/kgH_2, while the S–I cycle produces more than 1 kg. Energy use is highly variable (Sadeghi & Ghandehariun, 2023).	Exergy comparative analysis with LCA. Methods must be standardized in order to ensure reproducibility across investigations (Ozbilen et al., 2014; Sadeghi & Ghandehariun, 2023).

(*continued on next page*)

Photocatalysis	Assess H_2 generation using solar catalysis (TiO_2, ZnS, graphene, etc.) (Chen et al., 2023).	Information on catalysts, irradiation surface area, system lifespan, and photonic efficiency (Chen et al., 2019; Maurya et al., 2023; Weidner et al., 2023)	GHG emissions from noble catalysts exceed 2.5 kg CO_2-eq/kgH_2. 1–1.5 kg for plentiful catalysts (Chen et al., 2019; Maurya et al., 2023; Weidner et al., 2023)	Low technological maturity. Recommendations: Increase quantum yield and recyclability (Chen et al., 2019; Maurya et al., 2023; Weidner et al., 2023)
Biomass	The goal is to measure the consequences of lignocellulosic biomass or biowaste gasification (Iribarren et al., 2014; Valente et al., 2021).	Analysis of biomass fluxes, gasification, syngas treatment, and potential CO_2 capture (Ajaj et al., 2024; Valente et al., 2021).	GHG emissions range from 2 to 4 kg CO_2-eq/kgH_2. Variability is high depending on biomass kind, transit method, and humidity level (Ajaj et al., 2024; Valente et al., 2021)	The sensitivity to conversion yields and CO_2 capture. Integration of coproducts (biochar) (Ajaj et al., 2024; Valente et al., 2021).

AEL, *alkaline electrolyzer;* CPV-T, *concentrated photovoltaic thermal;* LCA, *life cycle assessment;* PEM, *proton exchange membrane;* PV-T, *photovoltaic thermal.*

TABLE 7.2 Comparative summary of life cycle assessment results across hydrogen production methods.

Hydrogen production method		LCA study highlights
Electrolysis methods	PV-electrolysis	Multiple LCA studies (Ravilla et al., 2024) highlight that solar PV-powered electrolysis has relatively low GHG emissions (0.8–3.5 kg CO_2-eq/kgH_2), high scalability, and strong performance under optimized solar irradiance. Studies emphasize the role of PEM versus AEL electrolyzers and regional solar resource variability (Ravilla et al., 2024).
	PV-T-electrolysis	PV-T systems increase efficiency by combining thermal and electrical capture. Studies suggest a 10%–25% reduction in GHG emissions compared to PV alone (Alghool et al., 2025; Ravilla et al., 2024). Improved water efficiency and lower land usage are also mentioned (Alghool et al., 2025).
	CPV-T-electrolysis	CPV-T is efficient at high temperatures and has the potential to enhance LCA ratings when tuned. However, increased complexity and expense remain restrictive. GHG emissions are equivalent or slightly lower than PV-T (Boudries, 2016).
	Wind-electrolysis	Wind-based electrolysis provides the lowest emissions (<1.5 kg CO_2-eq/kgH_2). LCA studies stress increased sustainability for offshore wind and highlight the effects of electrolyzer type, energy mix, and infrastructure (Zhang et al., 2023).
	Hydro-electrolysis	Hydropower-driven electrolysis is one of the most sustainable. LCA investigations (Zhao et al., 2020) consistently indicate <1 kg CO_2-eq/kgH_2. The environmental effect varies depending on the kind of hydro source (run-of-river vs. reservoir).

(*continued on next page*)

	Geothermal-electrolysis	Geothermal-powered electrolysis enables continuous H_2 generation with stable GHG profiles (1–2.5 kg CO_2-eq/kgH_2). LCA highlights strong site dependence and infrastructure impact (Ghorbani et al. 2023; Molho et al., 2024).
Thermochemical splitting		Many studies such as Ghorbani et al. (2023), Ozbilen et al. (2014), and Sadeghi and Ghandehariun (2023) evaluate multiple cycles (Cu–Cl, Fe–Cl, S–I), reporting varied performance. Cu–Cl cycles show lowest GHG values (~0.5–1.2 kg CO_2-eq/kgH_2). LCA studies remain limited and models lack harmonization.
Photocatalysis		Photocatalytic approaches show potential, although they are not yet commercially available. LCA studies (Chen et al., 2023) show that noble-metal catalysts have higher GHG emissions (~3–6 kg CO_2-eq/kgH_2) than earth-abundant catalysts (~1–2 kg) (Maurya et al., 2023).
Biomass		Biomass-based hydrogen provides low emissions (~2.5–4.5 kg CO_2-eq/kgH_2) that vary depending on feedstock and method. Many LCA studies (Ajaj et al., 2024; Iribarren et al., 2014) (highlight the benefits of combining CCS or ORC. Methane and NOx emissions can be substantial.

CCS, *carbon capture and storage;* CPV, *concentrated photovoltaic-thermal;* GHG, *greenhouse gas;* LCA, *life cycle assessment,* PV-T, *photovoltaic thermal.*

Hunt & Nascimento, 2021) and biomass technologies (Ajaj et al., 2024; Valente et al., 2019) also show significant heterogeneity depending on whether collection, transport, and treatment phases are included. Finally, photocatalysis stands out for its detailed modeling of catalyst synthesis and recyclability (Zhao et al., 2020). Inventory analysis in LCA studies of green hydrogen production via electrolysis reveals significant methodological heterogeneity, driven by the type of energy vectors used, the LCA software applied, and the territorial scale of the projects. The most commonly used software tools are SimaPro and GaBi, which rely predominantly on the Ecoinvent database, particularly for solar PV and wind pathways (Shaya & Glöser-Chahoud, 2024; Zhao et al., 2020). Some studies adapt these tools to integrate local contexts; for example, Jovan et al. (2021) incorporate site-specific electricity production, transportation, and cogeneration inventories in their analysis of hydropower-based electrolysis in Slovenia. Regarding methodologies, impact assessment frameworks such as ReCiPe (2008/2016), CML 2001, ILCD 2011, and IMPACT 2002+ are most frequently employed, with the choice often guided by study objectives. Hunt & Nascimento, 2021 tailor their inventory to account for maritime transport in their assessment of an electrolysis ship powered by hydropower. Other studies highlight that the methodological framework selected can significantly influence outcomes, particularly for impact categories such as eutrophication, acidification, and human toxicity (Ghorbani et al., 2023; Sadeghi & Ghandehariun, 2023). At the technological level, PV-based electrolysis systems yield inventories sensitive to the carbon intensity of the electricity mix, land use, and PV module production processes (Gerhardt-Mörsdorf et al., 2024; Palmer et al., 2021). The integration of PV-T or CPV-T hybrids further complicates inventories due to the thermal components, requiring joint modeling of electrical and thermal flows (Farhani et al., 2024; Li et al., 2024). For wind-based systems, impacts stem mainly from turbine manufacturing and the variability of wind resources (Díaz-Motta et al., 2023; Kim & Kim, 2024). Hydropower-based electrolysis benefits from shared infrastructure, which helps reduce environmental burdens related to transport and storage (Jovan et al., 2021; Kelman et al., 2020). Thermochemical methods, although less mature, show inventories dominated by the use of specific materials (e.g., perovskites, rare metal oxides) and complex thermodynamic cycles (Ghorbani et al., 2023; Ozbilen et al., 2014). Photocatalytic processes, often at pilot scale, concentrate their impacts on the synthesis of rare-earth-based catalysts and the low durability of these systems (Acar, 2020; Al-Ghussain et al., 2024; Luo & Su, 2024). For biomass-based technologies, LCA studies must account for upstream logistics (collection, drying), combustion emissions, and, in some cases, CO_2 capture (Arfan et al., 2023; Valente et al., 2019). Finally, the quality of inventories depends strongly on their temporal and geographical relevance. Several recent studies emphasize the need to align input data with emerging technologies and future scenarios to ensure more accurate and forward-looking assessments (Al-Ghussain et al., 2024; Shaya & Glöser-Chahoud, 2024; Zhao et al., 2020).

7.5.2 Impact assessment

7.5.2.1 Greenhouse gas emissions

GHG emissions represent the most commonly assessed environmental indicator in LCA studies of green hydrogen production. Among various production technologies, electrolysis powered by renewable energy sources consistently demonstrates significantly lower GHG emissions compared to steam methane reforming (SMR). For instance, hydrogen production via PV-powered PEM (PV+PEM) electrolysis typically results in emissions ranging from 1.5 to 3.5 kg CO_2-eq per kg of H_2, depending on regional and system-specific parameters (Díaz-Motta et al., 2023). In integrated Californian systems characterized by a dominant renewable electricity mix, values as low as < 2 kg CO_2-eq/kgH_2 have been reported (Hunt & Nascimento, 2021; Li et al., 2020). Hybrid configurations such as PV-T and concentrated PV-T (CPV-T) offer additional reductions of 10%–15%, primarily due to the recovery and utilization of thermal energy. Wind energy, particularly when deployed with electrolyzers, remains one of the most environmentally favorable options, with reported emissions of 0.8–1.2 kg CO_2-eq/kgH_2 (Weidner et al., 2023). Hydropower-based electrolysis can yield even lower values, often below 0.5 kg CO_2-eq/kgH_2, especially in cascading hydropower systems (Jovan et al., 2021; Kelman et al., 2020). By contrast, geothermal hydrogen systems show greater variability, with emissions between 1 and 2 kg CO_2-eq/kgH_2, depending on site-specific geological conditions and potential fluid leakages (Ghorbani et al., 2023). Thermochemical processes, such as the Cu–Cl cycle, exhibit relatively low GHG footprints (approximately 0.3–0.9 kg CO_2-eq/kgH_2), though other cycles, such as sulfur–iodine (S–I), have higher emissions, typically exceeding 1 kg CO_2-eq/kgH_2 (Sadeghi & Ghandehariun, 2023). Finally, photocatalytic hydrogen production, especially when employing noble-metal catalysts, can generate GHG emissions above 2.5 kg CO_2-eq/kgH_2. However, using earth-abundant catalysts has been shown to lower emissions significantly, reaching levels between 1 and 1.5 kg CO_2-eq/kgH_2 (Iribarren et al., 2014).

7.5.2.2 Water consumption

Water consumption varies significantly across hydrogen production technologies. PEM electrolyzers typically require more water than AEL, especially in arid environments where water scarcity may exacerbate sustainability concerns (Gerhardt-Mörsdorf et al., 2024; Hake et al., 2017). In addition, PV-based systems operating in desert regions may require substantial water volumes for panel cleaning, with estimates reaching up to 0.1 m^3/kg of H_2 produced (Abdelkareem et al., 2023). In contrast, wind and hydropower-based systems exhibit negligible direct water consumption, making them favorable in water-stressed regions (Jovan et al., 2021; Zhao and Pedersen, 2018). Geothermal hydrogen systems show moderate water use, primarily associated with drilling fluids and subsurface fluid circulation (Ghorbani et al., 2023). Finally, photocatalytic hydrogen production pathways generally involve low direct water consumption, but their overall water-related impacts are influenced by feedwater quality and the durability of the photocatalysts, which may require frequent replacement or regeneration (Maurya et al., 2023; Yang et al., 2022).

7.5.2.3 Depletion of critical resources and materials

Impacts related to the consumption of critical resources are particularly concerning for technologies reliant on rare metals. PEM electrolyzers powered by PV systems are heavily dependent on platinum and iridium, both of which are scarce and environmentally intensive to extract (Cetinkaya et al., 2012; Wang et al., 2025). Similarly, certain photocatalytic systems using noble-metal-based catalysts also contribute significantly to critical resource depletion (Al-Ghussain et al., 2024). By contrast, AELs and catalysts based on earth-abundant metals exhibit better performance in this impact category (Hake et al., 2017). The manufacturing of PV and CPV modules may also require rare earth elements, thereby increasing the mineral resource scarcity indicators associated with these technologies (Ravilla et al., 2024). Geothermal systems, depending on the type of working fluid and equipment used, can also show moderate to high impacts in terms of resource extraction, especially when advanced drilling technologies or rare materials are involved (Ghorbani et al., 2023).

7.5.2.4 Acidification, eutrophication, and ecotoxicity

Acidification impacts are amplified in scenarios relying on carbon-intensive energy mixes, even when the hydrogen production technology itself is considered clean (Hunt & Nascimento, 2021). The regional electricity mix plays a crucial role in shaping the environmental burdens related to both acidification and eutrophication. For instance, hydrogen production via hydropower in countries such as Austria or Slovenia exhibits a favorable environmental profile, in contrast to regions with a high share of fossil-based electricity (Kelman et al., 2020). Thermochemical processes, particularly those involving chlorine-based cycles or solvent use, can lead to indirect emissions that significantly contribute to these impact categories (Ghorbani et al., 2023). Additionally, the manufacturing of PV and CPV modules, which often involves solvents and critical metals, further increases the ecotoxicity-related burdens (Boudries, 2016; Tripathi & Subramanian, 2022).

7.5.2.5 Socioeconomic impact

From a socioeconomic perspective, green hydrogen production presents notable opportunities at the local level, particularly in terms of job creation related to the deployment of PV arrays, wind turbines, and the operation of electrolysis systems (Jovan et al., 2021; Valente et al., 2021). Air quality improvements are also frequently highlighted, especially in scenarios where hydrogen replaces coal-based fuels in transportation or industrial processes (Acar, 2020). Significant energy autonomy gains are anticipated in renewable-rich countries such as Brazil and India, where local production could reduce dependence on fossil fuel imports (Kelman et al., 2020). However, emerging technologies like photocatalysis and thermochemical water-splitting cycles still require high capital investments (exceeding 7–10 \$/kgH_2), which currently hinders their large-scale commercial deployment (Sadeghi & Ghandehariun, 2023).

7.5.3 Interpretation and comparison

First, the electricity source emerges as the most critical driver of the overall environmental profile. Systems based on PV-electrolysis powered by highly renewable grids (Alghool et al., 2025; Palmer et al., 2021; Tripathi & Subramanian, 2022) achieve significantly lower GHG emissions than those in regions where electricity is fossil-dominated. Table 7.3 represents the key parameters of inventory analysis for different green H_2 production routes. In this context, wind- and hydropower-based configurations (wind electrolysis, hydropower-electrolysis) demonstrate the most favorable climate performance, often below 1 kg CO_2-eq/kgH_2, underscoring the importance not only of the production technology but also of its integration into the local energy context (Gallandat et al., 2017). Second, the technological choice, whether PEM electrolyzers, alkaline systems, or emerging technologies like PEC or CPV-T, introduces clear trade-offs between efficiency, water consumption, and critical materials use. For instance, comparative studies of PV + electrolysis show that PEM electrolyzers deliver higher efficiencies (Gerhardt-Mörsdorf et al., 2024; Ravilla et al., 2024) but require valuable metals such as platinum and iridium, increasing their resource scarcity impacts (Gerhardt-Mörsdorf et al., 2024; Tripathi & Subramanian, 2022). Conversely, AELs are typically more robust and less reliant on rare materials, offering better performance in the resource use category despite slightly lower efficiencies (Gerhardt-Mörsdorf et al., 2024). Hybrid systems like PV-T or CPV-T offer gains in surface and thermal efficiency, reducing energy losses; however, their lower technological maturity still limits large-scale deployment (Al-Ghussain et al., 2024; Boudries, 2016). Third, regional and contextual factors play a decisive role. Studies conducted in high solar irradiation zones (e.g., Bangladesh, California) or renewables-rich regions (e.g., hydropower basins) exhibit marked advantages, not only in GHG emissions, but also in water consumption and land use (Al-Ghussain et al., 2024; Mazumder et al., 2023; Valente et al., 2021). However, in arid regions or areas with low renewable density, impacts such as PV panel cleaning water use, intermittency-related losses, or infrastructure constraints can reduce the overall benefits (Abdelkareem et al., 2023; Zhao & Pedersen, 2018). Moreover, scenarios that explicitly include logistics, transport, end-of-life, and recycling (Abdelkareem et al., 2023) reveal

TABLE 7.3 Key parameters of inventory analysis for different green H_2 production routes.

H_2 production method	GHG emissions	Water consumption	Energy consumption	Land use
Solar PV-electrolysis	Low emissions (<1.5—3.5 kg CO_2-eq/kgH_2), strongly depending on irradiance and electrolyzer performance (Díaz-Motta et al., 2023).	High water usage (~20–30 L/kgH_2), mostly for electrolysis and PV panel maintenance (Osman et al., 2024).	High electricity input; overall energy demand is determined by inverter and electrolyzer losses (Díaz-Motta et al., 2023; Osman et al., 2024).	Significant land footprint for PV arrays (Díaz-Motta et al., 2023).
Wind-electrolysis	Offshore wind generates very low emissions (~0.8–1.2 kg CO_2-eq/kgH_2), making it extremely beneficial (Weidner et al., 2023)	More efficient water consumption than solar, but still considerable. (Weidner et al., 2023)	Energy-efficient owing to the high wind capacity factor; reduced transmission losses when co-located (Weidner et al., 2023)	Lower land use than PV, offshore wind further reduces land use (Weidner et al., 2023).
PV-T hybrids	PV-T systems reduce GHG emissions by ~10%–20% versus PV-electrolysis, due to heat recovery (Alghool et al., 2025).	PV-T water use is similar to PV, driven by electrolysis and thermal loop needs (Alghool et al., 2025).	Thermal recovery in PV-T systems improves energy use efficiency, lowering electricity needs per kgH_2 (Alghool et al., 2025).	Land use is comparable to PV, with minor increases due to thermal components (Alghool et al., 2025).

(*continued on next page*)

CPV-T hybrids	Depending on solar concentration and location, performance may be comparable or superior to PV-T (Zhao et al., 2020).	Additional water is required to cool CPV modules, which increases demand above PV (Zhao et al., 2020).	Higher energy consumption for monitoring and cooling, although greater conversion rates offset some of it (Zhao et al., 2020).	Solar tracking equipment and CPV concentrators contribute to a larger land footprint (Zhao et al., 2020).
Hydropower-electrolysis	Low emissions (<0.5 kg CO_2-eq/kgH_2), particularly for cascade or run-of-river systems (Jovan et al., 2021; Kelman et al., 2020).	Stable water use, frequently coupled with existing reservoirs (Kelman et al., 2020).	Because the electricity comes from clean hydro sources, the upstream energy demand is low (Jovan et al., 2021).	Minimal additional land use; electrolyzer can be colocated with existing hydro infrastructure (Kelman et al., 2020).
Geothermal-electrolysis	Moderate emissions (1–2 kg CO_2-eq/kgH_2), sensitivity to site geology, and risk of fluid leakage (Ghorbani et al., 2023).	Significant water is required for reinjection and cooling (Ghorbani et al., 2023).	Depending on the type of geothermal heat and the electrolyzer used, the process is energy-intensive (Ghorbani et al., 2023).	Needs land for geothermal wells, pipes, and surface systems (Ghorbani et al., 2023).
Thermochemical splitting	The Cu–Cl cycle emits 0.3–0.9 kg CO_2-eq/kgH_2, while the S–I cycle produces more than 1 kg CO_2-eq/kgH_2 owing to increased energy input (Ozbilen et al., 2014).	Water requirements vary each cycle, mostly for the reaction and cooling phases (Sadeghi & Ghandehariun, 2023).	High thermal input required; some compensation achieved through waste heat recovery (Ghorbani et al., 2023).	Compact process; little land footprint compared to biomass and PV (Ozbilen et al., 2014).

(continued on next page)

TABLE 7.3 Key parameters of inventory analysis for different green H_2 production routes—cont'd

H_2 production method	GHG emissions	Water consumption	Energy consumption	Land use
Photocatalysis	Depending on the substance utilized, the CO_2-eq/kgH_2 can range from 1 to 2.5 kg; earth-abundant catalysts can reduce this amount (Chen et al., 2023).	Water is reactive, although it is normally utilized in little quantities; just modest cleaning water is necessary (Chen et al., 2023).	Mainly powered by solar input; little supplementary energy requirements (Iribarren et al., 2014).	Low land use, yet dependent on solar intensity and catalyst deployment density (Chen et al., 2023).
Biomass	Depending on the biomass supply and system performance, it may be carbon neutral or even negative (Chen et al., 2023).	Varies dependent on feedstock drying and syngas cooling; water reuse lowers the effect (Valente et al., 2021).	Pretreatment and gas cleaning need a lot of energy, which can be somewhat mitigated by energy recovery (Li et al., 2020).	Depends on the biomass type; agricultural wastes have a smaller impact than specific crops (Valente et al., 2021).

CPV-T, *concentrated photovoltaic thermal;* GHG, *greenhouse gas;* PV-T, *photovoltaic thermal.*

that these life cycle stages can account for a nonnegligible share of total impacts, especially in systems using critical metals or distributed modular units. Fourth, although co-benefits such as local job creation, air quality improvement, and energy security are often cited (Abdelkareem et al., 2023; Jovan et al., 2021; Valente et al., 2019), quantitative assessments of these social dimensions are still much rarer than those of environmental metrics. This gap limits the ability to fully evaluate their comparative role in the overall sustainability of green hydrogen systems. In summary, the interpretation of these studies converges on the notion that renewable-based electrolysis is among the most promising pathways, provided that it is supported by a clean electricity mix, optimized electrolyzer technologies, and a system-wide life cycle approach. The trade-offs between GHG emissions, water use, and critical resource consumption must be carefully managed according to regional conditions, and emerging technologies like PV-T, CPV-T, or PEC deserve further exploration to improve technological readiness and reduce material-related impacts.

7.6 Life cycle assessment of green hydrogen in Morocco

7.6.1 General context

Despite the diversity of LCA studies conducted globally on green hydrogen pathways, key trends have emerged, particularly the superior environmental performance of wind and hydro-powered electrolysis, which tend to exhibit the lowest life cycle GHG and water footprints (Kodgire, 2025). In contrast, solar and biomass-based hydrogen pathways reveal broader impact variability due to region-specific resource availability and technological efficiency (Ajeeb et al., 2024; Valente et al., 2021). However, these conclusions are highly sensitive to geographic and infrastructural contexts. Against this backdrop, Morocco stands out as a promising case, given its exceptional wind and solar potential, its strategic intent to export green hydrogen, and the current methodological limitations in existing domestic LCA studies (Nasser & Hassan, 2024). Recent works, such as Kanz et al. (2023), have started to address these gaps, estimating emissions for PV-powered hydrogen exports from Morocco at approximately 3.4 kg CO_2-eq/kgH_2, confirming the country's competitive advantage. Nevertheless, to ensure scientific credibility and inform sustainable policy, future Moroccan LCA studies must adopt more robust and regionally adapted methodological frameworks (Díaz-Motta et al., 2023).

7.6.2 Life cycle assessment of green hydrogen in Morocco: Status, analysis, and perspectives

In Morocco, the growing academic interest in green hydrogen is reflected in an expanding body of literature. However, most national contributions remain focused on TEAs, spatial planning (GIS/AHP), or hybrid system optimization, while quantitative LCA applied directly to domestic hydrogen production is still relatively scarce and fragmented (Amrani et al., 2023; Bahou, 2023; Bouramdane, 2024; Ennassiri et al., 2019; Ourya et al., 2023). These TEA studies typically model solar, wind, or hybrid (PV+CSP) power supplying PEM or AWE electrolyzers, with boundaries restricted to partial cradle-to-gate scopes, excluding downstream processes such as hydrogen storage, transport, or conversion for export (Amrani et al., 2023; Azzaoui & Alami Merrouni, 2025; Bahou, 2023; Bouramdane, 2024; Ourya et al., 2023). Cost estimates show promising competitiveness in optimized configurations: Ourya et al. (2023) report LCOH values around 2.54 \$/kgH_2 in Dakhla under hybrid PV-wind setups, while Bahou (2023) reports higher values (up to 12.6 \$/kg) in urban refueling contexts. Similar patterns are observed in El Hassani et al. (2023) and Touili et al. (2020), who compare solar technologies across Moroccan climatic zones. They further underline Morocco's cost advantage compared to Southern Europe for solar hydrogen. However, full LCAs remain underutilized, with only a few studies quantifying GHG emissions, resource consumption, or water use. Early work by Ennassiri et al. (2019) offers an energy and emission-based comparison of hybrid systems, whereas Fernández-Lobato et al. (2025) and Ozturk et al. (2023) apply gate-to-gate LCA to biomass valorization via gasification and also evaluate fuel cell technologies (PEMFC, SOFC, and AFC) using cradle-to-gate LCA, revealing critical insights on material intensity and electricity mix. Additional LCA-adjacent studies offer translatable insights: Azzaoui & Alami Merrouni (2025) and El-Maaroufi et al. (2025) assess the potential of hybrid solar systems for SDG-linked H_2 deployment, and Nasser and Hassan (2024) map regional LCOH and GWP variations across MENA, positioning Morocco as a key low-emission producer. At the strategic level, Bouramdane (2024) applies multicriteria decision frameworks to identify optimal hydrogen portfolios, while Caillard et al. (2024) explore production "hotspots" and export potential, though with limited environmental metrics. Across the board, life-cycle inventory completeness remains a challenge, with most studies overlooking stack degradation, desalination, or critical material flows. By contrast, international reviews like Osman et al. (2024) offer a clear framework for hydrogen LCAs, based on ISO 14040/44, using tools such as SimaPro and Ecoinvent, and calling for inclusion of multiple impact categories beyond GWP: water use, acidification, toxicity, and land occupation. Best-case renewable-based systems globally converge toward 1–2 kg CO_2e/kgH_2, with total water demand estimated around 20–30 L/kgH_2, indicators mostly missing in Moroccan literature. In conclusion, while Morocco demonstrates a clear techno-economic advantage for green hydrogen production, the LCA dimension remains underdeveloped, as shown in Table 7.4 . Future studies must adopt multiimpact ISO-compliant LCAs, integrate uncertainty propagation, model renewable grid dynamics, and assess export pathways such as ammonia, LOHC, or pipelines (Azzaoui & Alami Merrouni, 2025). Building a

TABLE 7.4 Comparative synthesis of the results of the life cycle analysis for hydrogen production in Morocco.

Studies	Energy source	Technology	CO_2eq (kg/kgH_2)	Water (L/kgH_2)	LCOH ($/kg)	GHG emission	Water consumption	Energy consumption	Waste production	Location
Ennassiri et al. (2019)	PV, wind	Alkaline electrolysis PEM/AWE	~1.5-3	~20-30	~3–5	Moderate	No water flows modeled	Initial estimates without energy breakdown	Preliminary modeling stage	Laâyoune , Ouarzazate, Midelt
Amrani et al. (2023)	PV and wind	PEM	~1.5-2.5	~20-30	~3	High variability depending on mix	Dependence on local desalination	High dependence on the power grid	Not discussed	Rabat Eastern Morocco
Bahou et al. (2023)	PV and Grid	PEM	~3–12	108>30	9.2–12.6	High emissions (urban network case)	Water not included in inventories	electricity demand	Not addressed	Urban Morocco Casablanca
Nasser and Hassan (2024)	Grid Mix	Fuel cell technologies (PEMFC, SOFC, AFC)	2–4	30–40	Not specified	PEMFC > SOFC > AFC in GHG PEMFC	consumes more water	Depends on the technology	Catalyst-dependent waste	Morocco

(*continued on next page*)

TABLE 7.4 Comparative synthesis of the results of the life cycle analysis for hydrogen production in Morocco—cont'd

Studies	Energy source	Technology	CO_2eq (kg/kgH_2)	Water (L/kgH_2)	LCOH ($/kg)	GHG emission	Water consumption	Energy consumption	Waste production	Location
Ourya et al. (2023)	PV and Wind	PEM/AWE	~1.5–2.5	~20–30	~2.54	Good GHG performance	Cooling water not specified	Optimized hourly profile, low cost	Not specified	Dakhla
Ozturk et al. (2023)	Biomass	Gasification	3–5	~30	Not specified	Local biomass LCA; variable based on process efficiency	Biomass drying/treatment likely increases use	Energy-intensive for biomass pretreatment	Biomass ash, biochar waste modeled	Morocco
Fernández-Lobato et al. (2025)	Biomass (olive pomace)	Gasification + engine-generator	Not specified	Low	Not specified	Moderate emissions, depending on co-products	Evaluated within the LCA, but not per kg H_2	Partially supplied through olive-pomace gasification	Olive pomace valorized; renewable electricity and biochar produced	Morocco
Bouramdane (2024)	Mixed ENR	Portfolio Modeling	Not specified	Not specified	Portfolio Costs	GHG hotspots identified in the South	Water consumption not detailed	LCA not systematic	No quantification	Southern Morocco

(*continued on next page*)

Nasser and Hassan (2024)	PV and Wind electrolysis (MENA atlas)	Electrolysis (MENA atlas)	Not specified	Not specified	Not specified	Favorable Morocco vs. MENA	Mapped water stress	High regional variability	Not integrated	MENA (including Morocco)
Osman et al. (2024)	Various	Production, storage, and usage	~1–2 (global)	20–30	Not specified	Benchmarks ~1–2 kg CO_2eq/kgH_2 for RE-based	Benchmarks 20–30 L/kgH_2 for RE electrolysis	Electrolysis energy demand central in modeling	Includes e-toxicity/solid waste among metrics	Global
Azzaoui and Alami Merrouni (2025), Touili et al. (2020)	CSP and PV	Hybrid Electrolysis	~1.5	~25	~2.5	Lower emissions than CSP alone	Water critical for CSP	Good energy hybridization	Not modeled	Ouarzazate
El-Maaroufi et al. (2025)	Hybrid PV-wind	Hybrid Electrolysis	~1–2	~20	~2.5–3.5	Reduced emissions if 100% renewable energy	Moderate water use, integrated via cooling	Optimized production in the South	Few waste products modeled	South Morocco
El Hassani et al. (2023), Touili et al. (2020)	PV and wind (off-grid hybrid)	Electrolysis + H_2 storage	~1.5–2	~30	~3	Significant reduction in GHG emissions	Not discussed	Storage increases consumption	Little waste treatment	Remote areas Morocco

AWE, *alkaline water electrolysis;* GHG, *greenhouse gas;* PEM, *proton exchange membrane;* PV, *photovoltaic;* CSP, *concentrating solar power;* ENR, *renewable energy;* ORC, *Organic Rankine Cycle.*

credible environmental case alongside cost competitiveness is key to supporting Morocco's role as a sustainable hydrogen exporter.

7.6.2.1 Perspectives

The integration of LCA in green hydrogen (GH_2) development represents a cornerstone for aligning technological innovation with climate objectives. Nonetheless, the maturity and standardization of LCA methodologies for H_2 production remain limited at the international scale, and even more so in emerging hydrogen markets such as Morocco. Although significant advancements have been achieved in modeling environmental impacts of major H_2 production routes, electrolysis powered by solar, wind, or hydro (Osman et al., 2024; Zhao et al., 2020); biomass gasification (Iribarren et al., 2014; Valente et al., 2021); and thermochemical or photocatalytic approaches (Ozbilen et al., 2014; Sadeghi & Ghandehariun, 2023), many studies still lack methodological consistency and context-sensitive indicators. At the global level, LCA studies frequently diverge in terms of functional unit definition, system boundary selection, and handling of co-products. These disparities hinder comparability and obstruct the creation of actionable environmental benchmarks (Ajaj et al., 2024; Weidner et al., 2023). The adoption of harmonized protocols, such as cradle-to-gate assessments with dynamic system modeling and multi-impact categories, is necessary to reflect the true environmental performance of H_2 systems under real-world variability. Particularly, the co-integration of LCA with TEA and circular economy principles, such as valorization of waste heat or biochar, can provide a holistic basis for sustainable technology development (Arfan et al., 2023; Iribarren et al., 2014). For Morocco, the situation is even more pressing. Despite its strong solar and wind resources, strategic geographical position, and growing international partnerships (IRESEN, ONEE), there is a notable lack of LCA and TEA studies adapted to local conditions. Most assessments of green Moroccan hydrogen potential rely on extrapolated or generic datasets, which undermines precision in environmental decision-making. To address this, we propose the creation of a national LCA database for energy systems, integration of site-specific environmental indicators (e.g., water scarcity, land-use sensitivity), and development of training programs for LCA practitioners in the region. Moreover, a regional platform bringing together policymakers, industry stakeholders, and researchers could serve as a foundation for standardizing LCA protocols adapted to North African contexts. Additionally, aligning green hydrogen pathways with infrastructure readiness and socioeconomic needs is essential. As highlighted by recent global studies (The Future of Hydrogen, 2019), deployment at scale will depend not only on lowering the production cost but also on enhancing storage, transport, and distribution infrastructure. For Morocco, a roadmap integrating distributed renewable microgrids, desalinated water use, and hybridization with existing gas systems could enable scalable, cost-effective deployment. Policy frameworks should thus prioritize not just

R&D funding but also support mechanisms for LCA-TEA integration, regulatory guidelines, and regional pilot projects. In conclusion, the future success of LCA for green hydrogen depends on three axes: (1) harmonization of methods at the international level, (2) localization of data, especially in emerging markets like Morocco, and (3) integration with economic and circular metrics to foster resilient and scalable systems. Through coordinated efforts among academia, industry, and government, LCA can become not just an evaluative tool but a strategic lever for the global hydrogen transition.

7.7 Conclusion

This chapter highlights the critical role of LCA in guiding the sustainable development of green hydrogen technologies. Despite global progress, methodological inconsistencies, especially in system boundaries, functional units, and co-product handling, continue to limit cross-study comparability and policy relevance. In emerging hydrogen markets like Morocco, the lack of localized LCA frameworks and data hinders accurate sustainability assessments. Addressing this gap requires harmonized LCA methodologies, region-specific indicators, and integration with techno-economic and circuleconomy analyses. Strengthening. This chapter highlights the critical role of LCA in guiding the sustainable development of green hydrogen technologies. Despite global progress, methodological inconsistencies, especially in system boundaries, functional units, and coproduct handling, continue to limit cross-study comparability and policy relevance. In emerging hydrogen markets like Morocco, the lack of localized LCA frameworks and data hinders accurate sustainability assessments. Addressing this gap requires harmonized LCA methodologies, region-specific indicators, and integration with techno-economic and circular economy analyses. Strengthening these foundations will enable LCA to evolve from a diagnostic tool into a strategic driver of the global green hydrogen transition.

Subscript and nomenclature

AEM	Anion exchange membrane electrolysis
AWE	Alkaline water electrolysis
CPV-T	Concentrated photovoltaic-thermal
GH_2	Green hydrogen
GHG	Greenhouse gas
IEA	International Energy Agency
IRENA	International Renewable Energy Agency
IRASEN	Institute for Research in Solar Energy and New Energies
LCIA	Life cycle impact assessment
LCA	Life cycle assessment
LCI	Life cycle inventory

MPPT	Maximum power point tracking
ONEE	Office National de l'Électricité et de l'Eau Potable
PEM	Proton exchange membrane
PV-T	Photovoltaic-thermal
SOEC	Solid oxide electrolysis
STH	Solar-to-hydrogen
TEA	Techno-economic assessment
TWSC	Thermochemical water splitting

References

Abdelkareem, M. A., Abdelghafar, A. A., Mahmoud, M., Sayed, E. T., Mahmoud, M. S., Alami, A. H., Agha, M. M. A., & Olabi, A. G. (2023). Optimized solar photovoltaic-powered green hydrogen: Current status, recent advancements, and barriers. *Solar Energy, 265*, 112072. doi:10.1016/j.solener.2023.112072.

Abdelsalam, R. A., Mohamed, M., Farag, H. E. Z., & El-Saadany, E. F. (2024). Green hydrogen production plants: A techno-economic review. *Energy Conversion and Management, 319*, 118907. doi:10.1016/j.enconman.2024.118907.

Acar, C. (2020). Solar hydrogen's role for a sustainable future. In T. S. Uyar (Ed.). In *Accelerating the transition to a 100% renewable energy era: 74* (pp. 309–331). Springer International Publishing. doi:10.1007/978-3-030-40738-4_14.

Ajaj, Y., Ali, M. B., Kumar, A., Jasim, D. J., Mancheno, C., Jabbar, A. H., Islam, S., Lasisi, A., & Zainul, R. (2024). Innovative biomass waste heat utilization for green hydrogen production: A comparative and optimization study of steam and organic Rankine cycles. *Process Safety and Environmental Protection, 190*, 148–174. doi:10.1016/j.psep.2024.08.006.

Ajeeb, W., Costa Neto, R., & Baptista, P. (2024). Life cycle assessment of green hydrogen production through electrolysis: A literature review. *Sustainable Energy Technologies and Assessments, 69*, 103923. doi:10.1016/j.seta.2024.103923.

Alghool, D., Bouden, C., Haouari, M., & Trucco, P. (2025). Harnessing enhanced solar efficiency for green hydrogen production: A comparative analysis of PV and PV-T systems. *International Journal of Hydrogen Energy, 98*, 394–406. doi:10.1016/j.ijhydene.2024.11.489.

Al-Ghussain, L., Alrbai, M., Al-Dahidi, S., & Lu, Z. (2024). Integrated assessment of green hydrogen production in California: Life cycle greenhouse gas emissions, techno-economic feasibility, and resource variability. *Energy Conversion and Management, 311*, 118514. doi:10.1016/j.enconman.2024.118514.

Ali, M. E. M., Almutairi, M. M., Mahmoud, M. S., Ismail, E., Khawassek, Y. M., & Atrees, M. S. (2022). Solar-driven photocatalyzed hydrogen production from water cleavage: Evaluation and process life cycle assessment. *Journal of Physics: Conference Series, 2305*(1), 012040. doi:10.1088/1742-6596/2305/1/012040.

Amrani, S.-E., Alami Merrouni, A., Touili, S., & Dekhissi, H. (2023). A multi-scenario site suitability analysis to assess the installation of large scale photovoltaic-hydrogen production units. Case study: Eastern Morocco. *Energy Conversion and Management, 295*, 117615. doi:10.1016/j.enconman.2023.117615.

Arfan, M., Eriksson, O., Wang, Z., & Soam, S. (2023). Life cycle assessment and life cycle costing of hydrogen production from biowaste and biomass in Sweden. *Energy Conversion and Management, 291*, 117262. doi:10.1016/j.enconman.2023.117262.

Azzaoui, A., & Alami Merrouni, A. (2025). Hybrid CSP-PV combination to enhance the green hydrogen production in Morocco: Solar technologies evaluation and techno-economic analysis. *Processes, 13*(3), 769. doi:10.3390/pr13030769.

Bahou, S. (2023). Techno-economic assessment of a hydrogen refuelling station powered by an on-grid photovoltaic solar system: A case study in Morocco. *International Journal of Hydrogen Energy, 48*(61), 23363–23372. doi:10.1016/j.ijhydene.2023.03.220.

Baker, J. W., & Lepech, M. D. (2009). September). Treatment of uncertainties in life cycle assessment. In Intl. Congress on Structral Safety and Reliability.

Boudries, R. (2016). Techno-economic assessment of solar hydrogen production using CPV-electrolysis systems. *Energy Procedia, 93*, 96–101. doi:10.1016/j.egypro.2016.07.155.

Bouramdane, A.-A. (2024). Crafting an optimal portfolio for sustainable hydrogen production choices in Morocco. *Fuel, 358*, 130292. doi:10.1016/j.fuel.2023.130292.

Bühler, L., & Möst, D. (2025). Projecting technological advancement of electrolyzers and the impact on the competitiveness of hydrogen. *International Journal of Hydrogen Energy, 98*, 1174–1184. doi:10.1016/j.ijhydene.2024.12.078.

Caillard, A., Yeganyan, R., Cannone, C., Plazas-Niño, F., & Howells, M. (2024). A critical analysis of Morocco's green hydrogen roadmap: A modelling approach to assess country readiness from the energy trilemma perspective. *Climate, 12*(5), 61. doi:10.3390/cli12050061.

Cetinkaya, E., Dincer, I., & Naterer, G. F. (2012). Life cycle assessment of various hydrogen production methods. *International Journal of Hydrogen Energy, 37*(3), 2071–2080. doi:10.1016/j.ijhydene.2011.10.064.

Chen, J., He, H., Quan, S., Zhang, Z., & Han, R (2023). Adaptive energy management for fuel cell hybrid power system with power slope constraint and variable horizon speed prediction. *International Journal of Hydrogen Energy, 48*(43), 16392–16405. doi:10.1016/j.ijhydene.2023.01.160.

Chen, Y., Hu, X., & Liu, J. (2019). Life cycle assessment of fuel cell vehicles considering the detailed vehicle components: Comparison and scenario analysis in china based on different hydrogen production schemes. *Energies, 12*(15), 3031. doi:10.3390/en12153031.

Chidire, A., Schifflechner, C., & Massier, T. (2023). Life cycle assessment of green hydrogen production via geothermal energy-driven electrolysis. In *2023 IEEE PES Innovative Smart Grid Technologies - Asia (ISGT Asia)* (pp. 1–5). doi:10.1109/ISGTAsia54891.2023.10372686.

Choi, B. B., Jo, J. H., Lee, T., Jeon, S.-Y., Kim, J., & Yoo, Y.-S. (2021). Operational characteristics of high-performance kW class alkaline electrolyzer stack for green hydrogen production. *Journal of Electrochemical Science and Technology, 12*(3), 302–307. doi:10.33961/jecst.2021.00031.

Curran, M. A. (2017). Overview of goal and scope definition in life cycle assessment. In M. A. Curran (Ed.), *Goal and scope definition in life cycle assessment* (pp. 1–62). Springer. doi:10.1007/978-94-024-0855-3_1.

Díaz-Motta, A., Díaz-González, F., & Villa-Arrieta, M. (2023). Energy sustainability assessment of offshore wind-powered ammonia. *Journal of Cleaner Production, 420*, 138419. doi:10.1016/j.jclepro.2023.138419.

Dincer, I., & Acar, C. (2015). Review and evaluation of hydrogen production methods for better sustainability. *International Journal of Hydrogen Energy, 40*(34), 11094–11111. doi:10.1016/j.ijhydene.2014.12.035.

Elgammal, A., & Ramlal, T. (2023). Adaptive demand-side management algorithm for grid-integrated PV-wind-battery-hydrogen systems using model predictive control. *European Journal of Electrical Engineering and Computer Science, 7*(6), 11–18. doi:10.24018/ejece.2023.7.6.578.

El Hassani, S., Oueslati, F., Horma, O., Santana, D., Moussaoui, M. A., & Mezrhab, A. (2023). Techno-economic feasibility and performance analysis of an islanded hybrid renewable energy system with hydrogen storage in Morocco. *Journal of Energy Storage, 68*, 107853. doi:10.1016/j.est.2023.107853.

El-Maaroufi, A., Daoudi, M., & Laamara, R. A. (2025). Hydrogen production for SDG 13 using hybrid renewables energies in southern Morocco. *Energy, 319*, 134986. doi:10.1016/j.energy.2025.134986.

El-Shafie, M. (2023). Hydrogen production by water electrolysis technologies: A review. *Results in Engineering, 20*, 101426. doi:10.1016/j.rineng.2023.101426.

Ennassiri, Y., Belhaj, I., & Bouzekri, H. (2019). Techno-economic assessment of hydrogen production from vRE in Morocco case study: Laayoune, Ouarzazate, midelt. In *2019 7th International Renewable and Sustainable Energy Conference (IRSEC)* (pp. 1–6). doi:10.1109/IRSEC48032.2019.9078331.

Farhani, S., Kaddachi, C., & Bacha, F. (2024). Study and simulation of green hydrogen production systems coupled with concentrated photovoltaic generators. In *2024 IEEE International Conference on Artificial Intelligence & Green Energy (ICAIGE)* (pp. 1–5). doi:10.1109/ICAIGE62696.2024.10776730.

Fernández-Lobato, L., Aguado, R., Escámez, A., Sánchez-Lozano, D., Oulbi, S., Hadria, R., & Vera, D. (2025). Gate-to-gate life cycle and techno-economic assessment of olive pomace gasification for renewable electricity and biochar production in Morocco. *Industrial Crops and Products, 234*, 121548. doi:10.1016/j.indcrop.2025.121548.

Ferrero, D., Lanzini, A., Santarelli, M., & Leone, P. (2013). A comparative assessment on hydrogen production from low- and high-temperature electrolysis. *International Journal of Hydrogen Energy, 38*(9), 3523–3536. doi:10.1016/j.ijhydene.2013.01.065.

Finnveden, G., Hauschild, M. Z., Ekvall, T., Guinée, J., Heijungs, R., Hellweg, S., Koehler, A., Pennington, D., & Suh, S. (2009). Recent developments in life cycle assessment. *Journal of Environmental Management, 91*(1), 1–21. doi:10.1016/j.jenvman.2009.06.018.

Gallandat, N., Romanowicz, K., & Züttel, A. (2017). An analytical model for the electrolyser performance derived from materials parameters. *Journal of Power and Energy Engineering, 05*(10), 34–49. doi:10.4236/jpee.2017.510003.

Ganesh, G. S., Dasappa,, S., Patil, B., & Shivapuji, A. M. (2025). Comparative economic and environmental performance assessment of biomass gasification pathway for green H2 production. *Detritus*, (30), 75–83. doi:10.31025/2611-4135/2025.19470.

Gebauer, C., Marić, R., Bulic, M., Ashley, M., & Swisher, J. (2024). Novel Ru-based PEM electrolysis catalyst – development from the nano- to the industrial-scale. *ECS Meeting Abstracts, MA2024-01*, (34), 1777. doi:10.1149/MA2024-01341777mtgabs.

Gerhardt-Mörsdorf, J., Incer-Valverde, J., Morosuk, T., & Minke, C. (2024). Exergetic life cycle assessment for green hydrogen production. *Energy, 299*, 131553. doi:10.1016/j.energy.2024.131553.

Ghorbani, B., Zendehboudi, S., Zhang, Y., Zarrin, H., & Chatzis, I. (2023). Thermochemical water-splitting structures for hydrogen production: Thermodynamic, economic, and environmental impacts. *Energy Conversion and Management, 297*, 117599. doi:10.1016/j.enconman.2023.117599.

Hake, J.-F., Koj, J. C., Kuckshinrichs, W., Schlör, H., Schreiber, A., Wulf, C., Zapp, P., & Ketelaer, T. (2017). Towards a life cycle sustainability assessment of alkaline water electrolysis. *Energy Procedia, 105*, 3403–3410. doi:10.1016/j.egypro.2017.03.779.

Hofrichter, A., Rank, D., Heberl, M., & Sterner, M. (2023). Determination of the optimal power ratio between electrolysis and renewable energy to investigate the effects on the hydrogen production costs. *International Journal of Hydrogen Energy, 48*(5), 1651–1663. doi:10.1016/j.ijhydene.2022.09.263.

IEA, Global hydrogen review 2022. (2022), IEA. https://www.iea.org/reports/global-hydrogen-review-2022

IRENA, Green hydrogen cost reduction: Scaling up electrolysers to meet the 1.5°C climate goal. (2020). International Renewable Energy Agency, https://www.irena.org/-/media/Files/IRENA/Agency/Publication/2020/Dec/IRENA_Green_hydrogen_cost_2020.pdf

Hunt, J., & Nascimento, A. (2021). Electrolysis ship for green hydrogen production and possible applications. *Int J Energy Environ Eng, 15*, 42e6.

Iribarren, D., Susmozas, A., Petrakopoulou, F., & Dufour, J. (2014). Environmental and exergetic evaluation of hydrogen production via lignocellulosic biomass gasification. *Journal of Cleaner Production, 69*, 165–175. doi:10.1016/j.jclepro.2014.01.068.

Ishaq, H., & Dincer, I. (2021). Comparative assessment of renewable energy-based hydrogen production methods. *Renewable and Sustainable Energy Reviews, 135*, 110192. doi:10.1016/j.rser.2020.110192.

Jang, D., Kim, J., Kim, D., Han, W.-B., & Kang, S. (2022). Techno-economic analysis and Monte Carlo simulation of green hydrogen production technology through various water electrolysis technologies. *Energy Conversion and Management, 258*, 115499. doi:10.1016/j.enconman.2022.115499.

Jónsson, V. K., Gunnarsson, R. L., Árnason, B., & Sigfússon, T. I. (1992). The feasibility of using geothermal energy in hydrogen production. *Geothermics, 21*(5-6), 673–681. doi:10.1016/0375-6505(92)90020-A.

Jovan, D. J., Dolanc, G., & Pregelj, B. (2021). Cogeneration of green hydrogen in a cascade hydropower plant. *Energy Conversion and Management: X, 10*, 100081. doi:10.1016/j.ecmx.2021.100081.

Kalinci, Y., Hepbasli, A., & Dincer, I. (2012). Life cycle assessment of hydrogen production from biomass gasification systems. *International Journal of Hydrogen Energy, 37*(19), 14026–14039. doi:10.1016/j.ijhydene.2012.06.015.

Kanz, O., Brüggemann, F., Ding, K., Bittkau, K., Rau, U., & Reinders, A. (2023). Life-cycle global warming impact of hydrogen transport through pipelines from Africa to Germany. *Sustainable Energy & Fuels, 7*(13), 3014–3024. doi:10.1039/D3SE00281K.

Kelman, R., Gaspar, L. D. S., Geyer, F. S., Barroso, L. A. N., & Pereira, M. V. F. (2020). Can Brazil become a green hydrogen powerhouse? *Journal of Power and Energy Engineering, 08*(11), 21–32. doi:10.4236/jpee.2020.811003.

Kim, K.-M., & Kim, D. (2024). Life cycle assessment of greenhouse gas emissions in hydrogen production via water electrolysis in South Korea. *Sustainability, 16*(24), 11010. doi:10.3390/su162411010.

Kim, S.-D., Choi, Y., & Park, H. (2024). Advances in solid oxide electrolysis cell (SOEC) technology for sustainable, efficient, and durable hydrogen production. *ECS Meeting Abstracts, MA2024-02*, (48), 3437. doi:10.1149/MA2024-02483437mtgabs.

(2014). *Life cycle assessment (LCA): A guide to best practice.* In Klöpffer, W., & Grahl, B. (Eds.) (1st ed.). Wiley. doi:10.1002/9783527655625.

Kodgire, P. (2025). Hydrogen - imminent clean and green energy: Hydrogen production technologies life cycle assessment review. *Process Safety and Environmental Protection, 193*, 483–500. doi:10.1016/j.psep.2024.11.019.

Li, G., Song, C., Han, G., Chen, Z., & Huang, J. (2024). Optimization of hydrogen production system performance using photovoltaic/thermal-coupled PEM. *Energies, 17*(21), 5405. doi:10.3390/en17215405.

Li, Q., Song, G., Xiao, J., Hao, J., Li, H., & Yuan, Y. (2020). Exergetic life cycle assessment of hydrogen production from biomass staged-gasification. *Energy, 190*, 116416. doi:10.1016/j.energy.2019.116416.

Li, X., Sun, X., Song, Q., Yang, Z., Wang, H., & Duan, Y. (2022). A critical review on integrated system design of solar thermochemical water-splitting cycle for hydrogen production. *International Journal of Hydrogen Energy, 47*(79), 33619–33642. doi:10.1016/j.ijhydene.2022.07.249.

Luo, Y., & Su, R. (2024). Environmental Impact of waste treatment and synchronous hydrogen production: Based on life cycle assessment method. *Toxics, 12*(9), 652. doi:10.3390/toxics12090652.

Mahfoud, R. J., Alkayem, N. F., Zhang, Y., Zheng, Y., Sun, Y., & Alhelou, H. H. (2023). Optimal operation of pumped hydro storage-based energy systems: A compendium of current challenges and future perspectives. *Renewable and Sustainable Energy Reviews, 178*, 113267. doi:10.1016/j.rser.2023.113267.

Maniscalco, M. P., Longo, S., Cellura, M., Miccichè, G., & Ferraro, M. (2024). Critical review of life cycle assessment of hydrogen production pathways. *Environments, 11*(6), 108. doi:10.3390/environments11060108.

Maurya, J., Gemechu, E., & Kumar, A. (2023). Life cycle assessment of earth-abundant photocatalysts for enhanced photocatalytic hydrogen production. *International Journal of Hydrogen Energy, 48*(52), 20077–20095. doi:10.1016/j.ijhydene.2023.01.345.

Mazumder, G. C., Hasan, S., Rahman Zishan, M. S., & Rahman, M. H. (2023). Solar PV- and PEM electrolyzer-based green hydrogen production cost for a selected location in Bangladesh. *Academia Green Energy, 1*(1). doi:10.20935/AcadEnergy6160.

Mulk, W. U., Aziz, A. R. A., Ismael, M. A., Ghoto, A. A., Ali, S. A., Younas, M., & Gallucci, F. (2024). Electrochemical hydrogen production through anion exchange membrane water electrolysis (AEMWE): Recent progress and associated challenges in hydrogen production. *International Journal of Hydrogen Energy, 94*, 1174–1211. doi:10.1016/j.ijhydene.2024.11.143.

Molho, S., Roccaro, E., Lainati, A., & Stefanucci, M. (2024). Techno-Economic Evaluation of the Integration of High-Temperature Electrolysis With Geothermal Energy for Low-Carbon Hydrogen Production. SPE Europe Energy Conference and Exhibition, D011S008R003. https://doi.org/10.2118/220050-MS.

Nasser, M., & Hassan, H. (2024). Feasibility analysis and Atlas for green hydrogen project in MENA region: Production, cost, and environmental maps. *Solar Energy, 268*, 112326. doi:10.1016/j.solener.2024.112326.

Osman, A. I., Nasr, M., Mohamed, A. R., Abdelhaleem, A., Ayati, A., Farghali, M., Al-Muhtaseb, A. H., Al-Fatesh, A. S., & Rooney, D. W. (2024). Life cycle assessment of hydrogen production, storage, and utilization toward sustainability. *WIREs Energy and Environment, 13*(3), e526. doi:10.1002/wene.526.

Ourya, I., Nabil, N., Abderafi, S., Boutammachte, N., & Rachidi, S. (2023). Assessment of green hydrogen production in Morocco, using hybrid renewable sources (PV and wind). *International Journal of Hydrogen Energy, 48*(96), 37428–37442. doi:10.1016/j.ijhydene.2022.12.362.

Ozbilen, A., Dincer, I., & Rosen, M. A. (2013). Comparative environmental impact and efficiency assessment of selected hydrogen production methods. *Environmental Impact Assessment Review, 42*, 1–9. doi:10.1016/j.eiar.2013.03.003.

Ozbilen, A., Dincer, I., & Rosen, M. A. (2014). Energy and exergy analyses of copper–chlorine (Cu–Cl) based integrated systems for hydrogen production. In I. Dincer, A. Midilli, & H. Kucuk (Eds.), *Progress in exergy, energy, and the environment* (pp. 111–119). Springer International Publishing. doi:10.1007/978-3-319-04681-5_10.

Ozturk, M. U., Ayol, A., & Tezer, O. (2023). Life cycle assessment of olive pomace gasification for an up-draft fixed bed gasifier system. *International Journal of Hydrogen Energy, 48*(60), 23339–23347. doi:10.1016/j.ijhydene.2023.01.206.

Palmer, G., Roberts, A., Hoadley, A., Dargaville, R., & Honnery, D. (2021). Life-cycle greenhouse gas emissions and net energy assessment of large-scale hydrogen production via electrolysis and solar PV. *Energy & Environmental Science, 14*(10), 5113–5131. doi:10.1039/D1EE01288F.

Panigrahy, B., Narayan, K., & Ramachandra Rao, B. (2022). Green hydrogen production by water electrolysis: A renewable energy perspective. *Materials Today: Proceedings, 67*, 1310–1314. doi:10.1016/j.matpr.2022.09.254.

Pitchaimuthu, S., Sridharan, K., Nagarajan, S., Ananthraj, S., Robertson, P., Kuehnel, M. F., Irabien, Á., & Maroto-Valer, M. (2022). Solar hydrogen fuel generation from wastewater—beyond photoelectrochemical water splitting: A perspective. *Energies, 15*(19), 7399. doi:10.3390/en15197399.

Ramadan, A., & Gabbar, H. A. (2024). Evaluation of hydrogen generation with hybrid renewable energy sources. *Applied Sciences, 14*(14), 6235. doi:10.3390/app14146235.

Ravilla, A., Song, Z., Yan, Y., & Celik, I. (2024). Comparative life cycle assessment of hydrogen production via photoelectrochemical (PEC) and photovoltaic-electrochemical (PV-EC) water splitting. In *2024 IEEE 52nd Photovoltaic Specialist Conference (PVSC)* (pp. 377–380). doi:10.1109/PVSC57443.2024.10748772.

Rieradevall, J., Domènech, X., & Fullana, P. (1997). Application of life cycle assessment to landfilling. *The International Journal of Life Cycle Assessment, 2*(3), 141. doi:10.1007/BF02978806.

Sadeghi, S., & Ghandehariun, S. (2023). Environmental impacts of a standalone solar water splitting system for sustainable hydrogen production: A life cycle assessment. *International Journal of Hydrogen Energy, 48*(50), 19326–19339. doi:10.1016/j.ijhydene.2023.01.234.

Scholz, R., & Tietje, O. (2002). *Embedded case study methods*. SAGE Publications, Inc. doi:10.4135/9781412984027.

Schropp, E., Naumann, G., & Gaderer, M. (2022). Prospective life cycle assessment : A case study of hydrogen production with water electrolysis. *Procedia CIRP, 105*, 92–97. doi:10.1016/j.procir.2022.02.016.

Sebbahi, S., Assila, A., Alaoui Belghiti, A., Laasri, S., Kaya, S., Hlil, E. K., Rachidi, S., & Hajjaji, A. (2024). A comprehensive review of recent advances in alkaline water electrolysis for hydrogen production. *International Journal of Hydrogen Energy, 82*, 583–599. doi:10.1016/j.ijhydene.2024.07.428.

Seghetta, M., & Goglio, P. (2018). Life Cycle assessment of seaweed cultivation systems. In K. Spilling (Ed.). In *Biofuels from algae: 1980* (pp. 103–119). Springer. doi:10.1007/7651_2018_203.

Sharafi Laleh, S., Gharamaleki, F. P., Alavi, S. F., Soltani, S., Mahmoudi, S. M. S., & Rosen, M. A. (2023). A novel sustainable biomass-fueled cogeneration cycle integrated with carbon dioxide capture utilizing LNG regasification and green hydrogen production via PEM electrolysis: Thermodynamic assessment. *Journal of Cleaner Production, 421*, 138529. doi:10.1016/j.jclepro.2023.138529.

Shaya, N., & Glöser-Chahoud, S. (2024). A review of life cycle assessment (LCA) studies for hydrogen production technologies through water electrolysis: Recent advances. *Energies, 17*(16), 3968. doi:10.3390/en17163968.

Sigurvinsson, J., Mansilla, C., Arnason, B., Bontemps, A., Maréchal, A., Sigfusson, T. I., & Werkoff, F. (2006). Heat transfer problems for the production of hydrogen from geothermal energy. *Energy Conversion and Management, 47*(20), 3543–3551. doi:10.1016/j.enconman.2006.03.012.

Siregar, K., Supriyanto, S., Saka Rani, D., Nurdiansyah, Y., & Wijayanto, F. (2021). Conceptual design of inventory analysis software to support the life cycle assessment in palm oil production. *E3S Web of Conferences, 226*, 00028. doi:10.1051/e3sconf/202122600028.

Sola, A., Rosa, R., & Ferrari, A. M. (2025). Green hydrogen and its supply chain. A critical assessment of the environmental impacts. *Advanced Sustainable Systems, 9*(2), 2400708. doi:10.1002/adsu.202400708.

Sonnemann, G., Tsang, M., & Castells, F. (2018). Life-cycle impact assessment. In G. Sonnemann, M. Tsang, & M. Schuhmacher (Eds.), *Integrated life-cycle and risk assessment for industrial processes and products* (pp. 79–118). CRC Press. doi:10.1201/9780429436949-3.

The Future of Hydrogen. (2019).

Touili, S., Alami Merrouni, A., El Hassouani, Y., Amrani, A., & Rachidi, S. (2020). Analysis of the yield and production cost of large-scale electrolytic hydrogen from different solar technologies and under several Moroccan climate zones. *International Journal of Hydrogen Energy, 45*(51), 26785–26799. doi:10.1016/j.ijhydene.2020.07.118.

Tripathi, A. K., & Subramanian, K. A. (2022). Comparative life cycle energy and GHG emissions assessment of green hydrogen production with different electrolysers and solar photovoltaics system. In *2022 IEEE North Karnataka Subsection Flagship International Conference (NKCon)* (pp. 1–9). doi:10.1109/NKCon56289.2022.10127022.

Valente, A., Iribarren, D., & Dufour, J. (2019). Life cycle sustainability assessment of hydrogen from biomass gasification: A comparison with conventional hydrogen. *International Journal of Hydrogen Energy, 44*(38), 21193–21203. doi:10.1016/j.ijhydene.2019.01.105.

Valente, A., Iribarren, D., & Dufour, J. (2021). Comparative life cycle sustainability assessment of renewable and conventional hydrogen. *Science of The Total Environment, 756*, 144132. doi:10.1016/j.scitotenv.2020.144132.

Vartiainen, E., Breyer, C., Moser, D., Román Medina, E., Busto, C., Masson, G., Bosch, E., & Jäger-Waldau, A. (2022). True cost of solar hydrogen. *Solar RRL, 6*(5), 2100487. doi:10.1002/solr.202100487.

Wang, J., Yang, J., Feng, Y., Hua, J., Chen, Z., Liao, M., Zhang, J., & Qin, J. (2025). Comparative experimental study of alkaline and proton exchange membrane water electrolysis for green hydrogen production. *Applied Energy, 379*, 124936. doi:10.1016/j.apenergy.2024.124936.

Weidner, T., Tulus, V., & Guillén-Gosálbez, G. (2023). Environmental sustainability assessment of large-scale hydrogen production using prospective life cycle analysis. *International Journal of Hydrogen Energy, 48*(22), 8310–8327. doi:10.1016/j.ijhydene.2022.11.044.

Yang, Y., Liu, S., Jing, D., & Zhao, L. (2022). Continuous hydrogen production characteristics and environmental analysis of a pilot-scale hydrogen production system by photocatalytic water splitting. *Energy Conversion and Management, 268*, 115988. doi:10.1016/j.enconman.2022.115988.

Yilmaz, & Kanoglu (2017). Investigation of hydrogen production cost by geothermal energy. *International Advanced Researches and Engineering Journal, 1*(1), 5-10. https://izlik.org/JA38KU62FR

Yilmaz, C., & Kanoglu, M. (2014). Thermodynamic evaluation of geothermal energy powered hydrogen production by PEM water electrolysis. *Energy, 69*, 592–602. doi:10.1016/j.energy.2014.03.054.

Zhang, J., Wang, Z., He, Y., Li, M., Wang, X., Wang, B., Zhu, Y., & Cen, K. (2023). Comparison of onshore/offshore wind power hydrogen production through water electrolysis by life cycle assessment. *Sustainable Energy Technologies and Assessments, 60*, 103515. doi:10.1016/j.seta.2023.103515.

Zhang, Q., Shan, Y., Pan, J., Kumar, P., Keevers, M. J., Lasich, J., Kour, G., Daiyan, R., Perez-Wurf, I., Thomsen, L., Cheong, S., Jiang, J., Wu, K.-H., Chiang, C.-L., Grayson, K., Green, M. A., Amal, R., & Lu, X. (2025). A photovoltaic-electrolysis system with high solar-to-hydrogen efficiency under practical current densities. *Science Advances, 11*(9), eads0836. doi:10.1126/sciadv.ads0836.

Zhao, G., Kraglund, M. R., Frandsen, H. L., Wulff, A. C., Jensen, S. H., Chen, M., & Graves, C. R. (2020). Life cycle assessment of H2O electrolysis technologies. *International Journal of Hydrogen Energy, 45*(43), 23765–23781. doi:10.1016/j.ijhydene.2020.05.282.

Zhao, G., & Pedersen, A. S. (2018). Life cycle assessment of hydrogen production and consumption in an isolated territory. *Procedia CIRP, 69*, 529–533. doi:10.1016/j.procir.2017.11.100.

Chapter 8

Weather-agnostic machine learning framework for forecasting offshore wind-to-hydrogen production

Salma Boujmiraz[a] and Asmae Berrada[b]
[a]*LRIT, Faculty of Science, Mohammed V University in Rabat, Rabat, Morocco,* [b]*LERMA Lab, College of Engineering and Architecture, International University of Rabat, Sala Al Jadida, Morocco*

8.1 Introduction

Green electricity from water electrolysis, known as "green hydrogen," produces hydrogen without direct carbon dioxide emissions. However, high production costs remain a barrier to widespread adoption. While hydrogen from steam methane reforming costs less than 2 $/kgH$_2$, electrolysis with renewable energy costs 8–12 $/kgH$_2$ (Safari and Dincer, 2020). Despite this, green hydrogen production has increased significantly in recent years, with offshore wind emerging as a particularly promising energy source.

Research on offshore wind-to-hydrogen systems has primarily focused on techno-economic optimization and system design. Studies have examined connection architectures (onshore, centralized offshore, decentralized offshore) (Calado et al., 2024; Ibrahim et al., 2022; Luo et al., 2022; Singlitico et al., 2021), cost comparisons between power and hydrogen export (Lucas et al., 2022; Lüth et al., 2023), and levelized cost of hydrogen (LCOH) calculations across different regions and configurations (d'Amore Domenech et al., 2021; Giampieri et al., 2023; Jiang et al., 2022; Komorowska et al., 2023; Miao et al., 2021). Dinh et al. (2023) developed geospatial models for Ireland showing LCOH values below 4 $/kg in offshore regions. Rogeau et al. (2023) projected European LCOH declining from 5–7.5 $/kg (2020) to 1.5–3.0 $/kg (2050) as technology costs decrease. Recent studies have assessed specific regional opportunities in the Iberian Peninsula (Calado et al., 2024), Mediterranean Sea (Bonacina et al., 2022), Norway (Meier, 2014), Ireland

Decarbonizing through Sustainable Hydrogen and its Derivatives.
DOI: https://doi.org/10.1016/B978-0-443-27632-3.00005-0

(Leahy et al., 2021), Poland (Komorowska et al., 2023), Chile and Korea (Ayeon et al., 2023), and China (Yamin et al., 2021), with reported LCOH values ranging from 1.64 to 12.6 $/kg depending on location, technology, and assumptions.

While techno-economic optimization dominates the literature, operational forecasting of hydrogen production remains underexplored. Accurate short-term forecasting is essential for production scheduling, storage management, delivery planning, and grid integration. However, conventional wind forecasting approaches create practical barriers: they require detailed meteorological inputs (wind speed, direction, temperature, pressure), individual turbine specifications, site-specific power curves, and access to numerical weather prediction (NWP) models. For offshore environments, obtaining high-quality meteorological data requires expensive infrastructure, including offshore met masts, LiDAR systems, or satellite data subscriptions. These data dependencies can delay forecasting system deployment, particularly for emerging offshore hydrogen projects.

Machine learning (ML) has increasingly been applied to renewable energy forecasting and optimization. Techniques including long short-term memory (LSTM), recurrent neural networks, and support vector regression have demonstrated success in power prediction, fault diagnosis, and resource evaluation (Allal et al., 2024; Altan et al., 2020; Cheng et al., 2023; Ren et al., 2024). Recent studies have applied ML to hydrogen production forecasting from renewable sources (Achour et al., 2023; Izadi et al., 2022; Javaid et al., 2022; Naveena et al., 2024; Salari et al., 2023; Shams et al., 2021; Shin et al., 2023). Kabir et al. (2023) used K-nearest neighbors and random forest to predict green hydrogen from dark fermentation and PEM technologies. Achour et al. (2025) employed deep learning for power and ammonia output forecasting. Javaid et al. (2022) utilized LSTM and SVR models to predict PEM electrolyzer hydrogen production from wind turbines. However, these studies focus on land-based systems with readily available meteorological data. The application of ML to offshore wind-hydrogen forecasting remains limited, particularly approaches that address offshore data accessibility challenges.

This study develops a weather-agnostic ML framework for forecasting offshore wind power and hydrogen production. Unlike conventional approaches requiring meteorological infrastructure, this framework operates exclusively on aggregated power generation measurements from transmission system operators (TSOs). The weather-agnostic design offers several advantages: require only TSO data that are often publicly available, eliminating meteorological infrastructure costs, naturally capture fleet-level operational realities including maintenance, curtailment, and control strategies that weather-based models miss, transfer easily to any offshore region with historical generation data without site-specific calibration, reduce computational requirements by avoiding complex physics-based transformations, and enable immediate deployment for emerging projects lacking meteorological data networks.

The research evaluates five deep learning architectures (LSTM, bidirectional long short-term memory [BiLSTM], gated recurrent unit [GRU], Transformer, TCN) using 3 years of operational data from Belgium's offshore wind fleet. The validated wind power forecasts are integrated with PEM electrolyzer modeling to predict hydrogen production at operationally relevant timescales (15–30 min). This forecasting capability supports real-time decision-making in electrolyzer control, storage management, and delivery scheduling without requiring weather forecasts or turbine-level monitoring.

This chapter is organized into seven sections. Section 8.1 introduces the topic and reviews relevant literature. Section 8.2 describes offshore wind-based hydrogen production systems. Section 8.3 presents mathematical modeling of wind turbines and electrolyzers. Section 8.4 details the weather-agnostic forecasting methodology, including data preparation, model architectures, and validation procedures. Section 8.5 presents results including cross-validation performance, test set evaluation, and hydrogen production forecasting. Section 8.6 concludes with key findings and future research directions.

8.2 Offshore wind-based hydrogen production

8.2.1 System overview

Green hydrogen is produced through electrolysis powered directly from renewables (e.g., wind, solar, or hydropower). The electrolyzer separates water into hydrogen and oxygen. Green hydrogen refers to hydrogen production without any CO_2 or other greenhouse gas emissions, unlike blue hydrogen made from fossil fuels with carbon capture.

Centralized offshore wind-to-hydrogen produces hydrogen in a single wind farm area. Depending on the size of the HPU, a 1 GW wind farm's installed HPU capacity might range from 900 to 1000 MW. A schematic representation of the centralized offshore wind-to-hydrogen setup is shown in Fig. 8.1. The flow lines between the wind turbines, interarray cables, electrical system, electrolyzers, BOP, and hydrogen pipelines are depicted in this image. Seawater, desalinated water, hydrogen, AC power, and DC power are among the flow lines. In the centralized offshore wind-to-hydrogen setup, the electrolyzers are installed on a single substation offshore. Due to the high electrolysis capability of a substation, a massive jacket substructure similar to the current oil and gas platforms is required to produce hydrogen.

8.2.2 System components

Desalination: Electrolyzers require 9–15 L of ultrapure water per kilogram of hydrogen produced (Ibrahim et al., 2022; Singlitico et al., 2021). Offshore

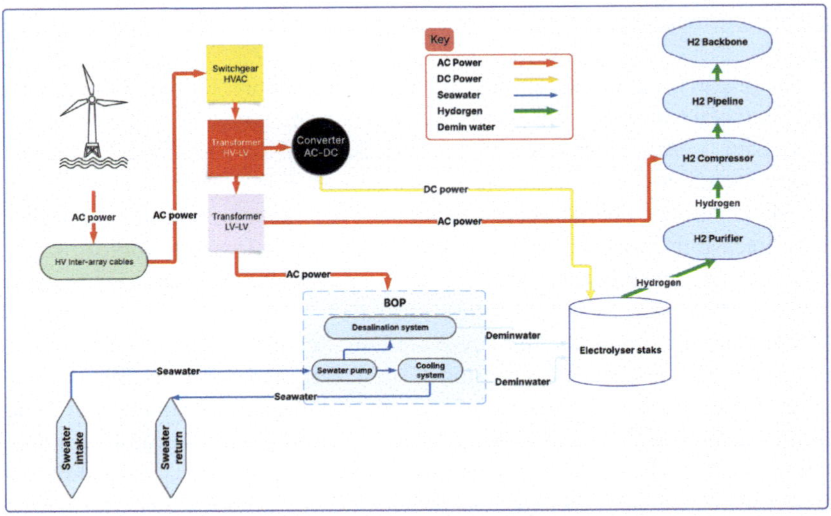

FIGURE 8.1 Schematic of offshore wind-based hydrogen production system.

systems use reverse osmosis (RO) desalination, consuming approximately 3–6 kWh/m^3 (Abdelkareem et al., 2018; Alkaisi et al., 2017; Ma & Lu, 2011). This represents less than 0.1% of total system energy use, as electrolysis requires about 60 kWh/kgH_2. Brine discharge requires site-specific equipment.

Electrolysis: Proton exchange membrane (PEM) electrolyzers are preferred for offshore applications due to compact design, fast response times, and the ability to handle variable power input from wind generation. Current PEM efficiency ranges from 65% to 70% with continuous improvement trends.

Hydrogen transport: Offshore-produced hydrogen is typically transported to shore via subsea pipelines, although ship-based transport in compressed or liquefied form is also being evaluated (Miao et al., 2021).

Substations: Hydrogen substations consist of topside structures housing electrolyzer equipment and substructures providing foundation support. Jacket structures are common for deeper waters, while monopiles suit shallower depths.

8.2.3 Operational challenges and forecasting needs

Several announced offshore hydrogen projects are in development (Table 8.1), with Sealhyfe being the first operational offshore electrolyzer deployed in June 2023 (Lhyfe 2023). Most projects remain in planning phases without accumulated operational experience.

Operational forecasting addresses critical challenges in these systems.

Variable power input: Offshore wind generation fluctuates on timescales from minutes to seasons. Electrolyzers operate most efficiently within 30%–90% capacity but can accommodate 10%–110% rated power

TABLE 8.1 Offshore hydrogen production projects.

Project name	Capacity (MW)	Expected completion	Project description	References
H2Res	2	2022	The Danish energy company Ørsted intends to build a 2 MW onshore facility in Denmark. The system will produce 1000 kg of hydrogen per day.	(H2RES, 2023)
Lhyfe	1	2023	June 2023 marked the first practical offshore hydrogen production off the coast of France. Every day, the 1 MW will generate up to 400 kg of hydrogen. The Sealhyfe platform occupies 200 square meters.	(Hill, 2023)
Oyster	>1	2024	The aim of the project is to develop and validate a marine electrolyzer for large-scale hydrogen production and export.	(Oyster, 2023)
Crosswind	2.5	2024	A 2.5 MW PEM will be used for a 759 MW offshore wind farm in a joint Shell and Eneco venture. When energy production is high, the project is intended to produce hydrogen.	
H2-Wind	5	2025	The project, led by the Fraunhofer Institute, aims to establish a technological and scientific foundation for producing hydrogen offshore. The project aims to have a 5 MW system operational by the end of the project.	(Thomas Schiestel, 2023)

(continued on next page)

TABLE 8.1 Offshore hydrogen production projects—cont'd

Project name	Capacity (MW)	Expected completion	Project description	References
PosHYdon	1	2025	400 kg of hydrogen will be produced daily by a Dutch initiative that will install a 1 MW system. It will be positioned 13 kilometers off the Dutch coast and has a desalination facility.	(Poshydon, 2023)
Salamander	5000	2030	The Dolphyn desalination module is being developed as part of a 100 MW floating wind energy project.	(Snieckus, 2021)
AquaVentus	300	2031	The project aims to set up a 300 MW electrolyzer to generate 20,000 tons of hydrogen a year. The project is broken down into three phases: decentralized hydrogen production, pipeline building, and production development.	(RWE, 2021)
Dolphyn	100	2032	Large-scale offshore floating wind-powered production of green hydrogen. The goal is to gradually build up production to GW-build by the early 2030s.	(Environmental Resources Management, 2021)

(Rezaei et al., 2024c). Forecasting upcoming power availability enables optimal load scheduling.

Storage management: Hydrogen storage buffers mismatches between production and delivery schedules. Short-term forecasts (15 min to hours) guide charging/ discharging decisions to maintain delivery commitments while operating electrolyzers efficiently.

Grid integration: Systems with hybrid power-hydrogen export require forecasts to optimize the power-hydrogen split based on electricity prices and hydrogen demand.

Maintenance planning: Predicting low-production periods enables strategic maintenance scheduling without disrupting hydrogen supply contracts. Despite growing interest in offshore hydrogen, real-time operational data remains scarce due to limited deployed systems. This creates challenges for developing and validating forecasting approaches. The weather-agnostic methodology developed in this study addresses this gap by using readily available wind generation data to predict production patterns relevant to these operational decisions.

8.3 Mathematical foundations for offshore wind

This section presents the mathematical foundations for offshore wind power generation and hydrogen production via electrolysis. While the forecasting approach developed in Section 8.4 is data-driven and weather-agnostic, understanding these physical relationships provides context for system behavior and electrolyzer integration.

8.3.1 Offshore wind turbine modeling

The mathematical modeling of offshore wind turbines is essential for predicting energy production and optimizing performance. The following equations provide a framework for analyzing mechanical power extraction, turbine efficiency, and wind power prediction (Achour et al., 2023; El Mrabet & Berrada, 2020).

Mechanical power extracted by the rotor is expressed in Eq. (8.1):

$$P = \frac{1}{2}\rho A C_P(\lambda, \beta) v^3 \tag{8.1}$$

Tip speed ratio (TSR) is expressed in Eq. (8.2):

$$\lambda = \frac{\omega R}{v} \tag{8.2}$$

Performance coefficient C_p is expressed in Eqs. (8.3) and (8.4):

$$C_p(\lambda, \beta) = 0,22\left(\frac{116}{\lambda_i} - 0.4\beta - 5\right)e^{\frac{-12,5}{\lambda_i}} \tag{8.3}$$

$$\frac{1}{\lambda_i} = \frac{1}{\lambda + 0.08\beta} - \frac{0.035}{\beta^3 + 1} \tag{8.4}$$

Wind power output is determined by Eq. (8.5) (Emrani et al., 2024):

$$P_w = \begin{cases} 0 v < v_{\text{cut-in}} \\ \frac{N_{wt} \times P_{\text{wind\_rate}} \times (v^3 - v_{\text{cut-in}}^3)}{v_{\text{rate}}^3 - v_{\text{cut-in}}^3} v_{\text{cut-in}} < v < v_{\text{rate}} \\ N_{wt} \times P_{\text{wind}_{\text{rate}}} v_{\text{rate}} \leq v \leq v_{\text{cut-out}} \\ 0 v_{\text{cut-out}} \leq v \end{cases} \tag{8.5}$$

$$v_{z1} = v_{z0} \left(\frac{z1}{z0} \right)^{\alpha} \tag{8.6}$$

8.3.2 Water electrolysis model

The electrochemical process of water electrolysis uses electricity to divide water into two gases: oxygen and hydrogen. Depending on the electrolyte utilized in the electrolysis cell, three primary technologies can be identified: alkaline electrolysis using an electrolyte that is liquid electrolyte, solid polymer electrolysis (SPE), also known as polymer electrolyte membrane (PEM) electrolysis with an acidic ionomer electrolyte, and solid oxide electrolyte-based high-temperature (HT) steam electrolysis. PEMs are used in this case study. The model has been modified to correspond with the required power The hourly hydrogen production is calculated using Eq. (8.7) (Achour et al., 2025)

$$M_t = \frac{E_{in_t}}{SEC_t} \tag{8.7}$$

where SEC_t is the specific energy consumption of the electrolysis per hour, M_t is the mass of hydrogen produced per hour, and E_{in_t} is the electrical power input. SEC fluctuates inversely with efficiency and is affected by changes in input power; in addition, it accounts for all hydrogen manufacturing losses. Rezaei et al. (2024c) have extensively investigated the overload capacity and power-dependent efficiency of PEM and alkaline electrolysis technologies. As a result of utilizing PEM technology and increasing production, lower LCOH was found (Rezaei et al., 2024b) A dual model has been developed, which links efficiency to input power. The model used in Kopp et al. (2017) incorporated a 0.25% increase in electrolysis efficiency annually. Eq. (8.8) provides an estimate of the efficiency of electrolysis based on the year 2024:

$$\eta_{PEM}^t = \begin{cases} \text{if} R_p < 15\% \rightarrow \left(5.0e - 05 \times R_p{}^5 - 6.1e - 03 \times R_p{}^4 + 0.2372 \times R_p{}^3 \right. \\ \quad -4.2014 \times R_p{}^2 + 36.675 \times R_p \\ \quad \left. -62.87\right) \times 1.0025^{\text{Basis year}-2020} \\ \text{if} R_p > 15\% \rightarrow \left(-0.149 \times R_p + 74.977\right) \times 1.0025^{\text{Basis year}-2020} \end{cases} \tag{8.8}$$

To ensure the plant's continuous operation, a bulk energy storage system is incorporated into the hydrogen production system. R_p represents the input power to the rated power of the electrolysis, and η_{PEM}^t reflects the electrolysis efficiency. As hydrogen production consumes most of the energy, a greater capacity for hydrogen storage might be necessary (Marocco et al., 2022).

TABLE 8.2 Dataset characteristics.

Attribute	Value
Time period	May 1, 2022 to May 31, 2025
Total observations	108,192
Sampling interval	15 min
Training samples (80%)	86,553
Test samples (20%)	21,639
Missing values (postprocessing)	0

8.4 Weather-agnostic production forecasting via machine learning

This section presents the ML framework developed to forecast offshore wind power and estimate hydrogen production. Unlike conventional approaches requiring meteorological forecasts and turbine-specific models, this framework learns temporal patterns directly from historical power generation measurements. The weather-agnostic design prioritizes data accessibility and deployment simplicity while maintaining forecast accuracy suitable for operational planning at short time horizons (15–30 min).

8.4.1 Data source and scope

Historical offshore wind power generation data was obtained from Elia, the Belgian transmission system operator. Table 8.2 summarizes the dataset characteristics. The covered period is from May 1, 2022 to May 31, 2025. It comprised 108,192 observations of measured and upscaled power output sampled at 15-min intervals, representing aggregate generation from Belgium's offshore wind fleet. This aggregated approach was deliberately chosen to develop a weather-agnostic forecasting system that captures operational patterns at the fleet level without requiring detailed meteorological inputs for individual turbine locations.

The preprocessing pipeline consisted of four key steps: temporal alignment and chronological sorting, resampling to a consistent 15-minute cadence, bidirectional linear interpolation for missing values, and nonnegativity constraint enforcement through clipping. The resulting dataset contained no missing values and maintained complete temporal consistency.

8.4.2 Weather-agnostic design rationale

The deliberate choice to use only aggregated power generation data, without meteorological inputs, addresses several practical challenges in offshore wind forecasting.

Data accessibility: Power generation data from TSOs is often publicly available through platforms like ENTSO-E Transparency Platform, while high-quality offshore meteorological data requires expensive infrastructure: offshore met masts, LiDAR systems, satellite data subscriptions, or commercial weather service contracts. For emerging offshore hydrogen projects, this data barrier can delay forecasting system deployment by months or years.
Operational realism: Aggregated generation inherently captures real-world operational factors that pure weather-based models miss: turbine availability due to maintenance, grid curtailment events, wake effects, control strategy changes, and performance degradation. These factors significantly affect actual production but are difficult to model from meteorological data alone.
Fleet-level forecasting: Offshore hydrogen production typically serves entire wind farm portfolios rather than individual turbines. Aggregated forecasting naturally operates at this decision-making level and benefits from smoothing effects that reduce relative forecast errors compared to individual turbine predictions.
Reduced complexity: Eliminating the meteorological transformation step (weather → power) simplifies the forecasting pipeline, reduces computational requirements, and removes error propagation from weather forecast uncertainty and power curve modeling inaccuracies.
Transferability: The approach can be replicated for any offshore wind fleet with historical generation records, without requiring site-specific meteorological characterization, turbine technical specifications, or layout information that may be proprietary.

The primary limitation of weather-agnostic approaches is reliance on temporal autocorrelation rather than causal meteorological relationships. This makes them most effective at short horizons (minutes to hours) where recent generation trends strongly predict near-term output. For the 15–30 min horizons addressed in this study—directly relevant to electrolyzer control, storage management, and real-time balancing—this approach excels while avoiding the data infrastructure burden of weather-based forecasting.

8.4.3 Forecasting problem formulation

The forecasting problem was formulated as supervised sequence-to-sequence learning using a sliding window approach. Each input consisted of a lookback window of w consecutive power measurements, while outputs predicted power at two future horizons: $t+1$ (15 min) and $t+2$ (30 min). Two window sizes were evaluated: 96 timesteps (24 h) and 192 timesteps (48 h). A chronological train–test split (80/20) preserved temporal dependencies, with min–max normalization [0,1] fitted exclusively on training data to prevent leakage. Fig. 8.1 illustrates the data flow from raw measurements through windowing to model inputs.

8.4.4 Neural network architecture

Five architectures were evaluated, representing different approaches to temporal modeling. Table 8.3 summarizes their key characteristics and computational properties. Additionally, a naive baseline model was implemented. This model simply predicts the last observed value of the input sequence as the next step. Although simplistic, it provides a meaningful reference to evaluate the added value of the deep learning models.

All models shared common elements: Adam optimizer (learning rate 0.001, gradient clipping 1.0), MSE loss, and a final dense layer with two neurons for dual-horizon output. Mixed precision training (float16) accelerated computation on GPU hardware.

8.4.5 Model validation and training

Model selection employed fivefold time series cross-validation with temporal gaps equal to the window size, preventing data leakage between training and validation sets. Each architecture-window combination was trained across all folds for up to 30 epochs with early stopping (patience = 3). Independent scalers were fitted per fold on training data only. Performance metrics (MAE, RMSE, R^2) were aggregated as mean ± standard deviation across folds, with the optimal window size selected by maximizing 15-min R^2.

Final models were trained on the full training set for up to 60 epochs (patience = 6, batch size = 512) using their optimal window sizes. A 10% temporal validation split with gap monitored convergence. Test set evaluation measured performance on completely held-out data using MAE (average error in MW), RMSE (penalizing large errors), and R^2 (variance explained). Two persistence baselines provided context: 15-min predictions used the last observed value, whereas 30-min predictions tested both the last value (t) and the penultimate value ($t - 1$).

8.4.6 Implementation details

All models were implemented using TensorFlow 2.19 with Keras API and trained on Google Colaboratory with NVIDIA GPU acceleration. Performance optimizations included mixed precision training (float16), XLA compilation, and dynamic GPU memory growth. Reproducibility was ensured by fixing random seeds at 42 across NumPy, Python, and TensorFlow, controlling weight initialization, dropout masks, and data shuffling.

8.5 Results and discussions

8.5.1 Cross-validation performance

The fivefold time series cross-validation revealed substantial differences in model performance and stability across architectures. Table 8.4 presents the

TABLE 8.3 Model architecture specifications.

Model	Type	Key components	Parameters[a]	Training Speed[b]
LSTM	Recurrent	128 units, dropout 0.2	~82K	Baseline
BiLSTM	Recurrent	2 × 128 units, dropout 0.2	~148K	1.8 × slower
GRU	Recurrent	128 units, dropout 0.2	~63K	1.2 × faster
TCN	Convolutional	5 layers, 64 filters, dilations [1,2,4,8,16]	~91K	2.3 × faster
Transformer	Attention	128-D, 4 heads, 2 layers	~105K	0.9 × slower

[a] *Approximate parameter count for window = 96.*
[b] *Relative to LSTM baseline on GPU.*

TABLE 8.4 Cross-validation results for 15-min ahead forecasting.

Model	Window	MAE ± Std (MW)	RMSE ± Std (MW)	R^2 ± Std
GRU	192	44.34 ± 3.05	73.15 ± 6.17	0.991 ± 0.001
BiLSTM	192	46.05 ± 5.59	75.13 ± 9.17	0.990 ± 0.002
LSTM	96	46.61 ± 5.77	76.07 ± 10.10	0.990 ± 0.002
Transformer	96	73.37 ± 16.71	100.50 ± 16.56	0.983 ± 0.003
TCN	96	96.46 ± 13.84	141.48 ± 14.63	0.964 ± 0.009

BiLSTM, *bidirectional long short-term memory;* GRU, *gated recurrent unit;* LSTM, *long short-term memory;* MAE, *mean absolute error;* RMSE, *root mean squared error.*

aggregated results for 15-min ahead forecasting using each model's optimal window size. Importantly, these results were achieved using only historical power generation data without any meteorological inputs. The weather-agnostic models matched or exceeded typical physics-based forecasting accuracy at these short horizons while avoiding the data infrastructure requirements of conventional approaches.

The three recurrent architectures demonstrated remarkably similar performance with R^2 values exceeding 0.990, indicating that all three captured over 99% of the variance in 15-min ahead power fluctuations. The GRU achieved the lowest mean absolute error (44.34 MW) with a notably small standard deviation (3.05 MW), suggesting both superior accuracy and consistent performance across temporal folds. Interestingly, both GRU and BiLSTM benefited from the longer 48-h lookback window (192 timesteps), whereas LSTM performed optimally with a 24-h context (96 timesteps). This suggests that bidirectional processing and the simpler gating structure of GRU may handle longer sequences more effectively than standard LSTM.

The Transformer architecture underperformed relative to recurrent models despite its recent prominence in natural language processing. With an MAE of 73.37 MW and an R^2 of 0.983, it lagged behind the best recurrent model by approximately 30 MW in average error. More concerning was the high variability across folds (standard deviation of 16.71 MW for MAE), indicating inconsistent generalization. This result aligns with the established understanding that Transformers require substantially larger datasets and longer sequences to leverage their self-attention mechanisms effectively. The relatively short sequences (96–192 timesteps) and moderate dataset size (~86,000 training samples) appear insufficient for the Transformer to match recurrent architectures that are explicitly designed for sequential processing.

TABLE 8.5 Cross-validation results for 30-min ahead forecasting.

Model	Window	MAE (MW)	RMSE (MW)	R^2
GRU	192	73.66 ± 4.93	119.11 ± 7.77	0.975 ± 0.003
BiLSTM	192	75.11 ± 5.72	121.03 ± 9.93	0.974 ± 0.003
LSTM	96	75.36 ± 5.22	120.98 ± 9.28	0.974 ± 0.003
Transformer	96	94.82 ± 19.42	138.77 ± 15.41	0.965 ± 0.011
TCN	96	115.98 ± 13.33	169.62 ± 12.78	0.949 ± 0.010

BiLSTM, *bidirectional long short-term memory;* GRU, *gated recurrent unit;* LSTM, *long short-term memory;* MAE, *mean absolute error;* RMSE, *root mean squared error.*

The TCN exhibited the poorest performance with an R^2 of 0.964 and an MAE nearly double that of the best models. The causal convolutional approach, while computationally efficient, struggled to capture the complex temporal dependencies in wind power generation. TCN performance with the 192-timestep window was low ($R^2 = 0.775 \pm 0.309$), suggesting fundamental instability with longer sequences. The extremely high standard deviation indicates that the model either converged to poor local minima or failed to train effectively in certain folds, making it unsuitable for operational deployment despite its theoretical advantages in parallel processing.

For 30-min ahead forecasting Table 8.5, similar patterns emerged with degraded absolute performance as expected for the more challenging longer horizon.

The 30-min horizon proved substantially more challenging, with all models experiencing approximately 60%–70% increases in MAE compared to 15-min forecasts. The GRU maintained its lead with an MAE of 73.66 MW and R^2 of 0.975, and the gap between top performers is narrowed. The consistent ranking across both horizons (GRU, BiLSTM, LSTM outperforming Transformer and TCN) validates the robustness of these findings and suggests that model selection based on 15-min performance generalizes effectively to the 30-min horizon.

8.5.2 Test set performance

Final models trained on the full training set were evaluated on the held-out test set representing genuinely unseen future data. Table 8.6 presents the results alongside persistence baselines.

The modest improvement over persistence at 15-min horizons (1.6–1.9 MW, ~5%) requires careful interpretation in the context of weather-agnostic forecasting. This result demonstrates that temporal autocorrelation in aggregated wind

TABLE 8.6 Test set performance for 15-min ahead forecasting.

Model	MAE (MW)	RMSE (MW)	R^2
GRU	31.22	52.93	0.994
LSTM	31.35	52.35	0.995
BiLSTM	31.78	52.36	0.995
Persistence	32.85	56.38	0.994
Transformer	77.51	98.56	0.982
TCN	76.18	97.86	0.982

BiLSTM, *bidirectional long short-term memory;* GRU, *gated recurrent unit;* LSTM, *long short-term memory;* MAE, *mean absolute error;* RMSE, *root mean squared error.*

power is sufficient for accurate short-term prediction without causal meteorological relationships. The weather-agnostic models achieved comparable accuracy to persistence using only learned patterns from historical data, a validation of the approach's viability rather than a limitation. The value emerges through capturing systematic patterns (diurnal cycles, seasonal trends, multihour ramps) that persistence cannot represent, particularly at longer horizons where the models' advantage increases substantially.

The near-identical performance of LSTM, BiLSTM, and GRU, despite their architectural differences, is itself an important finding. It suggests that for this specific forecasting task with aggregated fleet-level data, the particular gating mechanism (LSTM vs. GRU) and directionality (unidirectional vs. bidirectional) matter less than the fundamental capacity for sequential pattern learning. This result has practical implications: practitioners can select based on computational efficiency (GRU trains 20% faster than LSTM) or implementation simplicity rather than worrying about marginal accuracy differences.

At the 30-min horizon, the picture changes meaningfully (Table 8.7). The recurrent models now perform comparably to or slightly better than the persistence baseline using the last observed value, with GRU achieving an MAE of 54.99 MW versus 55.75 MW for persistence. However, all three recurrent models substantially outperform persistence using the penultimate value ($t-1$), which yields an MAE of 74.00 MW. This 25% error reduction demonstrates that the models have learned useful temporal patterns beyond simple autocorrelation.

The Transformer's improved relative standing at 30 min (MAE of 65.42 MW) compared to 15 min is noteworthy. While still trailing the recurrent models by approximately 10 MW, the Transformer demonstrates that attention mechanisms can capture longer-range dependencies more effectively as the forecast horizon extends. This suggests that for even longer horizons (1–2 h ahead), architectures emphasizing multiscale temporal attention might become competitive with or superior to recurrent approaches.

TABLE 8.7 Test set performance for 30-min ahead forecasting.

Model	MAE (MW)	RMSE (MW)	R^2
GRU	54.99	89.81	985
BiLSTM	55.78	89.26	985
LSTM	55.57	89.55	985
Persistence (t)	55.75	93.79	984
Persistence (t-1)	74.00	123.40	972
Transformer	65.42	103.27	980
TCN	90.83	124.13	971

BiLSTM, *bidirectional long short-term memory;* GRU, *gated recurrent unit;* LSTM, *long short-term memory;* MAE, *mean absolute error;* RMSE, *root mean squared error.*

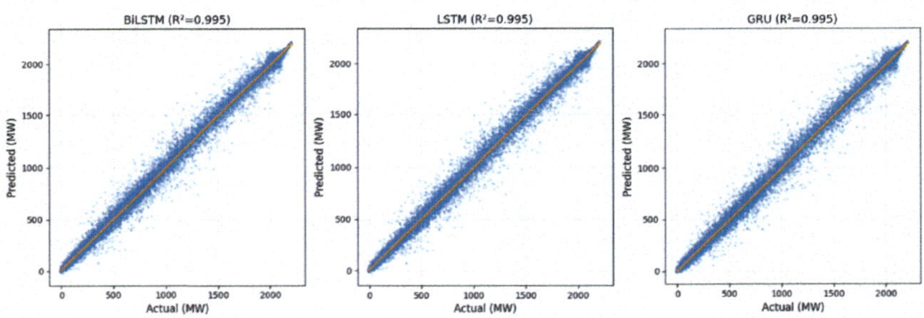

FIGURE 8.2 Predicted versus actual power for top-3 models (15-min ahead).

8.5.3 Forecast quality analysis

Fig. 8.2 presents scatter plots comparing predicted versus actual power for the three top-performing models on the 15-min horizon, revealing the nature of prediction errors. All three models exhibit tight clustering around the perfect prediction line, with R^2 values of 0.995 confirming that variance is almost entirely explained. However, subtle patterns emerge in the residuals. At very low power outputs (<100 MW), all models show slight underprediction, while at very high outputs (>2000 MW), slight overprediction occurs. This regression toward the mean is a common characteristic of statistical models and reflects the difficulty of predicting extreme values that occur rarely in the training distribution. The symmetry of errors around zero indicates no systematic bias, although the heteroscedasticity—greater absolute errors at higher power levels—suggests that percentage errors remain relatively constant across the output range.

Fig. 8.3 shows the time series comparison of actual versus predicted power for a representative 1000-timestep segment, illustrating the models' ability to

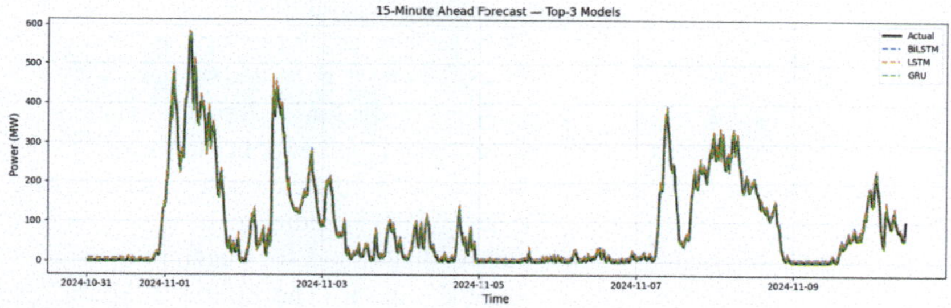

FIGURE 8.3 Time series forecast comparison (15-min ahead, 1000 samples).

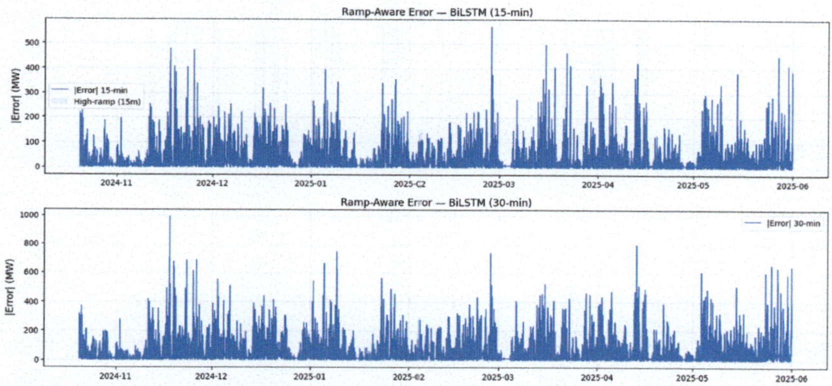

FIGURE 8.4 Error distribution for top-3 models.

track rapid fluctuations. The close alignment of all three model predictions with ground truth demonstrates effective learning of both slow trends and faster oscillations. Notably, the models track rapid ramp events, which means large power changes over short periods, with reasonable fidelity, although some lag is evident during the steepest ramps. This lag reflects the inherent smoothing effect of any predictive model that minimizes squared errors: the optimal prediction for highly uncertain ramp magnitudes regresses toward the mean, sacrificing perfect tracking of extreme excursions to minimize overall error.

Residual analysis (Fig. 8.4) reveals that prediction errors are approximately normally distributed with a mean near zero, validating the appropriateness of MSE loss for this problem.

The distributions show slight positive kurtosis (heavier tails than normal), indicating that large prediction errors occur more frequently than a Gaussian model would suggest. These outliers correspond to rapid, unpredictable ramp events driven by weather systems that move through the offshore region faster than the 24–48 h lookback window can anticipate. The similarity of error distributions across LSTM, BiLSTM, and GRU reinforces the conclusion that these architectures learn essentially equivalent representations of the underlying temporal patterns.

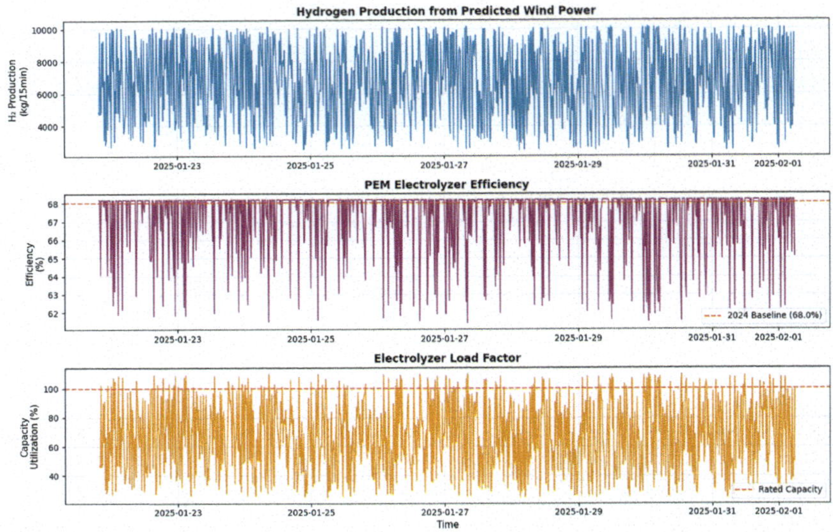

FIGURE 8.5 Hydrogen production, electrolyzer efficiency, and capacity utilization from 15-min wind power forecasts.

8.5.4 Hydrogen production forecasting from wind power

The validated GRU wind power forecasting model was integrated with a PEM electrolyzer system to translate predicted wind generation into hydrogen production estimates. This approach demonstrates the complete forecasting chain from renewable resource prediction to end-product quantification, essential for operational planning of integrated wind-hydrogen systems. The electrolyzer model assumed 2000 MW rated capacity with 68% baseline efficiency for 2024 PEM technology and 0.25% annual improvement. Efficiency was modeled as power-dependent, maintaining optimal performance between 30% and 90% capacity utilization but degrading approximately 15% below 10% load due to auxiliary system overhead.

Using 15-min ahead GRU wind power predictions as inputs, the electrolyzer model yielded average hydrogen production of 26,917 kg/h (640.9 tons/day). Individual 15-min intervals ranged from 2461 to 10,123 kg, which is a 4:1 dynamic range reflecting wind variability. Average electrolyzer efficiency of 67.3% matched design specifications, while 67% capacity utilization indicated operation at approximately 1340 MW mean power. This conservative sizing strategy avoids curtailment during generation peaks while accepting lower capital utilization during calm periods.

Time series analysis (Fig. 8.5) demonstrated tight coupling between predicted wind power fluctuations and hydrogen output on 15-min timescales. Efficiency remained stable during sustained operation within the optimal 30%–90% capacity range but varied during rapid wind ramps as the system traversed suboptimal regions. Capacity utilization showed pronounced variability, from

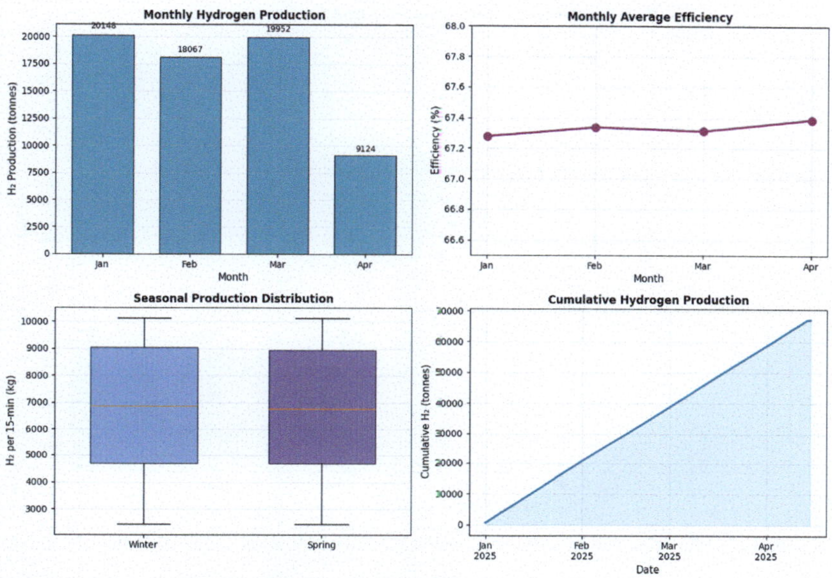

FIGURE 8.6 Monthly and seasonal hydrogen production patterns (Jan.–Apr. 2025).

near-zero during wind lulls to occasional excursions above 100% during storm events, validating the benefit of 110% overload capability.

Monthly aggregation over the 4-month test period (January–April 2025) revealed seasonal variation consistent with offshore wind climatology (Fig. 8.6). Winter months produced 18,000–20,000 tons monthly versus 15,000–17,000 tons during spring, approximately 15%–20% lower. Seasonal distribution analysis comparing winter and spring showed not only mean differences but distinct variability patterns, with winter exhibiting higher median values but larger interquartile ranges from variable storm passages.

The efficiency–power relationship (Fig. 8.7) confirmed that operating predominantly in 50%–80% capacity maintained near-optimal performance while providing headroom for peaks. Occasions exceeding 90% utilization resulted in only 1%–2% efficiency penalties, suggesting limited benefit from oversizing beyond 110% capacity. The cumulative production curve tracked smoothly despite individual timestep uncertainty, with monthly totals within approximately 5% of final values, indicating reliable planning guidance despite greater short-term uncertainty.

The validated wind power forecasts enable hydrogen production estimation through the integrated PEM electrolyzer model. The 15–30 min forecast horizons directly support operational decisions in electrolyzer scheduling, storage management, and delivery planning. The electrolyzer performance model is based on established PEM technology characteristics documented in recent techno-economic studies (Izadi et al., 2022; Lüth et al., 2023; Naveena et al., 2024). As offshore electrolysis projects transition from planning to operation in

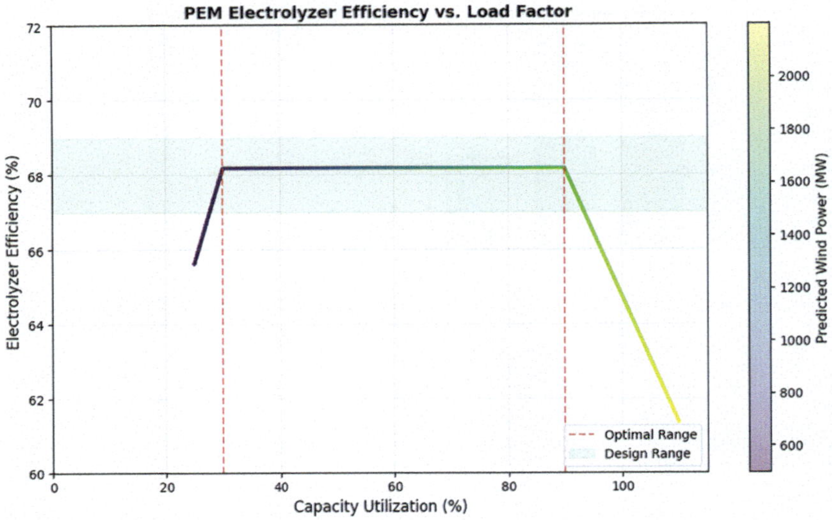

FIGURE 8.7 PEM electrolyzer efficiency versus capacity utilization. *PEM*, Polymer electrolyte membrane.

the coming years, the forecasting framework can incorporate actual performance data to enhance hydrogen-specific accuracy.

8.6 Conclusion

This study developed and validated a weather-agnostic ML framework for forecasting offshore wind power and hydrogen production. The approach operates exclusively on historical power generation measurements, eliminating requirements for meteorological data, turbine specifications, and site-specific parameters.

Five deep learning architectures were evaluated using 3 years of operational data from Belgium's offshore wind fleet (108,192 observations at 15-min intervals). Recurrent models (GRU, LSTM, BiLSTM) achieved nearly identical performance with R^2 exceeding 0.99 for 15-min forecasts. The GRU model delivered optimal results (MAE = 44.34 MW, R^2 = 0.995) while training 20% faster than LSTM, making it the recommended architecture for operational deployment. Transformer and TCN approaches underperformed, confirming that recurrent designs excel for this task at available dataset sizes and sequence lengths.

Integration with PEM electrolyzer modeling (2000 MW capacity, 68% efficiency) translated power forecasts into hydrogen production estimates averaging 640.9 tons/day with 67% capacity utilization. The system revealed 15%–20% seasonal variation between winter and spring, supporting production planning without meteorological modeling.

The weather-agnostic design addresses a critical deployment barrier: offshore meteorological data requires expensive infrastructure that delays forecasting system implementation. By using only transmission system operator data, the framework enables immediate deployment for emerging offshore hydrogen projects. The approach naturally captures operational realities, including maintenance, curtailment, and control strategies that pure weather-based models miss.

giThe modest 5% improvement over persistence at 15-min horizons reflects high temporal autocorrelation at ultra-short timescales rather than model limitations. The value emerges through capturing systematic patterns, such as diurnal cycles, seasonal trends, and ramp events, that persistence cannot represent, particularly at longer horizons where sophisticated models demonstrate clearer advantages.

Future work should extend forecast horizons to 1–24 h, where advanced models provide greater value over persistence, develop probabilistic forecasts with uncertainty quantification for risk management, and integrate operational data from commissioned offshore electrolyzers to validate end-to-end hydrogen production forecasts. As offshore wind-to-hydrogen projects transition from concept to commercial operation, data-driven forecasting tools requiring minimal infrastructure will be essential for reliable, economical production planning.

The demonstrated feasibility of accurate forecasting using only generation data removes a significant barrier to offshore hydrogen development and provides an immediately deployable solution for operational planning in this emerging sector.

Nomenclature

\$/kg	Cost per kilogram
\$/kWh	Cost per kilowatt-hour
TSR (tip	Ratio of the blade tip speed to the wind speed
ρ speed	Air density (kg/m^3)
A ratio)	Swept area by the wind turbine blades (m^2)
v	Wind speed (m/s)
ft	Forget gate
It	Input gate
H_2	Quantity of hydrogen produced (kg)
$P_{wind}(t)$	Power production from wind turbines at time t (kW)
C_{elec}	Capacity of the electrolyzer (kW)
$N_{turbines}$	Number of wind turbines
E_{elec}	Electricity consumption per kg of hydrogen (kWh/kgH_2)
E_{wind}	Efficiency of the wind turbine (kW)
H_{daily}	Daily hydrogen production target (20,000 kg/day)
T	Number of hours in a year (7860 h)

Abbreviations

MW	Megawatt
T/day	Tons per day
NWP	Numerical weather prediction
ML	Machine learning
AI	Artificial intelligence
LSTM	Long short-term memory
BiLSTM	Bidirectional long short-term memory
CV	Cross-validation
GRU	Gated recurrent unit
GPU	Graphics processing unit
PEM	Polymer electrolyte membrane (or proton exchange membrane)
R^2	Coefficient of determination
RMSE	Root mean squared error
MSE	Mean squared error
MAE	Mean absolute error
CAPEX	Capital expenditure
OPEX	Operational expenditure
LCOH	Levelized cost of hydrogen
LCOE	Levelized cost of energy
TSO	Transmission system operator

References

Abdelkareem, M. A., El Haj Assad, M., Sayed, E. T., & Soudan, B. (2018). Recent progress in the use of renewable energy sources to power water desalination plants. *Desalination using Renewable Energy, 435*, 97. https://www.journals.elsevier.com/desalination. doi:10.1016/j.desal.2017.11.018.

Achour, Y., Berrada, A., Arechkik, A., & El Mrabet, R. (May 2023). Techno-economic assessment of hydrogen production from three different solar photovoltaic technologies. *International Journal of Hydrogen Energy*. doi:10.1016/j.ijhydene.2023.05.017.

Achour, Y., El Mokrini, A., El Mrabet, R., & Berrada, A. (2025). Optimizing power-to-ammonia plant: Sizing, operation, and production forecasting using deep learning approach. *Renewable Energy, 240*, 122234. doi:10.1016/j.renene.2024.122234.

Alkaisi, A., Mossad, R., & Sharifian-Barforoush, A. (2017). A review of the water desalination systems integrated with renewable energy. *Energy Procedia, 110*, 268.

Allal, Z., Noura, H. N., Salman, O., & Chahine, K. (2024). Machine learning solutions for renewable energy systems: Applications, challenges, limitations, and future directions. *Journal of Environmental Management, 354*, 120392. doi:10.1016/j.jenvman.2024.120392.

Altan, S., Sajid, M., & Iqbal, S. (2020). Forecasting hydrogen production potential in Islamabad from solar energy using water electrolysis. *International Journal of Hydrogen Energy, 46*, 1671–1681. doi:10.1016/j.ijhydene.2020.10.059.

Ayeon, K., Heehyang, K., Changgwon, C., & Hankwon, L. (2023). Feasibility of offshore wind turbines for linkage with onshore green hydrogen demands: A comparative economic analysis. *Energy Conversion and Management, 277*, 116662. doi:10.1016/j.enconman.2023.116662.

Bonacina, C. N., Gaskare, N. B., Valenti, G. Assessment of offshore liquid hydrogen production from wind power for ship refueling. 2022;47:1279–1291.

Calado, G., Castro, R., Pires, A. J., & Marques Miguel, J. (2024). Assessment of hydrogen-based solutions associated offshore wind farms: The case of the Iberian peninsula. *Renewable Sustainable Energy Reviews, 192*, 114268.

Cheng, G., Luo, E., Zhao, Y., Yang, Y., Chen, B., Cai, Y., et al. (2023). Analysis and prediction of green hydrogen production potential by photovoltaic-powered water electrolysis using machine learning in China. *Energy, 284*, 129302. doi:10.1016/j.energy.2023.129302.

d'Amore Domenech, R., Leo, T. J., & Pollet, B. G. (2021). Bulk power transmission at sea: Life cycle cost comparison of electricity and hydrogen as energy vectors. *Applied Energy, 288*, 116625. doi:10.1016/j.apenergy.2021.116625.

Dinh, Q. V., Dinh, V. N., Mosadeghi, H., Todesco Pereira, P. H., & Leahy, P. G. (2023). A geospatial method for estimating the levelised cost of hydrogen production from offshore wind. *International Journal of Hydrogen Energy, 48*, 15000–15013. doi:10.1016/j.ijhydene.2023.01.016.

El Mrabet, R., & Berrada, R. A (2020). Hydrogen production and derivatives from renewable energy systems for a best valorization of sustainable resources. In *Hybrid energy system models* (pp. 343–363). Academic Press.

Emrani, A., Achour, Y., Sanjari, M. J., & Berrada, A. (2024). Adaptive energy management strategy for optimal integration of wind/PV system with hybrid gravity/battery energy storage using forecast models. *Journal of Energy Storage, 96*, 112613. doi:10.1016/J.EST.2024.112613.

Environmental Resources Management, ERM dolphyn hydrogen: Phase 2 final report (2021), Environmental Resources Management.

Ghandehariun, S., Ghandehariun, A. M., & Bahrami, N. (2023). Performance prediction and optimization of a hybrid renewable-energy-based multigeneration system using machine learning. *Energy, 282*, 128908. doi:10.1016/j.energy.2023.128908.

Giampieri, A., Ling-Chin, J., & Roskilly, A. P. (2023). Techno-economic assessment of offshore wind-to-hydrogen scenarios: A UK case study. *International Journal of Hydrogen Energy*. doi:10.1016/j.ijhydene.2023.01.346.

H2RES, H2RES Sustainable hydrogen production, (2023), State of Green, https://h2res.dk/?lang=en.

Hill, J. S., Sealhyfe produces first green hydrogen at sea from offshore wind farm, (2023) Renew Economy.

Ibrahim, O. S., Singlitico, A., Proskovics, R., McDonagh, S., Desmond, C., & Murphy, J. D. (2022). Dedicated large-scale floating offshore wind to hydrogen: Assessing design variables in proposed typologies. *Renewable Sustainable Energy Reviews, 160*, 112310. doi:10.1016/j.rser.2022.112310.

Izadi, A., Shahafve, M., & Ahmadi, P. (2022). Neural network genetic algorithm optimization of a transient hybrid renewable energy system with solar/wind and hydrogen storage system for zero energy buildings at various climate conditions. *Energy Conversion and Management, 260*, 115593. doi:10.1016/j.enconman.2022.115593.

Javaid, A., Umer, J., Muhammad, S., Muhammad, R., Emad, U., Yasar, A., et al. (2022). Suburban environment using machine learning. *Energies, 15*, 1–13.

Jiang, Y., Huang, W., & Yang, G. (2022). Electrolysis plant size optimization and benefit analysis of a far offshore wind-hydrogen system based on information gap decision theory and chance constraints programming. *International Journal of Hydrogen Energy, 47*, 5720–5732. doi:10.1016/j.ijhydene.2021.11.211.

Kabir, M. M., Roy, S. K, Alam, F., Nam, S. Y., Im, K. S., Tijing, L, & Shon, H. K. (2023). Machine learning-based prediction and optimization of green hydrogen production technologies from water industries for a circular economy. *Desalination, 567*, 116992. doi:10.1016/j.desal. 2023.116992.

Komorowska, A., Benalcazar, P., & Kamiński, J. (2023). Evaluating the competitiveness and uncertainty of offshore wind-to-hydrogen production: A case study of Poland. *International Journal of Hydrogen Energy, 48*, 14577–14590. doi:10.1016/j. ijhydene.2023.01.015.

Kopp, M., Coleman, D., Stiller, C., Scheffer, K., Aichinger, J., & Scheppat, B. (2017). Energiepark Mainz: Technical and economic analysis of the worldwide largest power-to-gas plant with PEM electrolysis. *International Journal of Hydrogen Energy, 42*(19), 13311–13320. doi:10.1016/j. ijhydene.2016.12.145.

Lhyfe, Sealhyfe produces its first kilos of green hydrogen in the Atlantic, (2023), Lhyfe,https://www.lhyfe.com/press/lhyfe-announces-that-sealhyfe-the-worlds-first-offshore-hydrogen-production-pilot-produces-its-first-kilos-of-green-hydrogen-in-the-atlantic-ocean/.

Leahy, P., McKeogh, E., Murphy, J., & Cummins, V. (2021). Development of a viability assessment model for hydrogen production from dedicated offshore wind farms. *International Journal of Hydrogen Energy, 46*, 24620–24631.

Lucas, T. R., Ferreira, A. F., Santos Pereira, R., & Alves, M. (2022). Hydrogen production from the WindFloat Atlantic offshore wind farm: A techno-economic analysis. *Applied Energy, 310*, 118481. doi:10.1016/j.apenergy.2021.118481.

Luo, Z., Wang, X., Wen, H., & Pei, A. (2022). Hydrogen production from offshore wind power in South China. *International Journal of Hydrogen Energy, 47*, 24558–24568. doi:10.1016/j. ijhydene.2022.03.162.

Lüth, A., Seifert, P. E., Egging-Bratseth, R., & Weibezahn, J. (2023). How to connect energy islands: Trade-offs between hydrogen and electricity infrastructure. *Applied Energy., 341*, 121045. doi:10.1016/j.apenergy.2023.121045.

Ma, Q., & Lu, H. (2011). Wind energy technologies integrated with desalination systems: Review and state-of- the-art. *Desalination, 277*, 274.

Marocco, P., Ferrero, D., Lanzini, A., & Santarelli, M. (2022). The role of hydrogen in the optimal design of off-grid hybrid renewable energy systems. *Journal of Energy Storage, 46*, 103893. doi:10.1016/j.est.2021.103893.

Meier, K. (2014). Hydrogen production with sea water electrolysis using Norwegian offshore wind energy potentials: Techno-economic assessment for an offshore-based hydrogen production approach with state-of-the-art technology. *International Journal of Energy and Environmental Engineering, 5*, 1–12.

Miao, B., Giordano, L., & Chan, S. H. (2021). Long-distance renewable hydrogen transmission via cables and pipelines. *International Journal of Hydrogen Energy, 46*, 18699–18718. doi:10.1016/ j.ijhydene.2021.03.067.

Naveena, K., Krishnamoorthy, M., Karuppiah, N., Gouda, P. K., Hariharan, S., Saravanan, K., & Kumar, A. (2024). Elevating sustainability with a multi-renewable hydrogen generation system empowered by machine learning and multi-objective optimization. *Measurement: Sensors, 33*, 101192. doi:10.1016/j.measen.2024.101192.

Oyster, Oyster Project, Oyster (2023), https://oysterh2.eu/about/project/.

Poshydon. Process (2023), Poshydon, https://poshydon.com/nl/home/proces-2/.

Ren, B., Chi, Y., Zhou, N., Wang, Q., Wang, T., & Luo, Y. (2024). Machine learning applications in health monitoring of renewable energy systems. *Renewable Sustainable Energy Reviews, 189*, 114039. doi:10.1016/j.rser.2023.114039.

Rezaei, M., Akimov, A., & Gray, E. M. A. (2024b). Techno-economics of renewable hydrogen export: A case study for Australia-Japan. *Applied Energy, 374*, 124015. doi:10.1016/J.APENERGY.2024.124015.

Rezaei, M., Akimov, A., & Gray, E. M. A. (2024c). Levelised cost of dynamic green hydrogen production: A case study for Australia's hydrogen hubs. *Applied Energy, 370*, 123645. doi:10.1016/J.APENERGY.2024.123645.

Rogeau, A., Vieubled, J., de Coatpont, M., Affonso Nobrega, P., Erbs, G., & Girard, R. (2023). Techno-economic evaluation and resource assessment of hydrogen production through offshore wind farms: A European perspective. *Renewable Sustainable Energy Reviews, 187*, 113699. doi:10.1016/j.rser.2023.113699.

RWE, AquaVentus – Hydrogen production in the North Sea, RWE, (2021), https://www.rwe.com/en/research-and-development/hydrogen-projects/aquaventus/.

Safari, F., & Dincer, I. (2020). A review and comparative evaluation of thermochemical water splitting cycles for hydrogen production. *Energy Conversion and Management, 205*, 112182. doi:10.1016/j.enconman.2019.112182.

Salari, A., Shakibi, H., Habibi, A., & Hakkaki-fard, A. (2023). Optimization of a solar-based PEM methanol/water electrolyzer using machine learning and animal-inspired algorithms. *Energy Conversion and Management, 283*, 116876. doi:10.1016/j.enconman.2023.116876.

Shams, M. H., Niaz, H., Na, J., & Anvari-moghaddam, A. (2021). Machine learning-based utilization of renewable power curtailments under uncertainty by planning of hydrogen systems and battery storages. *Journal of Energy Storage, 41*, 103010. doi:10.1016/j.est.2021.103010.

Shin, H., Jang, D., Lee, S., Cho, H., Kim, K., & Kang, S. (2023). Techno-economic evaluation of green hydrogen production with low-temperature water electrolysis technologies directly coupled with renewable power sources. *Energy Conversation and Management, 286*, 117083. doi:10.1016/j.enconman.2023.117083.

Singlitico, A., Østergaard, J., & Chatzivasileiadis, S. (2021). Onshore, offshore or in-turbine electrolysis? Techno-economic overview of alternative integration designs for green hydrogen production into offshore wind power hubs. *Renewable and Sustainable Energy Reviews, 1*, 100005. doi:10.1016/j.rset.2021.100005.

Snieckus, D., Salamander floating wind project off Scotland gets gassed up with new-look hydrogen kit, (2021), Recharge News.

Thomas Schiestel, H2Wind – Self-sufficient offshore H2 electrolysis, Fraunhofer Institute for Wind Energy Systems (2023).

Yamin, Y., Haoran, Z., Qi, L., Yongtu, L., & Jinyue, Y. (2021). Roadmap to hybrid offshore system with hydrogen and power co-generation. *Energy Conversion and Management, 247*, 114690.

Chapter 9

Integrating policy, regulation, and artificial intelligence in the hydrogen economy

Abdellatif El Ghazi[a] and Asmae Berrada[b]

[a] *TIC Lab, College of Engineering and Architecture, International University of Rabat, Sala Al Jadida, Morocco,* [b] *LERMA Lab, College of Engineering and Architecture, International University of Rabat, Sala Al Jadida, Morocco*

9.1 Introduction

The energy transition would appear all but complete if government white papers were the only metric used to assess the global hydrogen economy. The number of countries with official national hydrogen strategies soared to nearly 60 by the end of 2024, covering over 84% of global CO_2 emissions related to energy. Hydrogen was the natural vector for decarbonizing the "hard-to-abate" sectors; the diplomatic consensus was absolute. In late 2025, however, a stark dissonance has emerged between policymakers' ambition and investors' appetite. Capital allocation has entered the era of the "great filter," where exuberant announcements are colliding with cold mechanics.

Recent research has well-documented this "ambition-implementation gap." In a global analysis of over 190 hydrogen projects, only about 7% of announced capacity has been realized on schedule (Odenweller & Ueckerdt, 2025). Even though the global project pipeline has tripled in 3 years, achieving these goals without massive subsidies, estimated at USD 1.3 trillion, is highly unlikely.

The mismatch between political enthusiasm and private capital can also be found elsewhere. The lack of operational benchmarks and environmental uncertainties, especially for large-scale projects, are major obstacles for investors seeking predictable returns. Compared to other clean energy projects, studies of green hydrogen equity portfolios indicate lower returns and higher volatility than other clean energy investments (Tudor, 2023).

Complexity is not limited to the policy side. There are wide differences between national strategies and market realities in emerging economies such

Decarbonizing through Sustainable Hydrogen and its Derivatives.
DOI: https://doi.org/10.1016/B978-0-443-27632-3.00009-8

as Brazil, Chile, and South Africa, with governments prioritizing exports and technology leadership.

Regional studies in North China have shown that current hydrogen-supporting policies often increase emissions due to the use of nonclean hydrogen (Zhang et al., 2025). The capital markets remain skeptical, despite governments proclaiming hydrogen to be the key to a carbon-free future. It is unclear whether the hydrogen economy will succeed based on the number of national strategies, but rather on whether these ambitions can endure the scrutiny of investors navigating the Great Filter between political vision and financial viability.

There was an enormous distortion field created between 2020 and 2023 by the volume of "paper hydrogen" generated. Among the pipeline of projects announced by the end of 2023, 49 million tons of hydrogen per annum (Mtpa) would be available by 2030. Initially, these announcements appeared to indicate a promising industry. However, a deeper scrutiny reveals a more fragile situation. Only 3.4 Mtpa of that massive 49 Mtpa potential has reached a final investment decision (FID) or entered construction (Odenweller & Ueckerdt, 2025).

In the electrolyzer market, the disparity is even greater.

Despite manufacturers' and developers' claims of nearly 520 GW of capacity by 2030, only 20 GW had secured firm financial commitments. 96 % of all projects were in the precarious stage of "feasibility" or "concept," susceptible to the slightest shift in interest rates or regulatory guidance. The market was dominated by press releases, not development; project developers announced gigawatt-scale hubs before formal offtake agreements had been signed (Odenweller et al., 2022).

9.2 Divergent policy architectures

An increasing number of governments are recognizing the significance of the hydrogen industry and are enacting policy support measures necessary for ensuring technological preparedness, improving market penetration, and fostering market growth. Policies jointly influence the framework for hydrogen technologies and infrastructures. They provide a regulatory framework, financial support, and strategic direction to promote the sustainable deployment of hydrogen. They offer a regulatory framework, financial assistance, and strategic guidance to facilitate the sustainable implementation of hydrogen.

The global hydrogen map has fractured into three distinct regulatory philosophies, each creating a different gravitational pull for capital.

9.2.1 The United States: The tax credit machine

US policymakers have placed their bets on simplicity, using Section 45V of the Inflation Reduction Act (IRA) to subsidize supply. There is a simple, yet effective mechanism: a tiered subsidy that offers up to USD 3/kg of clean hydrogen. In 2024, however, the "three pillars" of compliance: additionality, temporal matching, and geographical correlation, severely tested the efficacy of this approach (Baker Botts 2025; King & Spalding 2024).

9.2.2 The European Union: The regulatory superpower

European nations have formulated national hydrogen programs to facilitate the transition to a low-carbon economy and attain their climate objectives. Since hydrogen can be generated from renewable resources and utilized as a clean energy carrier in a variety of industries, including transportation, industry, and energy, it is thought to be a crucial part of this shift. With targets set for hydrogen deployment across sectors and plans in place to promote innovation and create economic opportunities within the hydrogen sector, these strategies seek to scale-up the production and use of renewable hydrogen while supporting the development of related infrastructure and technologies.

Since hydrogen is a global commodity that needs international cooperation to ensure its safe and effective production, transportation, and use, national hydrogen strategies also seek to foster international coordination and cooperation. National hydrogen policies are a crucial instrument for nations to meet their climate goals, facilitate the shift to a low-carbon economy, and foster economic expansion and innovation.

The European Union has built a different policy architecture. A complex web of mandates has emerged as part of the Fit for 55 package. According to the updated Renewable Energy Directive (RED III), 42% of hydrogen used in industry must be renewable by 2030 (Hydrogen Europe 2024). Furthermore, the ReFuelEU aviation initiative, for example, is forcing the use of synthetic fuel, creating an affordable market for hydrogen derivatives.

These mandates have created a cost gap that the EU has attempted to bridge through surgical financial instruments. A first pilot auction conducted by the European Hydrogen Bank in April 2024 awarded premiums ranging from €0.37 to €0.48 per kilogram to the most competitive producers, well below the threshold of €4.50. In contrast to the US model, which relies on uncapped tax credits, the EU model relies on competitive tension and binding quotas, creating a market with fewer entry barriers but greater long-term certainty for those able to navigate its regulatory maze.

In addition to their national hydrogen policies, a number of European nations, such as Germany, France, the Netherlands, Portugal, Spain, and Italy, have created hydrogen roadmaps. To make sure that they are on pace to reach their national and international climate targets, these roadmaps are a crucial tool for countries to provide a more thorough plan for the deployment of hydrogen technologies and infrastructure in the country, including precise targets and timetables.

The EU hydrogen regulatory framework is presented in Fig. 9.1.

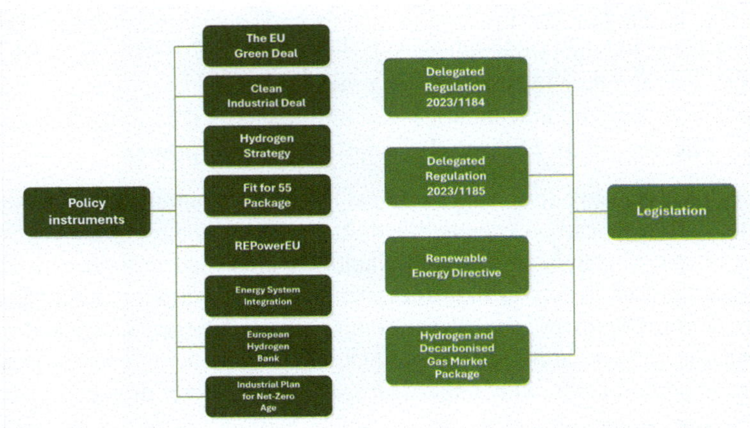

FIGURE 9.1 EU hydrogen regulatory framework. *EU*, European Union.

TABLE 9.1 Comparative capital expenditure (CAPEX) for electrolyzer systems (2024 estimates). China's industrial scaling provides a massive cost advantage (IEA, 2024).

Region	Alkaline (ALK) [USD/kW]	PEM [USD/kW]
China	300—450	800—1100
Europe	1100—1400	1600—2200
North America	1200—1500	1550—2100

9.2.3 China: The industrial juggernaut and the coal paradox

As the West debated definitions, China built factories. The Chinese government has leveraged state-owned enterprises and provincial mandates to ensure a dominant position in the supply chain, controlling approximately 60% of global electrolyzer capacity by 2023 (IEA 2024).

The Chinese approach puts strategic industrial scaling above the complex "certify-then-build" obsession of the West. A massive cost advantage has been created by China's focus on the mass production of alkaline electrolyzers. Western protectionist measures are attempting to close a gap where Chinese systems are 3 to 4 times cheaper than their European counterparts, as shown in Table 9.1.

There is, however, an environmental cost associated with this scaling speed. The Chinese hydrogen strategy faces a "coal dilemma" because a significant portion of their "green" hydrogen projects are dependent on grid connections. Zhang et al.'s (2025) recent modeling suggests that if hydrogen production policies are not strictly coupled with renewable capacity, they may inadvertently increase net emissions in regions like North China, where coal dependency continues to lead to high grid carbon intensity. A further concern is the high "embodied carbon" of the Chinese electrolyzers, which are produced using coal-heavy electricity, which contributes to their low CAPEX. This vulnerability may be addressed by future iterations of the EU's carbon border adjustment mechanism (CBAM).

9.2.4 Africa: Morocco's national roadmap for green hydrogen

Morocco has excellent renewable energy resources, including solar and wind, which enable high green energy performance and potential. The Kingdom's 2010 National Energy Strategy prioritizes the development of renewable energy sources to: (1) meet the country's energy needs with cheap, plentiful energy, thereby reducing its energy dependency (in 2009, Morocco imported more than 95% of its energy needs); and (2) realize the country's goal of becoming a regional leader in sustainable development.

Furthermore, Moroccan industry will be able to access global markets and become more competitive as a result of the country's decarbonization, particularly in light of the EU's impending implementation of a carbon price at Morocco's borders. Cars (USD 3.91 billion, of which 63% is exported to the EU), insulated wires (USD 3.75 billion, of which 80% is exported to Europe), mixed mineral or chemical fertilizers (USD 2.71 billion, of which 15% is exported to the EU), phosphoric acid (USD 1.48 billion, of which 23% is exported to the EU), and clothing (USD 1.39 billion, of which 80% is exported to the EU).

The production of Green H_2 in Morocco is an appealing prospect that has already drawn significant attention from both international public and private players, as well as Moroccan authorities. The nation is also a strong contender for the creation of new energy markets because of its extensive expertise with large-scale energy projects and infrastructure development in general (ports, high-speed line trains, industrial and mining infrastructure, ammonia manufacturing plants, etc.). Additionally, through the corporation office cheédrifien des phosphates (OCP), which currently imports over 2 million tons of ammonia annually for its transformation processes, Morocco is a major player in the extraction and valorization of phosphates. If hydrogen is produced locally, Morocco could generate ammonia, which is a derivative of hydrogen. Therefore, the expansion of Green H_2 production capacity may contribute to a more general goal of industrial and energy autonomy.

Morocco developed a national roadmap for green hydrogen, which was released in 2021, based on these suggestions and findings. According to the document, the nation can produce a very large quantity of green hydrogen to meet local needs, maximize the use of its potential, especially through exports, and develop an industrial and commercial sector centered around green molecules

FIGURE 9.2 The green hydrogen national roadmap's action plan.

like hydrogen, ammonia, and methanol. The national roadmap additionally specifies an implementation pathway that revolves around three main axes: market and demand, which refers to the realization of demand opportunities leading to the creation of new markets; investment and procurement, including the establishment of an industrial cluster and the development of a master plan for the corresponding infrastructure; and technological advancements and cost savings. Eight steps were outlined for the development of Green H_2 in Morocco in a comprehensive action plan for 2050 based on those axes: cost-cutting, innovation and research, local content, industrial cluster, domestic markets, storage, exports, and financing. The specifics of each step are displayed in Fig. 9.2.

A timeline from 2020 to 2050 is included in the roadmap, outlining the steps toward potential economic uses of green H_2. These latter exports, local use for domestic industries, particularly phosphates, and the discovery of natural hydrogen reserves are the top priority in the short term (2020–2030). Morocco will concentrate on developing hydrogen projects and employing greenhouse gas as a vector for energy storage in the medium term (2030–2040). The strategy predicts increased exports and ammonia production in the long run (2040–2050). Conversely, it is uncertain how the natural gas roadmap and the Green H_2 roadmap will interact. Long-term financial unsustainability and duplication of expenses and investments could result from this lack of coordination. Furthermore, a clear policy and regulatory framework, as well as actions to boost demand and incentives to encourage the switch to Green H_2, are not mentioned in the national roadmap. Implementing the roadmap's vision and methodically integrating it into current energy, climatic, economic, and geopolitical policies is a critical problem.

TABLE 9.2 The compliance revolution: Comparison of traditional auditing versus AI-enabled digital compliance.

Dimension	Traditional approach	AI-enabled digital compliance
Granularity verification	Monthly or annual averages, expost manual audits (high lag)	Hourly or subhourly (15-min) Real-time algorithmic validation
Traceability	Paper-based Guarantees of Origin	Blockchain/digital product passports
Cost structure	High OpEx (Consultants/Auditors)	High initial CapEx, near-zero marginal cost
Risk management	Reactive (fines after violation)	Predictive (adjustment before violation)

AI, *artificial intelligence.*

9.3 The digital hydrogen economy: Artificial intelligence as the compliance engine

The accelerating regulatory complexity of the global hydrogen economy, spanning carbon intensity thresholds, lifecycle emission accounting, renewable energy traceability, and evolving certification schemes, has created a "data tsunami" that far exceeds human administrative capacity. Meeting requirements under frameworks like the US Section 45V clean hydrogen production tax credit now demands the processing of millions of data points annually per facility.

Artificial intelligence (AI) is transitioning from an optional optimization tool to the critical "compliance engine" of this sector. Recent reviews underscore that AI significantly enhances precision in production optimization and lifecycle assessments (LCAs) by integrating real-time process data, IoT sensor streams, and dynamic emissions modeling (Bhuiyan & Chowdhury, 2025). Unlike static spreadsheets, machine learning (ML) models enable the continuous verification of renewable electricity inputs, forecasting emissions intensity under varying production scenarios to ensure adherence to the strict threshold of 3.38 kg CO_2e/kg H_2 (EU Renewable Fuel of Non-Biological Origin (RFNBO) standard) or 4 kg CO_2e/kg H_2 (US IRA standard).

9.3.1 From static certificates to digital product passports

When coupled with blockchain-based traceability systems, AI facilitates the deployment of digital product passports (DPP). These immutable provenance records ensure end-to-end transparency. As summarized in Table 9.2, the shift to AI-driven compliance represents a fundamental structural change in how the industry operates, moving from expostauditing to real-time validation.

AI's predictive capabilities now enable "dynamic compliance": reinforcement learning and Digital Twin systems can autonomously adjust operational parameters to maintain carbon intensity below thresholds, minimizing regulatory risks while optimizing resource use. Such systems transform compliance from a static reporting function into an active, real-time optimization process.

9.3.2 The hourly matching challenge: Beyond mere verification

The shift toward "hourly correlation" in the European Union and the United States, requiring proof that renewable electricity was generated in the same hour that hydrogen was produced, represents the most demanding compliance mechanism in the sector. Ricks et al. (2023) demonstrated that without such strict hourly matching, grid-connected hydrogen production could inadvertently increase system-wide emissions. Under the EU Delegated Acts, producers must link every megawatt-hour of renewable input to each kilogram of hydrogen produced. This creates a complex multiobjective optimization problem involving weather forecasting, spot market electricity prices, and storage constraints.

AI as an arbitrage tool: Managing this volume requires decision-making AI. Advanced algorithms do not merely record the match; they predict it. By forecasting wind/solar availability 24 h ahead, AI systems determine the optimal production schedule to maximize the capture of "green" electrons while avoiding hours where grid carbon intensity is high. This allows producers to perform regulatory arbitrage, ensuring economic viability.

9.3.3 Permitting acceleration via generative artificial intelligence

Project developers consistently cite permitting delays as a major bottleneck (IEA 2024). In response, emerging "PermitAI" systems using retrieval-augmented generation (RAG) can autonomously scan thousands of pages of legal text to identify compliance risks, effectively "de-bottlenecking" the administrative pipeline. Unlike generic models, RAG systems retrieve specific clauses from local zoning laws or EU directives to draft prefilled compliance documents. These tools improve consistency by cross-referencing past precedent data, identifying potential rejection risks before submission. Early pilot programs demonstrate that such digital permitting systems can reduce documentation review times by up to 40%, effectively accelerating deployment.

9.3.4 Critical perspective: The digital paradox

Strictly relying on AI and digitalization introduces new vulnerabilities that must be scrutinized to maintain an objective perspective:

- *Algorithmic bias and exclusion*: ML models used for carbon verification are only as good as their training data. There is a risk that algorithms trained primarily on stable, interconnected Western grids may incorrectly flag legitimate operations in emerging markets—where grid intermittency profiles differ—as "noncompliant anomalies." Without transparent, explainable AI (XAI) frameworks, these automated verification tools could inadvertently reinforce technical barriers to entry for producers in the global south.
- *Cyber-physical risks*: Connecting gigawatt-scale electrolyzers to the grid via digital control systems introduces significant attack vectors. A compromised "Digital Twin" could be manipulated to cause physical damage or grid instability.
- *Energy parasitism*: The computational power required for blockchain consensus mechanisms and continuous high-frequency AI optimization consumes significant energy. If not managed, this "digital parasitism" could erode the net energy return on investment (EROI) of the clean hydrogen produced.

9.4 The demand dilemma: You can build it, but will they buy it?

9.4.1 The offtake bottleneck

The defining flaw of the 2020–2023 policy cycle was a form of supply-side myopia. Governments aggressively subsidized electrolyzers (CAPEX) through instruments such as the US IRA, assuming that cheap supply would automatically create its own demand. However, it did not. By 2024, the market faced a glaring asymmetry: while global production targets for 2030 aimed for nearly 40 Mtpa, firm demand commitments barely scraped 11 Mtpa (Zhang et al., 2025).

This gap is not merely a scheduling issue; it is a structural failure of price formation. Industrial off-takers in steel, shipping, and ammonia operate in commoditized markets with thin margins. Without mechanisms to cover the "green premium"—often double or triple the price of fossil hydrogen—no CEO can sign a 15-year offtake agreement without risking bankruptcy. Consequently, the "bankability" of projects has evaporated. Financiers, unable to see secured revenue streams, refuse to unlock capital for construction, leaving gigawatt-scale projects stranded in the feasibility phase.

The paralysis is only now breaking as policy shifts focus from "building machines" to "bridging costs." We are seeing the emergence of contracts for difference (CfDs) and carbon contracts for difference (CCfDs), notably in Germany and the Netherlands. These instruments do not just subsidize the factory; they subsidize the difference between the strike price of green hydrogen and the fluctuating market price of gray hydrogen (or carbon), effectively de-risking the operational expenditure (OPEX) for the buyer rather than just the capital expenditure for the seller.

9.4.2 Mandates versus market-makers: Two philosophies

To force this reluctant market into existence, two distinct policy architectures have crystallized, creating a fascinating divergence between the European and global approaches. The coercive approach ("the stick"): the European Union has opted for direct obligation. The RED III directive effectively legislates demand by mandating that 42% of hydrogen used in industry must be renewable by 2030 (Clean Hydrogen JU 2025). This forces refiners and fertilizer producers to absorb the green premium or face penalties. While effective at creating a guaranteed market volume, this approach carries a high risk of "carbon leakage"—where industries simply relocate production to jurisdictions without such mandates to avoid the cost burden.

The intermediary approach ("the bridge"): alternatively, mechanisms like H_2Global act as a state-backed market maker. Operating as a double-sided auction, this entity buys hydrogen from global producers at the lowest offer price (e.g., USD 5/kg) and sells it to regional industries at the highest bid price (e.g., USD 3/kg), with the German government absorbing the delta. This disconnects the high cost of production from the willingness to pay, allowing the market to function before price parity is achieved.

Interestingly, AI is finding a niche here, not in "governance," but in contractual simulation. New algorithmic tools allow developers and banks to stress-test these CfD structures against thousands of electricity price and carbon tax scenarios. Rather than relying on static Excel models, these simulations help define the "strike price" probability bands needed to make a project bankable, effectively translating policy complexity into financial risk metrics that credit committees can accept.

9.5 Standards and certification

The industry has moved beyond the simplistic color wheel (green vs. blue vs. pink) to a rigorous, math-based standard: carbon intensity (CI). The ISO 19870 methodology and the International Partnership for Hydrogen and Fuel Cells in the Economy (IPHE's) work have established a common grammar for calculating lifecycle emissions (kg CO_2e/kg H_2). However, this shift from qualitative labels to quantitative thresholds has exposed deep methodological fractures.

A "clean" molecule in Texas is not necessarily "clean" in Rotterdam. Discrepancies in how jurisdictions account for upstream methane leakage rates (GWP100 vs. GWP20), grid intermittency, and nuclear inputs create friction. For instance, the EU's CBAM will soon penalize imports that cannot prove adherence to the specific RFNBO methodology, effectively functioning as a nontariff trade barrier.

In this contested landscape, AI serves as the necessary interoperability layer. Advanced "DPP" systems are being piloted to track molecules from production to end-use. Unlike static certificates, these AI-driven passports continuously ingest data, satellite methane monitoring, real-time grid intensity, and liquefaction energy, to calculate a dynamic CI score. This allows for algorithmic recertification in transit: a cargo can be digitally "translated" from US 45V compliance to EU RED III compliance automatically, reducing the friction that otherwise threatens to paralyze global trade.

9.6 Conclusion and policy roadmap

The hydrogen economy has left the era of "irrational exuberance" to enter the "hard" decade of industrial execution. The "great filter" described in this chapter is not merely a financial hurdle but a structural stress-test for global energy policy. The survivors of this shakeout will not be the projects with the grandest visions, but those with the most rigorous compliance architectures and secured offtake.

To bridge the ambition-implementation gap and operationalize AI as the sector's "central nervous system," three concrete recommendations are proposed:

- *Harmonize digital certification by 2027*: The fragmentation of standards (US 45V vs. EU RED III) is a trade barrier. International bodies (ISO, IPHE) must prioritize the interoperability of "DPP". We recommend establishing a unified ISO-standardized digital API for CI verification, allowing AI systems to automatically adjudicate compliance across borders.
- *Shift from CAPEX subsidies to OPEX certainty*: Building electrolyzers is insufficient if the molecule is too expensive to buy. Governments must pivot funding toward demand-side mechanisms. Specifically, the expansion of CfDs and double-auction schemes (like H_2Global) should be prioritized over one-off grants to cover the "green premium" for early adopters.
- *Institutionalize algorithmic de-risking*: To avoid a "two-speed" transition where the global south is priced out by high capital costs (WACC), Multilateral Development Banks should adopt AI-driven risk modeling as a standard for loan underwriting. By 2026, guarantees should be linked to "Digital Twin" performance data to lower interest rates for projects in high-potential regions.

Hydrogen is not a miracle cure; it is a complex industrial machine. Without a digital brain to manage its complexity and a synchronized policy framework to support its cost, it will remain a niche solution. The technologies are ready; the regulatory synchronization must now follow.

References

Baker Botts. (2025). Final section 45V clean hydrogen production tax credit regulations: A closer look. bakerbotts.com.

Bhuiyan, M., & Chowdhury, T. (2025). AI-driven optimization in renewable hydrogen production: A review. *American Journal of Interdisciplinary Studies*, 76–94.

Clean Hydrogen JU. (2025). *The European hydrogen policy landscape - January 2025*. European Hydrogen Observatory.

Hydrogen Europe. (2024). Hydrogen Europe Position paper. https://hydrogeneurope.eu/wp-content/uploads/2024/06/H2Europe_RED-III_Position-Paper.pdf.

IEA. (2024). *Global hydrogen review 2024*. International Energy Agency.

King & Spalding. (2024). *Guidance on section 45V clean hydrogen production tax credit*. kslaw.com.

Odenweller, A., & Ueckerdt, F. (2025). The green hydrogen ambition and implementation gap. *Nature Energy, 10*, 110–123. doi:10.1038/s41560-024-01684-7.

Odenweller, A., Ueckerdt, F., Nemet, G. F., Jensterle, M., & Luderer, G. (2022). Probabilistic feasibility space of scaling up green hydrogen supply. *Nature Energy, 7*, 854–865. doi:10.1038/s41560-022-01097-4.

Ricks, W., Xu, Q., & Jenkins, J. D. (2023). Minimizing emissions from grid-based hydrogen production in the United States. *Environmental Research Letters, 18*(1), 014025.

Tudor, C. (2023). Enhancing sustainable finance through green hydrogen equity investments: A multifaceted risk-return analysis. *Risks, 11*, 212. doi:10.3390/risks11120212.

Zhang, Z., Zhou, W., Hui, G., Hunt, J., Zhang, X. B., Ren, H., & Qin, P. (2025). Evaluating the CO_2 emissions implications of hydrogen energy policies in North China based on a dedicated hydrogen planning model. *Climate Policy, 25*(8), 1131–1143. doi:10.1080/14693062.2024.2443451.

Chapter 10

Future research directions in sustainable hydrogen: Production, storage, and utilization

Samson Olaitan Jeje[a], **Eseoghene Helen Umukoro**[b] **and Mxolisi Brendon Shongwe**[a]
[a] *Department of Chemical, Metallurgical and Materials Engineering, Faculty of Engineering and Built Environment, Tshwane University of Technology, Pretoria, South Africa,* [b] *Department of Chemistry, Obafemi Awolowo University, Ile Ife, Nigeria*

10.1 Introduction

The global climate crisis demands urgent, systemic decarbonization across all sectors. While renewable energy and electrification have reduced power generation and light transport emissions, heavy industries, shipping, aviation, and chemical manufacturing remain stubbornly reliant on fossil fuels. Green hydrogen—produced via renewable-powered electrolysis—has emerged as a critical solution for these hard-to-abate sectors (Jeje et al., 2024). Its versatility as an energy carrier, chemical feedstock, and high-temperature heat source positions it uniquely to displace fossil fuels where alternatives fall short. However, the hydrogen economy faces significant barriers, including technological immaturity, infrastructure gaps, and cost disparities with fossil-based hydrogen (Tzschöckel, 2024). This chapter focuses on transformative research priorities to accelerate hydrogen's role in achieving net-zero targets. Unlike earlier chapters, which detail current applications in steelmaking, ammonia production, and transportation, this section explores underexplored innovations capable of redefining hydrogen's scalability, affordability, and sustainability. For instance, while established methods, such as proton-exchange membrane (PEM) electrolysis and LOHCs, are well-documented, emerging approaches—such as

Decarbonizing through Sustainable Hydrogen and its Derivatives.
DOI: https://doi.org/10.1016/B978-0-443-27632-3.00010-4

plasma-driven water splitting, biohybrid systems, and AI-optimized material discovery—remain nascent yet pivotal to unlocking hydrogen's full potential.

The urgency of hydrogen innovation cannot be overstated. Driven by climate imperatives and global commitments to net zero, the International Renewable Energy Agency (IRENA) projects that hydrogen could account for up to 12% of global energy use by 2050 (International Renewable Energy Agency, 2022). Achieving this vision requires a drastic reduction in production costs, with initiatives such as the U.S. Department of Energy's (U.S. DOE) Hydrogen Shot targeting $1 per kilogram within the next decade (by 2031) (U.S. Department of Energy DOE, 2021). To scale hydrogen infrastructure effectively, challenges related to freshwater consumption, critical mineral dependence, and system integration must also be addressed. This chapter bridges these gaps by proposing actionable research pathways, emphasizing interdisciplinary collaboration between engineers, data scientists, and policymakers. Three key themes underpin this analysis:

- Disruptive production methods: Moving beyond incremental electrolysis improvements to explore low-cost, resource-efficient alternatives.
- Next-generation storage and utilization: Leveraging advanced materials and digital tools to enhance energy density, safety, and cross-sector applicability.
- Systemic integration: Aligning technological advances with circular economy principles, equitable deployment, and policy frameworks.

This chapter aims to shift hydrogen from a promising concept to a cornerstone of global decarbonization by prioritizing high-impact innovations. The following sections outline these opportunities in detail, providing a roadmap for researchers, industries, and governments to accelerate the transition to a sustainable hydrogen economy.

10.2 Innovative approaches to hydrogen production

The transition to a sustainable hydrogen economy hinges on developing efficient, cost-effective, and scalable hydrogen production methods. While conventional electrolysis and steam methane reforming (SMR) currently dominate the landscape, these approaches either face high energy demands or emit significant carbon dioxide (Sikiru et al., 2024). To address these challenges, researchers are exploring novel production techniques that leverage advanced materials, biological systems, and artificial intelligence (AI) to enhance efficiency and sustainability. This section delves into emerging hydrogen production technologies, including plasma-driven methods, biohybrid systems, and thermochemical solar-assisted processes. These advancements represent critical pathways toward a low-carbon hydrogen future, ensuring that production methods remain both environmentally and economically viable.

10.2.1 Emerging production technologies

10.2.1.1 Plasma-based hydrogen production

Plasma-assisted hydrogen production is a promising avenue in sustainable energy, leveraging high-energy ionized gases to drive chemical transformations. Unlike conventional electrolysis, plasma-based methods facilitate the dissociation of water and hydrocarbon molecules through reactive species such as radicals, ions, and photons, enabling selective and efficient hydrogen generation. Plasma-driven solution electrochemistry (PDSE) replaces conventional electrodes with plasma-liquid interfaces, inducing highly reactive interfacial layers that enhance reaction kinetics (Bruggeman et al., 2021). This approach, though historically inspired by electrolysis, presents unique challenges in understanding reaction mechanisms and optimizing process parameters. Recent advancements in plasma modeling and diagnostics have shed light on the fundamental differences between PDSE and traditional electrochemical processes, highlighting the need for predictive control to maximize efficiency. Among emerging techniques, microwave-excited steam plasma offers a novel route for hydrogen production. Computational models have demonstrated that under controlled conditions—such as 800 W microwave power at 1 Torr pressure—water molecules dissociate with a hydrogen conversion efficiency of approximately 24% (Oner & Dincer, 2023). However, the complex plasma kinetics and energy-exergy relationships necessitate further optimization to improve overall system efficiency.

Hydrogen production via plasma varies with feedstock type, reactor configuration, and energy input. Plasma activation pathways, including electronic, ionic, and vibrational excitations, play a crucial role in reaction effectiveness (Lee, 2023). Key innovations in this space include:

- Plasma catalysis, where nonthermal plasma enhances reaction kinetics while reducing energy consumption, enabling hydrogen production with lower energy penalties compared to traditional electrolysis.
- Plasma pyrolysis of methane (PPM) is a carbon-neutral method that decomposes methane into hydrogen and solid carbon without CO_2 emissions. Microwave plasma pyrolysis, in particular, has demonstrated high efficiency and could serve as a transitional technology for regions still reliant on natural gas.
- Nonthermal plasma electrolysis, which dissociates water molecules under ambient conditions, achieving efficiencies comparable to alkaline electrolyzers while operating on intermittent renewable power sources.

Plasma-assisted hydrogen production is gaining traction through initiatives like the EU's Plasmalysis project and other global efforts in the USA, Canada, and Europe (Clean Hydrogen JOINT UNDERTAKING, 2024; Hydrogen Europe, 2022). Technologies such as high-temperature plasma heating, plasmalysis, and microwave-driven processes are being explored for large-scale hydrogen generation, with some projects targeting production capacities of several tons per day. However, challenges such as electrode degradation, energy efficiency, and reactor scalability remain. Advancing reactor designs and material innovations will be critical to optimizing these processes for widespread adoption in hydrogen production and energy transition efforts.

10.2.1.2 Biohybrid systems for hydrogen generation

Biohybrid hydrogen production integrates biological processes with synthetic catalysts to enhance sustainable hydrogen generation. Recent advances in nanobiohybrid systems demonstrate the potential of combining bacterial metabolism with semiconductor materials to harness solar energy for hydrogen production (Ramprakash & Incharoensakdi, 2023). Photocatalytic bio-hybrids incorporating cadmium sulfide, metal oxides, and quantum dots have achieved significant improvements in hydrogen yield, efficiency, and stability under both anaerobic and ambient conditions (Cui, 2022; Ramprakash & Incharoensakdi, 2023; Wang et al., 2022). These systems offer a promising pathway for scalable hydrogen production while contributing to waste valorization and renewable energy solutions. Key approaches include:

- Microbial electrolysis cells (MECs): Microorganisms catalyze the breakdown of organic waste (e.g., agricultural residues, wastewater) into hydrogen at a low applied voltage, making it an energy-efficient alternative to conventional electrolysis. U.S. DOE projects' recent advancements demonstrated scalable reactor designs that can achieve hydrogen production rates exceeding 30 L/L/day employing raw wastewater while maintaining cost-effectiveness (Department of Energy, 2023).
- Photobiological hydrogen production: Photosynthetic microorganisms, such as cyanobacteria and algae, produce hydrogen through enzymatic activity under specific light conditions. Photobioelectrochemical systems further integrate algae-based photosynthesis with electrolysis, using sunlight to split water while simultaneously sequestering CO_2.
- Enzyme-assisted hydrogen production: Synthetic enzyme systems mimic natural hydrogenase enzymes to catalyze water splitting at ambient temperatures, offering a biologically inspired route to green hydrogen production.

Despite their promise, industrial deployment of biohybrid systems hinges on improving microbial stability, optimizing bioreactor designs, and reducing operational costs. Research efforts are currently focused on engineering robust microbial consortia and miniaturizing bioreactors to enhance scalability.

10.2.1.3 Thermochemical and solar-driven hydrogen production

Thermochemical and solar-driven hydrogen production offer promising pathways for large-scale hydrogen generation while reducing reliance on fossil fuels (Jeje et al., 2024).

- Thermochemical water splitting: This method involves high-temperature chemical cycles utilizing metal oxides to generate hydrogen without electricity. Advanced materials such as perovskite oxides and doped ceria are being explored to enhance reaction efficiency, stability, and reversibility. These cycles can be integrated with industrial waste heat sources to improve overall energy efficiency and lower production costs.
- Photoelectrochemical (PEC) water splitting: Semiconductor-based photocatalysts harness solar energy to drive direct hydrogen production from water. Emerging research focuses on nanostructured electrodes, co-catalysts, and hybrid PEC-PV (photovoltaic) systems to improve conversion efficiency, operational stability, and overall hydrogen yield. Materials such as metal-organic frameworks (MOFs) and 2D transition metal dichalcogenides (TMDs) are being investigated to enhance charge transport and light absorption properties.

Future research directions in thermochemical and solar-driven hydrogen production include developing cost-effective, corrosion-resistant materials and integrating AI-based predictive modeling for process optimization. By leveraging these advancements, hydrogen production can become more energy-efficient and commercially viable, accelerating the transition to a sustainable hydrogen economy. Table 10.1 compares emerging and conventional hydrogen production technologies—including plasma catalysis, biohybrid systems (MECs), PEM electrolysis, and SMR with carbon capture—based on their efficiency, feedstock requirements, scalability prospects, and key challenges. The analysis emphasizes qualitative performance trends and barriers to adoption, such as material durability, feedstock dependency, and system stability, providing a holistic view of technological readiness for sustainable energy transition.

10.2.2 Integration of artificial intelligence and machine learning

AI and machine learning (ML) are revolutionizing hydrogen production by accelerating material discovery, optimizing process efficiency, and enabling predictive analytics. AI can improve electrolyzer efficiency from 60%–80% to over 90%, reduce energy consumption by 10%, and lower operating costs by 15% (Saxena, 2024). These technologies address the challenges of high energy demands and costs, advancing the hydrogen economy, especially in regions like Europe, Asia, and North America. ML methods also enhance hydrogen electrochemical devices, such as electrolyzers and fuel cells, by predicting performance, detecting faults, and optimizing efficiency (Franić et al., 2025). Advanced deep learning techniques, despite challenges in interpretability, outperform traditional methods, and when combined with optimization algorithms, can revolutionize the design and diagnostics of these systems. Key applications include:

- AI-driven catalyst discovery: ML algorithms analyze vast datasets of material properties to predict novel compositions for oxygen and hydrogen evolution

TABLE 10.1 Comparative analysis of hydrogen production methods (Aiken et al., 2019; Lim et al., 2023; Loenders et al., 2023; Muddasar et al., 2022; Peng et al., 2023; Zang et al., 2024).

Method	Efficiency	Feedstock	Scalability	Key challenges
Plasma catalysis	Moderate-high (energy-intensive)	Methane/seawater	High (pending electrode R&D)	Electrode degradation, reactor scale-up, energy optimization
PEM electrolysis	High (water-splitting optimized)	Freshwater	Moderate (iridium scarcity)	High capital costs, strict water purity requirements, catalyst durability
Biohybrid (MECs)	Moderate-high (substrate-dependent)	Organic waste/wastewater	Moderate (pending cost/material R&D)	Microbial stability, methanogen competition, material costs, and low energy density
SMR with carbon capture and storage (CCS)	High (fossil-optimized)	Natural gas	Limited (methane leakage concerns)	CO_2 sequestration risks, fossil dependency, and infrastructure inertia

reactions. Google DeepMind's discovery of low-iridium catalysts for PEM electrolyzers exemplifies AI's potential to mitigate supply chain constraints.

- Predictive process analytics: Digital twin models simulate electrolyzer performance under fluctuating renewable inputs, improving efficiency and reducing downtime. A data-driven approach for high-temperature proton-exchange membrane electrolyzers (HT-PEMEC) combines multiphysics simulations and nonlinear system identification, achieving a 96% accuracy in predicting power consumption, hydrogen production, and temperature. This model helps design control strategies, such as fuzzy and neural network predictive controls, to reduce power overshoot and improve durability, with neural network control reducing overshoot by 92% (Zhao et al., 2022).
- Automated process control: AI-integrated electrolysis and plasma reactors dynamically adjust input parameters, enhancing cost-effectiveness and system longevity.
- Demand-response optimization: ML algorithms align hydrogen production with real-time grid renewable availability, minimizing energy costs and maximizing sustainability.

Despite these advancements, challenges remain in data availability, particularly for emerging methods such as plasma catalysis. Standardized benchmarking and collaborative frameworks for data sharing are essential to train robust AI models. Additionally, balancing model complexity with interpretability is critical to gaining industry acceptance and ensuring transparency in AI-driven decision-making.

10.2.3 Addressing feedstock sustainability and renewable energy integration

The sustainability of hydrogen production hinges on feedstock availability and the seamless integration of renewable energy sources. Key strategies include:

- Seawater electrolysis: Innovations in chloride-resistant PEM membranes and redox-mediated processes are making seawater electrolysis a viable alternative to freshwater-based systems, thus increasing efficiency (Liu et al., 2024; Mishra et al., 2025).
- Circular hydrogen production from biomass: Utilizing agricultural residues, organic waste, and industrial byproducts can create carbon-negative hydrogen. For example, rice straw is an inexpensive and abundant biomass feedstock that can mitigate greenhouse gas emissions. By converting rice straw into hydrogen, emissions from burning or decomposing the straw are avoided, and the hydrogen produced serves as a clean-burning fuel, further reducing emissions (Morya, 2023). Its renewability and versatility in conversion processes make rice straw a promising alternative to traditional hydrogen production methods.

- Hybrid renewable-powered hydrogen systems: Cointegration of solar, wind, and hydropower with electrolysis maximizes efficiency and ensures grid stability. A PV wind-tidal fuel cell system reduces net present costs while maintaining reliability, highlighting the economic viability of hybrid renewable energy (Naderipour et al., 2021). Similarly, a geothermal-hydro-biomass-solar-wind system improves efficiency, lowers energy costs, and offers substantial environmental benefits (Tafavogh & Zahedi, 2021).
- Advanced energy buffering: Flow batteries integrated with electrolyzers mitigate power variability and prevent renewable energy curtailment.
- Water resource management: Emerging technologies such as brine-splitting and wastewater electrolysis address freshwater scarcity while expanding hydrogen production potential.

Policy reforms, including streamlined zoning for co-located facilities and updated grid codes for electrolyzer–grid interaction, are essential to fully harness these synergies. By advancing sustainable feedstocks and optimizing renewable energy integration, hydrogen production pathways can achieve carbon neutrality while maintaining economic viability.

10.3 Advanced hydrogen storage materials and methods

Efficient hydrogen storage remains a fundamental challenge in realizing a sustainable hydrogen economy. The success of hydrogen as a clean energy carrier depends not only on its production but also on safe, cost-effective, and high-capacity storage solutions that enable its widespread adoption across transportation, industry, and grid applications. Conventional storage methods, such as high-pressure gas cylinders (350–700 bar) and cryogenic liquid hydrogen (−253°C), suffer from safety concerns, high-energy consumption, and economic limitations. Consequently, research is increasingly focused on next-generation storage materials and dynamic carrier systems that offer superior energy density, stability, and scalability.

10.3.1 Development of next-generation storage materials

Recent advancements in material science are unlocking new possibilities for hydrogen storage. Next-generation materials aim to overcome the limitations of conventional storage by enabling reversible hydrogen binding at near-ambient conditions. Key developments include:

- Metal hydrides: Solid-state hydrogen storage using metal hydrides (e.g., magnesium hydride, titanium-based alloys, and complex hydrides like lithium borohydride) has gained attention due to high volumetric hydrogen density (up to 150 kgH_2/m^3) and reversible absorption/desorption properties (Mehr et al., 2024). However, challenges such as slow kinetics, high desorption temperatures, and material degradation persist. Recent advances in nanostructuring (e.g., ball-milled MgH_2 doped with transition metals such as Ni and Fe) have reduced desorption temperatures to <250°C, improving their viability for industrial applications (Xu et al., 2024).

- Porous frameworks (MOFs/COFs): MOFs and covalent organic frameworks (COFs) offer ultrahigh surface areas and tunable pore structures that allow for hydrogen physisorption at lower pressures. Notable materials, NU-1501-Al, an ultra-porous MOF, achieves a high Brunauer–Emmett–Teller (BET) surface area (7310 m^2/g) and exceptional hydrogen storage capacity (14.0 wt.% at 160 K/5 bar and 2.9 wt.% at 100 bar/296 K) outperforming most reported MOFs (Chen et al., 2020). Its balance of gravimetric and volumetric storage makes it a promising material for onboard hydrogen storage. However, performance at ambient conditions remains limited, necessitating novel functionalization strategies like spillover catalysts (e.g., platinum nanoparticles) to enhance hydrogen binding.
- Hybrid composites: The integration of different storage materials, such as MOFs with metal hydrides or graphene-based nanostructures, presents promising hybrid solutions. These materials can leverage the strengths of each component, improving hydrogen uptake and release efficiency while maintaining stability. MOF-graphene hybrids, such as MOF-5/GO and UiO-66/graphene acid, improve stability, conductivity, and porosity, optimizing hydrogen uptake and release (Jayaramulu et al., 2022). Advanced synthesis techniques, including layer-by-layer deposition and nucleation-guided growth, can further refine these composites for practical hydrogen storage applications.

Building on recent advancements in next-generation hydrogen storage materials, various technologies are being explored to enhance efficiency, scalability, and practical implementation. Table 10.2 summarizes key hydrogen storage technologies, highlighting their characteristics, advantages, and challenges to provide insights into their potential for future applications.

10.3.2 Artificial intelligence–enabled discovery and optimization of storage materials

AI and ML are revolutionizing hydrogen storage by accelerating material discovery, optimizing system performance, and enhancing safety. These technologies enable rapid screening of potential materials, predictive modeling of performance, and real-time monitoring of storage systems. Robotics further enhances precision and efficiency in hydrogen production and storage, mitigating operational risks. Key advancements include:

- Computational modeling: AI-driven algorithms analyze vast datasets to identify materials with superior hydrogen adsorption and desorption properties. Generative adversarial networks (GANs) and ML models trained on materials

TABLE 10.2 Comparison of hydrogen storage technologies: Characteristics, advantages, and challenges (Chen et al., 2020; Chu et al., 2023; Li et al., 2024; Mehr et al., 2024; Xu et al., 2023).

Material/technology	Hydrogen storage characteristics	Key advantages	Key challenges
Metal hydrides	Solid-state hydrogen storage using metal alloys that absorb and release hydrogen	High volumetric density, safe storage under moderate pressures	High operating temperatures, slow hydrogen absorption/desorption rates
Metal-organic frameworks (MOFs)	Porous materials with high surface area for hydrogen adsorption	High storage capacity, tunable properties for optimization	Requires cryogenic temperatures and high-pressure operation is needed for efficiency
Liquid organic hydrogen carriers (LOHCs)	Hydrogen is stored in liquid organic molecules and released through catalytic dehydrogenation	High cycle life, ease of transport using existing fuel infrastructure	Requires high temperatures for hydrogen release, energy-intensive dehydrogenation
Cryo-compressed hydrogen	Hydrogen stored at cryogenic temperatures under high pressure	High gravimetric and volumetric storage densities, suitable for mobile applications	Requires specialized insulation and high-pressure containment, and high-energy consumption for cooling

databases predict novel compositions with enhanced stability and kinetics (Delpisheh et al., 2024).

- High-throughput experiments: ML-guided experimental design significantly reduces the time required for material testing. Molecular dynamics simulations optimize pore geometries in MOFs and COFs, improving volumetric storage capacity.
- Real-time performance monitoring: AI-integrated nanomaterial-based sensors ensure precise tracking of critical storage parameters, such as hydrogen purity, pressure, and temperature. Predictive maintenance models extend the lifespan of storage tanks and prevent failures under extreme conditions. Collaborative platforms like the HyMarc Data Hub's Hydrogen Materials Database (Hydrogen Tools, 2025) aggregate experimental and computational data, though standardization of testing protocols remains a hurdle.
- Robotic automation in storage and production: Robotics streamlines the assembly of proton-exchange membrane fuel cells (PEMFCs) and other hydrogen-related components, enhancing precision and reducing operational hazards (Quintanilla et al., 2025). Automated systems improve scalability and reduce manufacturing costs.

Despite its transformative potential, AI-driven hydrogen storage faces challenges, including the "simulation-to-reality" gap, limited high-quality datasets, and the high computational costs of multiscale modeling. Additionally, robotic systems must be optimized for cryogenic and high-pressure conditions, requiring specialized materials and AI-driven adaptive control. Potential solutions include developing physics-informed neural networks (PINNs) for enhanced model interpretability, reinforcement learning-based adaptive control for robotics, and energy-efficient "Green AI" to reduce computational demands. Standardization of AI-driven hydrogen storage testing protocols and robust cybersecurity measures are also essential for ensuring long-term viability (Quintanilla et al., 2025). By integrating AI and robotics, hydrogen storage technologies can achieve significant breakthroughs, accelerating the transition to sustainable and efficient energy solutions.

10.3.3 Dynamic hydrogen carriers and modular storage systems

To enhance the flexibility and mobility of hydrogen storage, innovative carrier technologies and modular systems are under development:

- Liquid organic hydrogen carriers (LOHCs): LOHCs, such as benzyl toluene and dibenzyl toluene, enable safe, low-pressure hydrogen transport using existing fuel infrastructure. Recent breakthroughs in catalyst design (e.g., Pt-Sn/Al_2O_3 for dehydrogenation) have reduced energy penalties by 30%, though challenges persist in minimizing side reactions (e.g., cracking) and catalyst fouling. To address these, the EU's sherLOHCk consortium (European Union, 2020) is developing innovative catalytic solutions aimed at improving process efficiency ($\leq$10 kWh/kgH_2), enhancing dehydrogenation kinetics (>3 g H_2/g catalyst/min), and ensuring long-term stability (<0.01% performance loss per cycle). Their efforts also focus on increasing hydrogen storage capacity (>5.0 wt%) and volumetric energy density (>1.6 MWh/m^3) while ensuring catalyst compatibility with mixed gas streams. These advancements seek to optimize LOHC technology for scalable hydrogen applications in the energy and mobility sectors.

- Modular solid-state storage systems: Systems including containerized hydrogen units and pilot-scale metal hydride tanks are emerging as decentralized solutions for industrial clusters and off-grid applications. While residential-scale metal hydrides remain experimental due to high desorption temperatures and cycle durability challenges, advancements in modular hydrogen systems continue. Technologies like Siemens' Silyzer 300 PEM electrolyzer are being piloted in remote mining and microgrid applications, leveraging hydrogen's energy density for long-duration storage (AG, 2018).

Advancing hydrogen storage technologies requires a focus on multifunctional materials that integrate self-monitoring capabilities, ensuring real-time performance tracking and enhanced durability. Circular design principles must be incorporated to improve recyclability and reduce reliance on critical minerals. Hybrid storage solutions that combine LOHCs with solid-state materials can address different application needs, particularly in high-demand sectors such as heavy transport. Additionally, optimizing the balance between energy density, storage safety, and material longevity will be crucial in meeting cost and efficiency targets set by global energy initiatives. Addressing these challenges will unlock hydrogen's full potential as a ubiquitous, safe, and cost-effective energy vector for the future.

10.4 Expanding hydrogen utilization pathways

The conventional applications of hydrogen, such as fuel cells for transportation and industrial heating, are well established. However, future advancements in the hydrogen economy will be driven by expanding its utilization into unconventional sectors, integrating it into renewable energy systems, and enhancing its role in distributed energy networks. These emerging pathways have the potential to increase hydrogen demand, improve grid stability, and create new industrial markets, thereby accelerating global decarbonization efforts.

10.4.1 Unconventional and emerging industrial applications

While hydrogen's role in steelmaking and ammonia production is well established, its potential in niche industries remains underexplored. In semiconductor manufacturing, ultra-high-purity hydrogen is critical for processes such as chemical vapor deposition (CVD) and as a reducing atmosphere in wafer fabrication.

A report by QY Research (2024) indicated that in 2023, the market for ultra-high-purity hydrogen used in semiconductors was valued at US$ 142 million. It is expected to expand to US$ 247 million by 2030, with a compound annual growth rate (CAGR) of 8.7% during the period from 2024 to 2030. A key driver of this growth is the adoption of high-pressure hydrogen annealing (HPHA), an innovative process pioneered by companies like HPSP, a South Korean semiconductor equipment provider (HPSP, 2024). HPHA enhances device reliability and performance by curing defects in advanced nodes, such as FinFET, gate-all-around (GAA), and 3D NAND memory devices. As semiconductor scaling intensifies and thermal budgets shrink, conventional high-temperature annealing methods face limitations. HPSP's HPHA technology enables defect reduction at lower temperatures using high-pressure hydrogen, making it indispensable for cutting-edge fabrication. However, contamination risks (e.g., moisture, hydrocarbons) necessitate advanced purification systems, increasing costs and complexity.

In advanced chemical processes, hydrogen is emerging as a reducing agent for novel pathways (Jeje et al., 2024). Johnson Matthey's HyCOgen technology leverages green hydrogen and captured CO_2 to produce syngas—a precursor for sustainable aviation fuel (SAF)—through a catalytic reverse water gas shift process (Matthey, 2022). Integrated with FT CANS Fischer–Tropsch technology (developed with bp), this solution converts over 95% of CO_2 into synthetic crude oil, which is refined into drop-in fuels like SAF, renewable diesel, and naphtha. Designed for scalability, the system supports projects ranging from single-electrolyzer setups to industrial-scale deployments, addressing aviation's urgent need to curb emissions. By bridging renewable hydrogen with CO_2 utilization, the HyCOgen/FT CANS platform enables cost-effective, large-scale SAF production, aligning with global decarbonization targets for air transport. Similarly, Mitsubishi Gas Chemical has pioneered hydrogen peroxide (H_2O_2) synthesis via direct hydrogen-oxygen reaction, eliminating hazardous anthraquinone solvents and reducing waste by 70%. Despite these promising applications, hurdles remain in catalyst longevity and reaction control, particularly under fluctuating hydrogen supply conditions.

In the metallurgical sector, hydrogen is gaining traction in direct reduction of iron (DRI) for steelmaking, reducing reliance on carbon-intensive coke (Basirat, 2024). Future research is exploring its application in aluminum smelting and rare earth element extraction, where hydrogen can replace conventional reductants to minimize environmental impact. While hydrogen's industrial applications dominate sectors like semiconductors and metallurgy, its potential in medical therapy emerges as a groundbreaking, unconventional frontier. Molecular H_2, long confined to industrial reductants and energy systems, is gaining applications in biomedicine for its unique anti-inflammatory and antioxidant properties (Boretti, 2024). Experimental therapies now explore inhaled or dissolved hydrogen for treating neurodegenerative disorders, chronic inflammation, and radiation-induced injuries—leveraging ultra-high-purity gas standards akin to semiconductor manufacturing. This nascent field bridges industrial gas infrastructure with clinical innovation, offering a sustainable pathway to address unmet medical needs. Though still experimental, hydrogen's dual role as an industrial staple and therapeutic agent underscores its versatility, positioning it as a high-potential offshoot in the evolving hydrogen economy. Critical challenges are as follows:

- Purity requirements: Semiconductor-grade hydrogen (99.9999% purity) requires energy-intensive purification, increasing the levelized cost of hydrogen (LCOH).
- Process integration: Retrofitting existing chemical plants for hydrogen compatibility demands capital-intensive upgrades to piping, valves, and safety systems.

10.4.2 Integration into renewable hybrid systems

As discussed briefly in Section 10.2.3, hydrogen's ability to stabilize intermittent renewable energy sources is being tested in hybrid systems that combine solar, wind, and hydrogen storage. The modeled green hydrogen system in Port Augusta integrates 870 MW PEM electrolyzers with 800 MW hydrogen-fired turbines, utilizing underground aquifers for seasonal storage (Currie et al., 2024). Excess renewable energy from local wind/solar projects is converted to hydrogen during low-price periods, providing grid stability and dispatchable power during demand peaks. This system addresses South Australia's intermittency challenges while enabling 63,840 MWh of seasonal storage capacity. Mitsubishi Power successfully conducted a grid-connected demonstration of its J-series air-cooled (JAC) gas turbine with a 30% hydrogen-natural gas fuel blend at the T-point 2 combined cycle power plant in Takasago Hydrogen Park, Japan (2023). The test verified stable combustion and low NOx emissions using a dry low NOx (DLN) combustor, demonstrating the feasibility of hydrogen cofiring in large-scale power generation.

Concentrated solar power (CSP) is emerging as a key technology for decarbonizing energy systems by integrating baseload electricity generation with industrial heat and hydrogen production. According to the U.S. DOE (Vijaykumar, 2024), 4.8 GW of CSP is deployed worldwide, including 1.8 GW in the U.S. When integrated with thermal storage, CSP enables continuous 24/7 operation, making it a reliable energy source. Its capability to generate high-temperature heat (>700°C) makes it particularly suited for solar-driven hydrogen production through high-temperature electrolysis (HTE). The U.S. Department of Energy's Solar Energy Technologies Office (SETO) is advancing CSP-SOEC (Solid Oxide Electrolysis Cell) hybrid systems, optimizing thermal efficiency and molten salt storage to reduce hydrogen production costs. Additional innovations, such as thermochemical redox cycles and proton-conducting membranes, aim to enhance scalability and lower reaction temperatures for efficient hydrogen synthesis. With CSP's capacity to address intermittency and reduce reliance on grid electricity, it is poised to play a pivotal role in achieving the DOE's hydrogen cost target of $1/kg by 2031. Critical challenges are as follows:

- Efficiency losses: Energy penalties from multiple conversion steps (e.g., electrolysis, reelectrification) reduce overall system competitiveness.
- Infrastructure costs: Colocating hydrogen storage with renewables demands land-use reforms and new safety protocols.

10.4.3 Hydrogen-based microgrid and off-grid solutions

Hydrogen-based microgrids and off-grid energy solutions offer a transformative approach to energy access, particularly in remote regions and industrial sites with limited grid infrastructure. Advancements in this area include decentralized hydrogen power generation, hydrogen-powered backup systems, and integration with distributed energy resources (DERs).

A group of researchers from the University of South Africa and the University of Ibadan, Nigeria (Ayodele et al., 2021) demonstrated the transformative potential of hydrogen-based microgrids for off-grid energy access through a case study in rural South Africa. Their work focused on designing a hybrid renewable energy system (40 kW solar PV, 30 kW wind turbines) paired with hydrogen storage (25 kW electrolyzer, 40 kg hydrogen tank, and 8.6 kW fuel cell) to power a rural health clinic requiring 18.67 kWh/day. The system achieved 100% renewable penetration, leveraging decentralized hydrogen production to address intermittency while eliminating reliance on short-lived battery storage. Key advancements highlighted include the system's ability to operate autonomously in remote regions, with a levelized energy cost of $2.34/kWh and a breakeven grid extension distance of 8.81 km—far shorter than the clinic's 21.35 km proximity to the nearest grid. This underscores hydrogen's viability as a resilient backup and long-term storage solution, particularly in areas with sparse populations or challenging terrain. Future priorities are as follows:

- Cold-climate adaptations: Development of frost-tolerant membranes and passive thermal management systems for Arctic deployments.
- Community engagement: Addressing safety perceptions and training local technicians for system maintenance.
- Circular design: Repurposing decommissioned electric vehicle (EV) batteries for hybrid hydrogen-battery storage, as demonstrated in Norway's HyBaT project.

As hydrogen utilization expands beyond conventional applications, ongoing research, and technological innovation will continue to unlock new opportunities. By leveraging its versatility, hydrogen can play a pivotal role in the future energy landscape, supporting sustainable industrial processes, enhancing grid stability, and providing reliable energy access worldwide. Fig. 10.1 crystallizes that vision, illustrating how hydrogen's multifaceted pathways interconnect to

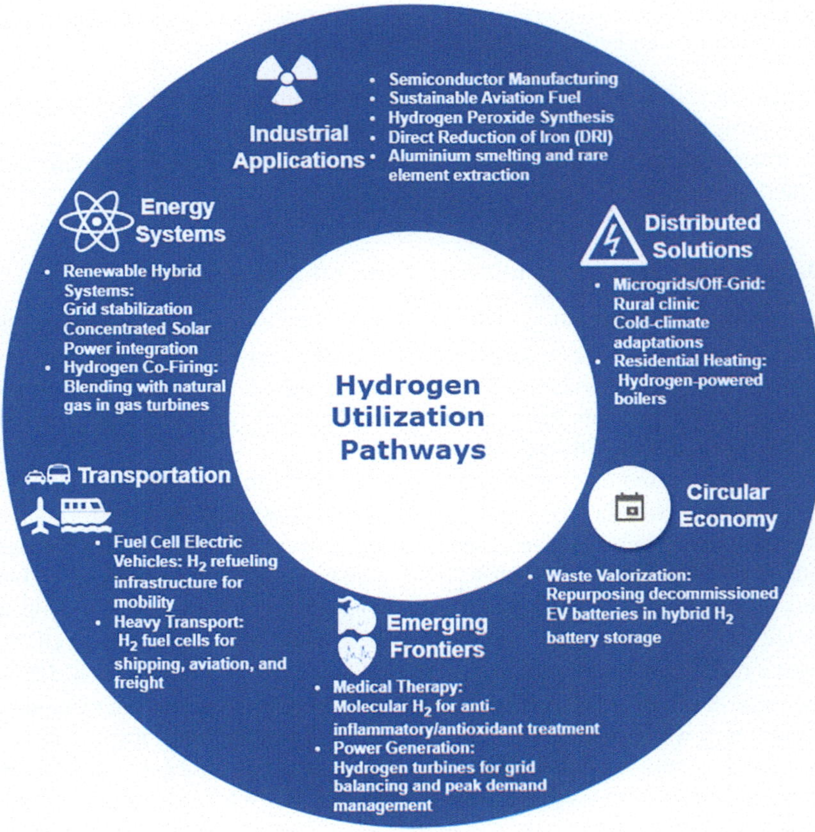

FIGURE 10.1 Hydrogen utilization pathways.

form a cohesive, adaptive ecosystem. It reinforces the idea that hydrogen's value lies not only in individual applications but also in its systemic potential to bridge sectors, amplify renewable integration, and address global energy inequities.

10.5 Digital transformation in hydrogen systems

As the hydrogen economy advances, digital transformation is playing a pivotal role in optimizing production, enhancing storage technologies, and ensuring efficient utilization. The integration of AI, ML, blockchain, the Internet of Things (IoT), and digital twin technology is revolutionizing the way hydrogen is produced, stored, and distributed. These technologies enable predictive analytics, enhance material discovery, improve operational efficiency, and provide real-time monitoring, ultimately driving cost reductions and accelerating adoption across industries.

10.5.1 Artificial intelligence and machine learning: Revolutionizing production and utilization

AI and ML are playing an increasingly vital role in improving the efficiency of hydrogen production, storage, and utilization. As discussed in Sections 10.2 and 10.3, their key applications can be summarized as follows:

- Predictive analytics for production optimization: AI-driven models analyze vast datasets from electrolysis, biomass conversion, and thermochemical production processes to optimize reaction conditions, reduce energy consumption, and enhance hydrogen yield.
- Demand forecasting: AI-based models predict hydrogen demand fluctuations based on market trends, energy grid dynamics, and industrial requirements. This helps optimize supply chain logistics, ensuring hydrogen availability while minimizing waste and inefficiencies.
- Accelerated design of advanced storage materials: ML is transforming the discovery of hydrogen storage materials by accelerating the identification of optimal candidates such as MOFs. A study by Wang et al. (2022) introduced an advanced ML framework that integrates structure decomposition, feature embedding, and predictive modeling to assess MOF adsorption capacities. By analyzing chemical and geometric descriptors, the model rapidly predicts hydrogen uptake and delivers accurate screening results, significantly reducing computational costs compared to traditional simulations. This approach enhances the development of next-generation hydrogen storage materials, optimizing energy efficiency and scalability.
- Process optimization in hydrogen fuel cells: AI enhances fuel cell performance by adjusting operational parameters in real time. For example, ML algorithms optimize plasma catalysis parameters (e.g., power density and gas flow rates) to maximize hydrogen yield while minimizing electrode wear.

Challenges are as follows:

- Data scarcity: Limited experimental data for novel processes (e.g., seawater electrolysis) restricts ML model accuracy.
- Model interpretability: Industrial stakeholders remain cautious of "black-box" AI solutions, demanding transparent decision-making frameworks.
- Cybersecurity risks: Increased connectivity exposes hydrogen infrastructure to cyberattacks targeting grid-integrated electrolyzers.

10.5.2 Blockchain and Internet of Things: Enhancing transparency and efficiency

Blockchain and IoT technologies are transforming the hydrogen supply chain by providing transparency, security, and efficiency in storage, distribution, and consumption. Key applications include:

- Enhancing supply chain transparency and efficiency: Blockchain ensures traceability across the hydrogen value chain by securely recording transactions, production sources, and carbon footprints. Initiatives like HydroChain document the clean credentials of hydrogen production, while H_2-Share leverages blockchain for decentralized hydrogen trading, enhancing market transparency and operational efficiency (Abdin, 2024).
- Real-time monitoring of hydrogen storage and distribution: IoT sensors embedded in hydrogen storage tanks, pipelines, and transportation systems provide real-time data on pressure, temperature, leakage detection, and overall system health. Hexagon Purus, for instance, integrates wireless sensors in composite storage tanks to monitor anomalies and transmit data via 5G networks.
- Automated smart contracts for hydrogen transactions: Blockchain-based smart contracts automate hydrogen trading, ensuring real-time settlements between producers, suppliers, and consumers. This increases efficiency, reduces transaction costs, and enhances trust in the hydrogen market.
- IoT-enabled hydrogen refueling stations: Hwang and Hwang (2024) developed an IoT system for hydrogen refueling stations using ESP32 microcontrollers, LoRa/Wi-Fi networks, and semiconductor sensors to enable real-time hydrogen leak detection. GPS pinpoints leak locations, while a Unity-based 3D digital twin visualizes data via InfluxDB, achieving a 646 ms transfer delay (under a 2-s sensor threshold). This mobile, cost-effective solution enhances safety by replacing fixed infrastructure and supports scalable monitoring, critical for advancing reliable hydrogen ecosystems.

  Key innovations are as follows:

- Smart contracts: Automating hydrogen transactions based on predefined sustainability thresholds.
- Edge computing: Processing sensor data locally to reduce latency in safety-critical applications (e.g., leak detection in pipelines).
- Interoperability standards: Addressing communication challenges between legacy industrial systems and modern IoT architectures.

10.5.3 Digital twins and predictive maintenance

Digital twins (virtual replicas of physical hydrogen systems) are becoming indispensable for predictive maintenance and scenario testing. Feng et al. (2024) demonstrate their role in hydrogen energy systems through real-time data integration and AI-driven simulations, enabling precise failure prediction and risk mitigation. For instance, digital twins of PEMFCs achieve >90% accuracy in forecasting remaining useful life, allowing proactive interventions. They also simulate scenarios like leaks or grid disruptions, enhancing safety and operational resilience in complex hydrogen infrastructure. By optimizing hydrogen storage, extending fuel cell lifetimes, and enabling robust

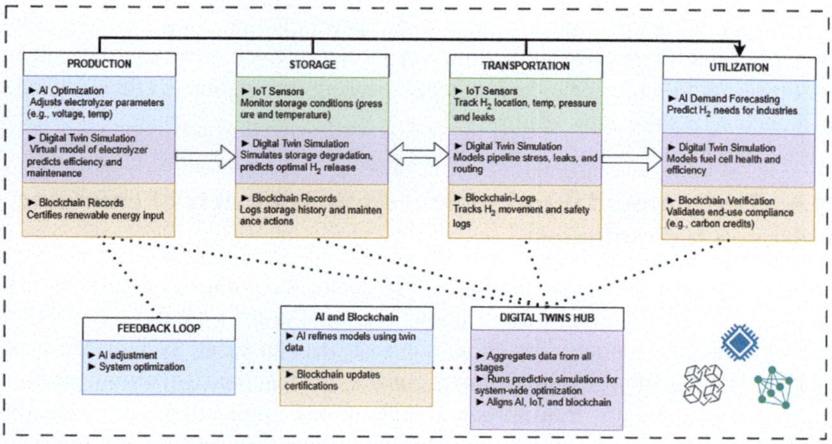

FIGURE 10.2 Integration of digital twins, AI, blockchain, and IoT in the hydrogen value chain. *AI*, Artificial intelligence; *IoT*, Internet of Things.

disaster recovery planning, digital twins bridge physical and digital realms to advance sustainable and resilient hydrogen ecosystems. The following are the barriers to adoption:

- Computational costs: High-fidelity simulations of complex systems require significant processing power.
- Skill gaps: Workforce training is needed to operationalize digital twin insights, particularly in developing economies.

Fig. 10.2 illustrates the integration of digital twins, AI, blockchain, and IoT across the hydrogen value chain, demonstrating their synergistic roles in advancing efficiency, transparency, and sustainability. Digital twins serve as virtual replicas of physical systems, enabling predictive simulations and real-time optimization of hydrogen production, storage, transportation, and utilization. AI leverages insights from these twins to dynamically refine processes, such as reducing energy consumption in electrolysis. IoT devices feed critical real-time data (e.g., pressure, temperature, leaks) into digital twins, ensuring accurate and adaptive modeling. Meanwhile, blockchain securely records twin-generated data—such as maintenance logs, carbon footprints, and certifications—to enhance accountability and traceability. Together, these technologies create a cohesive framework where predictive analytics, automated monitoring, and immutable record-keeping drive smarter, safer, and more sustainable hydrogen ecosystems. To fully harness digital tools, stakeholders must prioritize collaborative data ecosystems that enable anonymized operational data sharing across academia, industry, and governments. Standardization of IoT device communication, AI model benchmarking, and blockchain certification is crucial to ensure interoperability and efficiency. Additionally, implementing robust cybersecurity frameworks with advanced encryption and intrusion detection systems tailored to hydrogen infrastructure is essential to mitigate potential risks. By addressing these priorities, digital transformation can accelerate hydrogen's transition from pilot projects to large-scale deployment, ensuring transparency, efficiency, and resilience across the value chain.

10.6 Environmental and socioeconomic impacts of emerging hydrogen technologies

As the hydrogen economy continues to expand, assessing its environmental footprint and socioeconomic implications is critical to ensuring its sustainability and widespread adoption. The transition to hydrogen-based systems must be evaluated from a life cycle perspective, considering resource extraction, production processes, storage, transportation, and end-use applications. Additionally, societal acceptance, infrastructure readiness, and economic feasibility play significant roles in shaping the future of hydrogen technologies.

10.6.1 Environmental footprint of emerging hydrogen systems

While green hydrogen is often regarded as a zero-emission fuel, its true sustainability depends on the energy source used for electrolysis and the broader life cycle impacts of production, distribution, and use. Life cycle assessments (LCAs) provide a comprehensive framework for evaluating the environmental impact of emerging hydrogen technologies, considering factors such as energy consumption, greenhouse gas emissions, water usage, and land impact across the hydrogen value chain. Key considerations include the following:

- Water consumption and availability: Producing 1 kg of hydrogen via PEM electrolysis requires approximately 9 L of ultrapure water, raising concerns about freshwater availability as hydrogen production scales. Meeting projected hydrogen demand by 2030 could add 2.1 billion cubic meters to global freshwater consumption, making alternative sources essential. Seawater electrolysis and wastewater treatment offer viable pathways, though challenges such as electrolyzer corrosion, energy costs, and economic feasibility must be addressed. Reverse osmosis and membrane distillation are promising desalination techniques, with the latter offering potential integration with waste heat recovery to improve overall system efficiency (Kumar et al., 2024).
- Critical mineral demand and circular economy: Electrolyzers and fuel cells rely on scarce platinum group metals like iridium and platinum, posing supply chain risks as demand rises with Europe's 100 GW electrolyzer target by 2030. The EU's Critical Raw Materials Act (Ragonnaud, 2023) aims to boost domestic supply by covering 10% of extraction, 40% of processing, and 25% of recycling needs. However, dependence on PGM-rich regions such as South Africa and Russia remains a challenge, and scalable recycling solutions for degraded catalysts are still underdeveloped.

- Carbon footprint of hydrogen production: A Stanford and Cornell University study (Howarth & Jacobson, 2021) found that blue hydrogen, which relies on natural gas with carbon capture and storage (CCS), has a high greenhouse gas footprint due to methane leakage throughout the supply chain. At a 3.5% methane leakage rate, the study revealed that blue hydrogen's 20-year global warming impact is over 20% higher than burning coal or natural gas for heat, largely because CCS fails to capture all associated emissions. This raises concerns about blue hydrogen's viability as a truly low-emission alternative, emphasizing the need for stricter methane management and cleaner hydrogen pathways.
- Hydrogen leakage and climate impact: Hydrogen itself does not trap infrared radiation, but its leakage can indirectly impact the climate by influencing atmospheric chemistry. A study by Lakshmanan and Bhati (2024) highlights that hydrogen reacts with hydroxyl radicals (OH) and ozone (O_3), altering the balance of greenhouse gases such as methane and tropospheric ozone. The oxidation of hydrogen by O_3, in particular, can lead to ozone depletion rather than formation, challenging the assumption that hydrogen leakage always increases greenhouse gas levels.

Addressing the environmental footprint of hydrogen requires a holistic approach that balances resource efficiency, emissions reduction, and supply chain sustainability. Advancements in water sourcing, circular economy strategies for critical materials, and stricter methane management in blue hydrogen production are essential to ensuring hydrogen's role as a truly clean energy carrier. Additionally, mitigating hydrogen leakage and its atmospheric effects demands further research and regulatory oversight. By prioritizing these considerations, the hydrogen economy can progress toward greater sustainability and climate resilience.

10.6.2 Socioeconomic equity and infrastructure readiness

For hydrogen to be successfully integrated into global energy systems, equitable access and societal acceptance must be prioritized. However, disparities in hydrogen infrastructure and investment may exacerbate global inequalities, with industrialized nations leading adoption while emerging economies struggle with financing and technology deployment. Key challenges and opportunities include the following:

- Public perception and safety concerns: Hydrogen is often perceived as a high-risk fuel due to its flammability and storage complexities, with incidents like the 1937 Hindenburg disaster and the 2022 explosion at Equinor's refinery at Mongstad reinforcing public skepticism. However, from a technical standpoint, hydrogen's risks are comparable to conventional fuels like natural gas and petroleum, requiring different but manageable safety measures. Its wide flammable range and high storage pressures present challenges, yet hydrogen's buoyancy allows for rapid dispersion if properly ventilated. The key concern is not just actual risk but perceived risk—public unfamiliarity heightens fear, even though industrial hydrogen incidents remain infrequent. Effective education, transparent safety regulations, and advancements in storage and infrastructure design are essential to fostering public trust and enabling broader adoption.

- Equitable distribution of hydrogen infrastructure: Equitable hydrogen infrastructure deployment is crucial to ensuring that the benefits of hydrogen economies reach beyond industrialized regions. Namibia's Green Hydrogen and Derivatives Strategy outlines the development of three hydrogen valleys across key regions, with the Southern Corridor Development Initiative (SCDI) aiming to position the country as a global hydrogen producer (Altenburg & Kantel, 2024). Projects such as the Hyphen Hydrogen Energy initiative, expected to generate up to 300,000 tons of green hydrogen, integrate renewable power, desalination, and infrastructure expansion, with co-benefits like improved energy access and local development. To maximize socioeconomic benefits, measures such as revenue-sharing mechanisms and infrastructure investments in underserved areas are essential.
- Workforce development and reskilling: The hydrogen economy presents both opportunities and challenges for workforce dynamics, particularly in regions transitioning from fossil fuels. In South Africa, a research report by the Labour Market Intelligence (2024) research program, Department of Higher Education and Training, estimates that the hydrogen sector could create up to 650,000 jobs by 2050, while also posing a risk of displacing workers in the coal and oil industries. While high-skilled roles, such as engineers and electrochemical technicians, require specialized training, many elementary positions, like construction and maintenance, offer limited wage mobility. Addressing skills mismatches, initiatives such as hydrogen-focused vocational training, curriculum reforms, and industry-academic partnerships are essential to fostering an inclusive transition.
- Infrastructure lock-in and material compatibility: Repurposing natural gas infrastructure for hydrogen transport offers cost and time advantages but presents material challenges. Hydrogen's lower density and potential to cause embrittlement can reduce pipeline integrity, necessitating upgrades such as internal coatings, pressure regulation, and component replacements. Studies indicate that while conversion is feasible, long-term durability depends on pipeline material and operating conditions. Strategic repurposing, combined with new infrastructure, will be key to minimizing stranded assets and ensuring a resilient hydrogen transport network .

10.6.3 Economic opportunities and market challenges

Hydrogen presents significant economic opportunities, but several barriers must be addressed to enable its large-scale deployment. These include production costs, policy frameworks, and market integration challenges (Table 10.3) .

TABLE 10.3 SWOT analysis of hydrogen deployment (Basirat, 2024; Hossain Bhuiyan & Siddique, 2025; Jeje et al., 2024; International Renewable Energy Agency, 2019; Irena, 2022).

SWOT category	Key factors	Implications	Possible mitigation strategies
Strengths	Carbon-free energy potential	Supports decarbonization of industries, transport, and power sectors	Strengthen policies to integrate hydrogen into national energy strategies
	Versatility across multiple applications	Usable in fuel cells, industrial heat, and energy storage	Develop sector-specific deployment strategies
	Growing government and industry support	Incentives, subsidies, and R&D investments accelerating adoption	Increase global collaboration and funding mechanisms
Weaknesses	High production costs	Hydrogen is still more expensive than fossil fuels in many regions	Scale-up production, invest in low-cost renewable energy PPAs
	Infrastructure limitations	Lack of dedicated pipelines, storage, and refueling stations	Retrofit existing gas networks, develop hydrogen hubs
	Water and resource constraints	Electrolysis requires significant water and critical minerals	Advanced wastewater electrolysis and seawater desalination technologies

(*continued on next page*)

TABLE 10.3 SWOT analysis of hydrogen deployment (Basirat, 2024; Hossain Bhuiyan & Siddique, 2025; Jeje et al., 2024; International Renewable Energy Agency, 2019; Irena, 2022)—cont'd

SWOT category	Key factors	Implications	Possible mitigation strategies
Opportunities	Declining renewable energy costs	Makes green hydrogen more viable for large-scale deployment	Promote long-term PPAs and grid integration
	International trade potential	Hydrogen export markets emerging, especially in Europe and Asia	Build export-oriented infrastructure, harmonize certification standards
	Advancements in storage and transport	LOHCs, ammonia, and cryogenic storage improving hydrogen logistics	Invest in scalable storage solutions for global supply chains
Threats	Energy losses in conversion and transport	Efficiency concerns may slow adoption	Improve electrolyzer technology and hybrid storage approaches
	Policy and regulatory uncertainty	Inconsistent global policies hinder investment confidence	Standardize regulations and provide long-term policy frameworks
	Competing low-carbon alternatives	Batteries, biofuels, and synthetic fuels pose competition	Focus on hard-to-abate sectors where hydrogen has a unique advantage

LOHCs, *liquid organic hydrogen carriers;* PPAs, *power purchase agreement;* SWOT, *strengths, weaknesses, opportunities, and threats.*

- Cost reduction and market competitiveness: Green hydrogen remains significantly more expensive than gray hydrogen, with production costs ranging from $2.28 to $7.39/kg compared to gray hydrogen's $0.67–1.31/kg (Hossain Bhuiyan & Siddique, 2025). Achieving cost parity will require large-scale electrolyzer deployment, improved efficiency, and lower renewable electricity prices. Chile is actively leveraging power purchase agreements (PPAs) to secure affordable renewable energy and drive down hydrogen production costs. A recent long-term PPA between Abastible and Zelestra exemplifies this approach, integrating a 220 MWdc solar PV plant with a 1 GWh battery energy storage system to ensure a stable, carbon-free energy supply (Abastible , 2025). By enabling continuous renewable power availability, such agreements strengthen Chile's position in the global hydrogen market while supporting the country's broader decarbonization strategy.
- Policy incentives and investment strategies: Government policies, subsidies, and carbon pricing mechanisms play a crucial role in incentivizing hydrogen adoption. China's subsidies ($2.3/kg for domestic hydrogen) have led to market imbalances, while the EU's carbon border adjustment mechanism (CBAM) risks penalizing exporters lacking certified low-carbon hydrogen.
- Integration into existing energy markets: The compatibility of hydrogen with current energy systems, including gas pipelines, refineries, and industrial processes, will determine its market viability. Developing cost-effective retrofitting solutions and hybrid energy systems will be essential for seamless integration.

10.6.4 Public engagement and community participation

Public acceptance plays a pivotal role in the successful deployment of hydrogen technologies. However, community concerns, regulatory gaps, and sociopolitical dynamics shape attitudes toward hydrogen infrastructure.

- Community concerns and safety measures: Public perceptions of hydrogen safety remain a major barrier to adoption. Concerns over flammability and explosion risks often overshadow technical assessments showing hydrogen's risk profile is comparable to conventional fuels. For example, hydrogen refueling stations have faced public opposition in several regions, including Germany and Australia, due to perceived safety threats (Gordon et al., 2022). Strengthening public trust through transparent risk assessments, strict adherence to safety standards, and targeted education campaigns is crucial.
- Participatory decision-making: Integrating local communities into hydrogen project planning can significantly reduce resistance and improve project timelines. Evidence from Europe suggests that participatory approaches (where residents engage in co-designing energy infrastructure) enhance acceptance and reduce permitting delays. Lessons from offshore wind projects highlight

how early engagement fosters trust, mitigates concerns, and ensures smoother deployment (Bögel et al., 2018).

- Indigenous and marginalized community inclusion: Many large-scale hydrogen projects risk neglecting Indigenous and marginalized communities, exacerbating existing energy inequalities. Research highlights the importance of ensuring these communities have a voice in hydrogen initiatives, particularly in regions where land and resource rights intersect with energy infrastructure (Gordon et al., 2022). Transparent benefit-sharing agreements and community-led consultation frameworks are necessary to prevent socioeconomic displacement and ensure equitable participation.

Building public trust through proactive engagement, inclusive governance, and transparent communication is vital for the successful integration of hydrogen technologies. Ensuring that diverse stakeholders (ranging from local communities to policymakers and industry leaders) are actively involved in decision-making will help address concerns, accelerate project acceptance, and create equitable opportunities within the hydrogen economy.

10.7 Conclusion and research roadmap

The transition to a sustainable hydrogen economy presents a transformative opportunity to decarbonize industry, transportation, and power generation. However, achieving large-scale adoption requires overcoming key challenges in production scalability, storage efficiency, digital integration, and equitable infrastructure deployment. Advancements in plasma catalysis, biohybrid hydrogen systems, and AI-driven system optimization are paving the way for more efficient and cost-effective hydrogen technologies. Additionally, digital tools such as blockchain, IoT, and predictive analytics will enhance safety, streamline supply chains, and optimize operations. Ensuring the long-term success of hydrogen as a clean energy carrier will depend on proactive governance, circular economy principles, and inclusive stakeholder engagement to prevent socioeconomic disparities and environmental harm. Future research should prioritize scaling up hydrogen production through innovative methods like plasma-assisted and biohybrid systems, targeting cost reductions below $2/kg. Advanced storage solutions, including MOF-hydride composites and LOHCs, must be optimized for durability and efficiency. Digital transformation will play a crucial role in integrating hydrogen into existing energy markets, with AI-driven predictive maintenance and cybersecurity standards essential for infrastructure resilience. Hydrogen's integration into hybrid energy systems, microgrids, and industrial applications will further support grid stability and peak demand management. Global policy frameworks must accelerate hydrogen adoption through certification standards, carbon accounting mechanisms, and market incentives, while also ensuring that investment reaches developing nations to promote equitable access and job creation.

Collaboration among key stakeholders is essential to drive progress. Academia should lead fundamental research and interdisciplinary training programs, while the industry must invest in commercialization pathways and partnerships with startups to pilot emerging technologies. Policymakers must streamline regulations, facilitate cross-border agreements, and implement transparent carbon accounting systems. Civil society also plays a vital role in ensuring community-led decision-making and fostering public awareness of hydrogen's role in the energy transition. The urgency of advancing hydrogen technology cannot be overstated. Achieving net-zero emissions will require a fourfold increase in annual investments, reaching approximately $130 billion. Any delay in scaling up hydrogen deployment will lock in reliance on fossil fuels and slow climate progress. To avoid this, research, policy, and financing must align to make hydrogen a cornerstone of the global clean energy transition. However, rapid deployment must not come at the cost of environmental or social equity. The hydrogen economy must be built on bold, systemic change to ensure a just, sustainable, and resilient future.

References

Abastible, 2025. Abastible signs a long-term renewable energy PPA with Zelestra in Chile for a tailor-made energy solution that includes 220 MWdc of solar and 1 GWh of battery energy storage. Zelestra. https://zelestra.energy/abastible-signs-a-long-term-renewable-energy-ppa-with-zelestra-in-chile-for-a-tailor-made-energy-solution-that-includes-220-mwdc-of-solar-and-1-gwh-of-battery-energy-storage/.

Abdin, Z. (2024). Empowering the hydrogen economy: The transformative potential of blockchain technology. *Renewable and Sustainable Energy Reviews, 200*, 114572.

AG, S., SILYZER 300: The next paradigm of PEM electrolysis. (2018), Scribd Inc.

Aiken, D. C., Curtis, T. P., & Heidrich, E. S. (2019). Avenues to the financial viability of microbial electrolysis cells [MEC] for domestic wastewater treatment and hydrogen production. *International Journal of Hydrogen Energy, 44*(5), 2426–2434. http://www.journals.elsevier.com/international-journal-of-hydrogen-energy/. doi:10.1016/j.ijhydene.2018.12.029.

Altenburg, T., Kantel, A., Green hydrogen in Namibia: opportunities and risks. IDOS Discussion Paper. (2024), German Institute of Development and Sustainability (IDOS).

Ayodele, T. R., Mosetlhe, T. C., Yusuff, A. A., & Ogunjuyigbe, A. S. O. (2021). Off-grid hybrid renewable energy system with hydrogen storage for South African rural community health clinic. *International Journal of Hydrogen Energy, 46*(38), 19871–19885. doi:10.1016/j.ijhydene.2021.03.140.

Basirat, S. (2024). *Hydrogen unleashed: Opportunities and challenges in the evolving H2-DRI-EAF pathway beyond 2024*. Institute for Energy Economics and Financial Analysis.

Bögel, P., Oltra, C., Sala, R., Lores, M., Upham, P., Dütschke, E., Schneider, U., & Wiemann, P. (2018). The role of attitudes in technology acceptance management: Reflections on the case of hydrogen fuel cells in Europe. *Journal of Cleaner Production, 188*, 125–135. doi:10.1016/j.jclepro.2018.03.266.

Boretti, A. (2024). Therapeutic potential of hydrogen in sports orthopaedics and traumatology: A narrative review. *Sports Orthopaedics and Traumatology, 40*(2), 120–132. https://www.sciencedirect.com/science/journal/0949328X. doi:10.1016/j.orthtr.2024.02.008.

Bruggeman, P. J., Frontiera, R. R., Kortshagen, U. R., Kushner, M. J., Linic, S., Schatz, G. C., Andaraarachchi, H., Exarhos, S., Jones, L. O., Mueller, C. M., Rich, C. C., Xu, C., Yue, Y., & Zhang, Y. (2021). Plasma-driven solution electrolysis. *Journal of Applied Physics, 129*(20). doi:10.1063/5.0044261.

Chen, Z., Li, P., Anderson, R., Wang, X., Zhang, X., Robison, L., Redfern, L. R., Moribe, S., Islamoglu, T., Gómez-Gualdrón, D. A., Yildirim, T., Stoddart, J. F., & Farha, O. K. (2020). Balancing volumetric and gravimetric uptake in highly porous materials for clean energy. *Science, 368*(6488), 297–303. https://science.sciencemag.org/content/368/6488/297/tab-pdf. doi:10.1126/science.aaz8881.

Chu, C., Wu, K., Luo, B., Cao, Q., & Zhang, H. (2023). Hydrogen storage by liquid organic hydrogen carriers: Catalyst, renewable carrier, and technology – A review. *Carbon Resources Conversion, 6*(4), 334–351. www.keaipublishing.com/en/journals/carbon-resources-conversion/. doi:10.1016/j.crcon.2023.03.007.

Clean Hydrogen JOINT UNDERTAKINGWork Programme 2025. (2024), https://www.clean-hydrogen.europa.eu/document/download/e5846471-328c-48fd-b91f-837da07556b6_en?filename=Clean%20Hydrogen%20JU%20AWP%202025_FINAL_ADOPTED.pdf.

Cui, S. (2022). Light-assisted fermentative hydrogen production in an intimately-coupled inorganic-bio hybrid with self-assembled nanoparticles. *Chemical Engineering Journal, 428*, 131254.

Currie, G., Behrens, E., Bolitho, S., Coen, M., & Wilson, T. (2024). Green hydrogen driven by wind and solar—An Australian case study. *Wind., 4*(2), 111–134. https://www.mdpi.com/journal/wind. doi:10.3390/wind4020006.

Delpisheh, M., Ebrahimpour, B., Fattahi, A., Siavashi, M., Mir, H., Mashhadimoslem, H., Ali Abdol, M., Ghorbani, M., Shokri, J., Niblett, D., Khosravi, K., Rahimi, S., Mojtaba Alirahmi, S., Yu, H., Elkamel, A., Niasar, V., & Mamlouk, M. (2024). Leveraging machine learning in porous media. *Journal of Materials Chemistry A, 12*(32), 20717–20782. doi:10.1039/d4ta00251b.

Department of Energy, Scalable and highly-efficient microbial electrochemical reactor for hydrogen generation from lignocellulosic biomass and waste. Annual Merit Review and Peer Evaluation Meeting (2023), Oregon State University, Hong Liu, PI.https://www.hydrogen.energy.gov/docs/hydrogenprogramlibraries/pdfs/review23/p184_liu_2023_o-pdf.pdf?Status=Master.

European Union, Catalyst development for improved economic viability of LOHC technology. European Union. https://cordis.europa.eu/programme/id/H2020_FCH-02-1-2020.

Feng, Z., Eiubovi, I., Shao, Y., Fan, Z., & Tan, R. (2024). Review of digital twin technology applications in hydrogen energy. *Chain, 1*(1), 54–74. doi:10.23919/chain.2024.000001.

Franić, N., Pivac, I., & Barbir, F. (2025). A review of machine learning applications in hydrogen electrochemical devices. *International Journal of Hydrogen Energy, 102*, 523–544. https://www.sciencedirect.com/science/journal/03603199. doi:10.1016/j.ijhydene.2025.01.070.

Gordon, J. A., Balta-Ozkan, N., & Nabavi, S. A. (2022). Homes of the future: Unpacking public perceptions to power the domestic hydrogen transition. *Renewable and Sustainable Energy Reviews, 164*, 112481.

Hossain Bhuiyan, M. M., & Siddique, Z. (2025). Hydrogen as an alternative fuel: A comprehensive review of challenges and opportunities in production, storage, and transportation. *International Journal of Hydrogen Energy, 102*, 1026–1044. https://www.sciencedirect.com/science/journal/03603199. doi:10.1016/j.ijhydene.2025.01.033.

Howarth, R. W., & Jacobson, M. Z. (2021). How green is blue hydrogen? *Energy Science and Engineering, 9*(10), 1676–1687. http://onlinelibrary.wiley.com/journal/10.1002/(ISSN)2050-0505. doi:10.1002/ese3.956.

HPSP. HPSP expands its R&D capabilities and study on HPA and HPO. HPSP https://www.imec-int.com/en/press/hpsp-expands-its-rd-capabilities-and-study-hpa-and-hpo.

Hwang, H. J., & Hwang, S. H. (2024). Developing an internet of things system for hydrogen leak detection at hydrogen refueling stations integrating LoRa and global positioning system. *Edelweiss Applied Science and Technology, 8*(5), 1582–1593. https://learning-gate.com/index.php/2576-8484/article/download/1874/697. doi:10.55214/25768484.v8i5.1874.

Hydrogen Europe, Pyrolysis: Potential and possible applications of a climate friendly hydrogen production. (2022), Hydrogen Europe. https://hydrogeneurope.eu/wp-content/uploads/2022/10/ewp_kompakt_pyrolyse_english_web.pdf.

U.S. Department of Energy (DOE). (2021). Hydrogen Shot. U.S. Department of Energy. Available https://www.energy.gov/topics/hydrogen-shot. (Accessed: 11 March 2025)

Hydrogen Tools. Hydrogen storage materials database - data and resources. (2025), https://datahub.hymarc.org/dataset/hydrogen-storage-materials-db.

International Renewable Energy Agency, Hydrogen economy hints at new global power dynamics. (2022), International Renewable Energy Agency. Available https://www.irena.org/News/pressreleases/2022/Jan/Hydrogen.

International Renewable Energy Agency, The future of hydrogen, Report Prepared by the IEA for the G20, Japan. International Renewable Energy Agency (2019).

Jayaramulu, K., Mukherjee, S., Morales, D. M., Dubal, D. P., Kumar Nanjundan, A., Schneemann, A., Masa, J., Kment, S., Schuhmann, W., Otyepka, M., Zbořil, R., & Fischer, R. A. (2022). Graphene-based metal–organic framework hybrids for applications in catalysis, environmental, and energy technologies. *Chemical Reviews, 122*(24), 17241–17338. doi:10.1021/acs.chemrev.2c00270.

Jeje, S. O., Marazani, T., Obiko, J. O., & Shongwe, M. B. (2024). Advancing the hydrogen production economy: A comprehensive review of technologies, sustainability, and future prospects. *International Journal of Hydrogen Energy, 78*, 642–661. https://www.sciencedirect.com/science/journal/03603199. doi:10.1016/j.ijhydene.2024.06.344.

Kumar, P., Date, A., Mahmood, N., Das, R. K., & Shabani, B. (2024). Freshwater supply for hydrogen production: An underestimated challenge. *International Journal of Hydrogen Energy, 78*, 202–217. https://www.sciencedirect.com/science/journal/03603199. doi:10.1016/j.ijhydene.2024.06.257.

Labour Market Intelligence., Identification of skills needed for the hydrogen economy. (2024), Labour Market Intelligence.

Lakshmanan, S., & Bhati, M. (2024). Unravelling the atmospheric and climate implications of hydrogen leakage. *International Journal of Hydrogen Energy, 53*, 807–815. https://www.sciencedirect.com/science/journal/03603199. doi:10.1016/j.ijhydene.2023.12.010.

Lee, D. H. (2023). Plasma-assisted hydrogen generation: A mechanistic review. *Fuel Processing Technology, 247*, 107761.

Li, S., Han, F., Liu, Y., Xu, Z., Yan, Y., & Ni, Z. (2024). Evaluation criterion for filling process of cryo-compressed hydrogen storage vessel. *International Journal of Hydrogen Energy, 59*, 1459–1470. https://www.sciencedirect.com/science/journal/03603199. doi:10.1016/j.ijhydene.2024.01.086.

Lim, K. H., Yue, Y., Bella, N., Gao, X., Zhang, T., Hu, F., Das, S., & Kawi, S. (2023). Sustainable hydrogen and ammonia technologies with nonthermal plasma catalysis: Mechanistic insights and technoeconomic analysis. *ACS Sustainable Chemistry and Engineering., 11*(13), 4903–4933. http://pubs.acs.org/journal/ascecg. doi:10.1021/acssuschemeng.2c06515.

Liu, D., Cai, Y., Wang, X., Zhuo, Y., Sui, X., Pan, H., & Wang, Z. (2024). Innovations in electrocatalysts, hybrid anodic oxidation, and electrolyzers for enhanced direct seawater electrolysis. *Energy and Environmental Science, 17*(19), 6897–6942. http://pubs.rsc.org/en/journals/journal/ee. doi:10.1039/d4ee01693a.

Loenders, B., Michiels, R., & Bogaerts, A. (2023). Is a catalyst always beneficial in plasma catalysis? Insights from the many physical and chemical interactions. *Journal of Energy Chemistry, 85*, 501–533. elsevier.com/journals/journal-of-energy-chemistry/2095-4956. doi:10.1016/j.jechem.2023.06.016.

Matthey, J., JM launches HyCOgen, an enabling technology to efficiently convert CO2 and green hydrogen into sustainable aviation fuel. JM. 2022. https://matthey.com/news/2022/hycogen.

Mehr, A. S., Phillips, A. D., Brandon, M. P., Pryce, M. T., & Carton, J. G. (2024). Recent challenges and development of technical and technoeconomic aspects for hydrogen storage, insights at different scales; a state of art review. *International Journal of Hydrogen Energy, 70*, 786–815. https://www.sciencedirect.com/science/journal/03603199. doi:10.1016/j.ijhydene.2024.05.182.

Mishra, S., Shanbhag, M. M., Pollet, B. G., & Kalanur, S. S. (2025). Breaking down the barrier: The progress and promise of seawater splitting. *International Journal of Hydrogen Energy, 106*, 334–352. https://www.sciencedirect.com/science/journal/03603199. doi:10.1016/j.ijhydene.2025.01.447.

Mitsubishi, M. H. Power Successfully operates an advanced class gas turbine with 30% hydrogen fuel co-firing at grid-connected T-point 2. (2023), Fuel Cells Works.

Morya, R. (2023). Exploring rice straw as substrate for hydrogen production: Critical challenges and opportunities. *Environmental Technology & Innovation, 31*, 103153.

Muddasar, M., Liaquat, R., Aslam, A., Ur Rahman, M. Z., Abdullah, A., Khoja, A. H., Latif, K., & Bahadar, A. (2022). Performance efficiency comparison of microbial electrolysis cells for sustainable production of biohydrogen—A comprehensive review. *International Journal of Energy Research, 46*(5), 5625–5645. http://onlinelibrary.wiley.com/journal/10.1002/(ISSN)1099-114X. doi:10.1002/er.7606.

Naderipour, A., Abdul-Malek, Z., Nowdeh, S. A., Kamyab, H., Ramtin, A. R., Shahrokhi, S., & Klemeš, J. J. (2021). Comparative evaluation of hybrid photovoltaic, wind, tidal and fuel cell clean system design for different regions with remote application considering cost. *Journal of Cleaner Production, 283*, 124207. doi:10.1016/j.jclepro.2020.124207.

Oner, O., & Dincer, I. (2023). Numerical investigation of hydrogen production from low-pressure microwave steam plasma. *Computers & Chemical Engineering, 174*, 108230.

Peng, P., Su, J., & Breunig, H. (2023). Benchmarking plasma and electrolysis decomposition technologies for ammonia to power generation. *Energy Conversion and Management, 288*, 117166.

Quintanilla, P., Elhalwagy, A., Duan, L., Soltani, S. M., Lai, C. S., Foroudi, P., Huda, M. N., & Nandy, M. (2025). Artificial intelligence and robotics in the hydrogen lifecycle: A systematic review. *International Journal of Hydrogen Energy, 113*, 801–817. doi:10.1016/j.ijhydene.2025.03.016.

QY Research. Global Ultra-High-Purity Hydrogen for Semiconductors Market Insights, Forecast to 2030. 2024. QY Research. https://www.qyresearch.com/reports/2538843/ultra-high-purity-hydrogen-for-semiconductors.

Ragonnaud, G. (2023). Critical Raw Materials Act. *EPRS,*.

Ramprakash, B., & Incharoensakdi, A. (2023). Extracellular self-photosensitizer combined with metal oxide-based nano bio-hybrid system encapsulated by alginate improves hydrogen production in the presence of oxygen. *Bioresource Technology, 388*, 129703.

Saxena, A. (2024). AI-driven optimization for green hydrogen production efficiency. *The Journal of Scientific and Engineering Research, 11*, 145–155.

Sikiru, S., Adedayo, H. B., Olutoki, J. O., & Rehman, Z. u. (2024). Hydrogen integration in power grids, infrastructure demands and techno-economic assessment: A comprehensive review. *Journal of Energy Storage, 104*, 114520.

Tafavogh, M., & Zahedi, A. (2021). Design and production of a novel encapsulated nano phase change materials to improve thermal efficiency of a quintuple renewable geothermal/hydro/biomass/solar/wind hybrid system. *Renewable Energy, 169*, 358–378. http://www.journals.elsevier.com/renewable-and-sustainable-energy-reviews/. doi:10.1016/j.renene.2020.12.118.

Tzschöckel, T. S., A stakeholder-based assessment of the opportunities and hurdles to the development of the Icelandic green hydrogen economy. [Thesis, University of Iceland], Landsvirkjun. (2024).

Vijaykumar, R. SETO/CSP research on hydrogen and fuels. U.S. Department of Energy. (2024), https://www.hydrogen.energy.gov/docs/hydrogenprogramlibraries/pdfs/review24/seto000_bauer_2024_o.pdf?sfvrsn=fb1342db_3.

Wang, X. M., Chen, L., He, R. L., Cui, S., Li, J., Fu, X. Z., Wu, Q. Z., Liu, H. Q., Huang, T. Y., & Li, W. W. (2022). Anaerobic self-assembly of a regenerable bacteria-quantum dot hybrid for solar hydrogen production. *Nanoscale, 14*(23), 8409–8417. http://pubs.rsc.org/en/journals/journal/nr. doi:10.1039/d2nr01777f.

Wang, Z., Zhou, Y., Zhou, T., & Sundmacher, K. (2022). Identification of optimal metal-organic frameworks by machine learning: Structure decomposition, feature integration, and predictive modeling. *Computers & Chemical Engineering, 160*, 107739. doi:10.1016/j.compchemeng.2022.107739.

Xu, Y., Deng, Y., Liu, W., Zhao, X., Xu, J., & Yuan, Z. (2023). Research progress of hydrogen energy and metal hydrogen storage materials. *Sustainable Energy Technologies and Assessments, 55*, 102974. doi:10.1016/j.seta.2022.102974.

Xu, Y., Li, Y., Hou, Q., Hao, Y., & Ding, Z. (2024). Ball milling innovations advance Mg-based hydrogen storage materials towards practical applications. *Materials., 17*(11). http://www.mdpi.com/journal/materials. doi:10.3390/ma17112510.

Zang, G., Graham, E. J., & Mallapragada, D. (2024). H2 production through natural gas reforming and carbon capture: A techno-economic and life cycle analysis comparison. *International Journal of Hydrogen Energy, 49*, 1288–1303. https://www.sciencedirect.com/science/journal/03603199. doi:10.1016/j.ijhydene.2023.09.230.

Zhao, D., He, Q., Yu, J., Guo, M., Fu, J., Li, X., & Ni, M. (2022). A data-driven digital-twin model and control of high temperature proton exchange membrane electrolyzer cells. *International Journal of Hydrogen Energy, 47*(14), 8687–8699. http://www.journals.elsevier.com/international-journal-of-hydrogen-energy/. doi:10.1016/j.ijhydene.2021.12.233.

Index

Page numbers followed by "*f*" and "*t*" indicate, fig.ures and tables respectively.

Printed and bound by CPI Group (UK) Ltd, Croydon, CR0 4YY
17/09/2026
02218533-0001